ST(P) MATHEMATICS 3A

ST(P) MATHEMATICS 3A

ST(P) MATHEMATICS series:

ST(P) 1A
ST(P) 1B
ST(P) 1A Teacher's Notes and Answers
ST(P) 1B Teacher's Notes and Answers

ST(P) 2A
ST(P) 2B
ST(P) 2A Teacher's Notes and Answers
ST(P) 2B Teacher's Notes and Answers

ST(P) 3A
ST(P) 3B
ST(P) 3A Teacher's Notes and Answers
ST(P) 3B Teacher's Notes and Answers

ST(P) 4A
ST(P) 4B
ST(P) 4A Teacher's Notes and Answers
ST(P) 4B Teacher's Notes and Answers

ST(P) 5A (with answers)
ST(P) 5B (with answers)

ST(P) 5C
ST(P) 5C Copy Masters
ST(P) 5C Teacher's Note and Answers

ST(P) Resource Book

ST(P) Workbooks:
Drawing and Using Curved Graphs
Measuring Instruments
Symmetry and Transformation
Straight Line Graphs

ST(P) MATHEMATICS 3A

L. Bostock, B.Sc.

S. Chandler, B.Sc.

A. Shepherd, B.Sc.

E. Smith, M.Sc.

Second Edition

Stanley Thornes (Publishers) Ltd

First Published 1985 by:
Stanley Thornes (Publishers) Ltd,
Ellenborough House,
Wellington Street,
CHELTENHAM GL50 1YD
England

Reprinted 1985
Reprinted 1986
Reprinted 1987
Reprinted 1988
Reprinted 1989
Reprinted 1990
Second Edition 1992
Reprinted 1992
Reprinted 1993

British Library Cataloguing in Publication Data
ST(P) mathematics 3A.—2nd ed.—(ST(P)
mathematics)
   I. Bostock, L.   II. Series
   510

   ISBN 0-7487-1260-7

Typeset by Cotswold Typesetting Ltd, Gloucester
Printed and bound in Great Britain at The Bath Press, Avon

CONTENTS

INTRODUCTION

TO THE PUPIL

This book has been designed for those of you who are working at Levels 7 and 8 of the National Curriculum and who are hoping to attempt the highest level GCSE papers in Mathematics. The book contains a good deal of revision of earlier work and this makes it large. However you will find that you will not need to work through all the revision sections, and for those that you do, just a few questions will usually be sufficient to restore your confidence in that topic. As in the case with earlier books there are plenty of straightforward questions and the exercises are divided into three types of question:

The first type, identified by plain numbers, e.g. **12.**, helps you to see if you understand the work. These questions are considered necessary for every new topic that you attempt.

The second type, identified by a single underline, e.g. **12.**, are extra, but not harder, questions for extra practice or for later revision.

The third type, identified by a double underline, e.g. **12.**, are for those of you who manage the first type of question fairly easily and would like to tackle questions that are a little harder.

Most chapters end with mixed exercises. These can be used for revision, either when you have finished the chapter or at a later date.

At this stage you will find that you use your calculator a great deal. However, you should avoid getting into the habit of using it for calculations that can be done easily in your head. Using your brain is faster than using a calculator and your brain needs the exercise to keep it working efficiently. Whether you use a calculator or not, always estimate an answer and always ask yourself, "Is my answer a sensible one?"

TO THE TEACHER

A number of topics have been introduced into this new edition as a result of the National Curriculum. Some topics have been removed and two sections, the formal treatment of congruent triangles and matrix transformations, have been moved into the new edition of Book 4A.

Together with Books 1A and 2A, this book completes coverage of Levels 6 and 7 and most of Level 8. Some of the work in this book goes beyond Level 8. In particular, there is a large section on algebra which starts the preparation for the skills needed to cope with the algebra content of level 10.

1 MAKING SURE OF ARITHMETIC

NAMES FOR NUMBERS

As young children we start learning about numbers by counting objects: 1, 2, 3, ... 10, ...

The numbers that we use for counting are called the *natural numbers*.

We next learn that whole objects can be divided into parts, or fractions, for example: half an orange, one and a half bars of chocolate.

We use *common fractions*, e.g. $\frac{1}{2}$, $\frac{3}{4}$, $\frac{3}{2}$, ... to describe the size of the part. The *denominator* (bottom number) describes the number of equal parts of the object and the numerator (top number) tells us how many parts we have. For example $\frac{3}{4}$ means that we have 3 of the 4 equal parts of the whole object.

A fraction such as $\frac{3}{2}$, where the numerator is bigger than the denominator, is called an *improper fraction*. The improper fraction $\frac{3}{2}$ can be written as $1\frac{1}{2}$ and in this form it is called a *mixed number*. Note that $1\frac{1}{2}$ means $1 + \frac{1}{2}$.

ADDITION AND SUBTRACTION OF FRACTIONS

We cannot add apples to oranges unless we reclassify them both as, say, fruit. In much the same way, we cannot add tenths to quarters unless we change them both into the same kind of fraction, i.e. change them so that both have the same, or common, denominator. To do this we use the following fact.

> The value of a fraction is unaltered if both numerator and denominator are multiplied by the same number.

1

To find a common denominator for, say, $\frac{1}{4}$ and $\frac{3}{10}$, we look for the lowest number that both 4 and 10 divide into exactly. This number is called the *lowest common multiple* (LCM) of 4 and 10 and in this case it is 20.

Then, as $20 = 4 \times 5$ we have $\frac{1}{4} = \frac{1 \times 5}{4 \times 5} = \frac{5}{20}$

and as $20 = 10 \times 2$ we have $\frac{3}{10} = \frac{3 \times 2}{10 \times 2} = \frac{6}{20}$

Therefore $\frac{1}{4} + \frac{3}{10} = \frac{5}{20} + \frac{6}{20} = \frac{11}{20}$

EXERCISE 1a   Find the LCM of the following sets of numbers:

1. 3, 7	**5.** 2, 6, 3	**9.** 2, 3, 7
2. 2, 9	**6.** 5, 4	**10.** 18, 6, 9
3. 2, 8, 10	**7.** 4, 6	**11.** 24, 8, 6
4. 3, 4, 6	**8.** 3, 5, 12	**12.** 12, 36, 8

Find $\frac{5}{11} + \frac{2}{5} + \frac{7}{10}$

(110 is the LCM of 11, 5 and 10)

$$\frac{5}{11} + \frac{2}{5} + \frac{7}{10} = \frac{50 + 44 + 77}{110}$$

$$= \frac{171}{110}$$

$$= 1\frac{61}{110}$$

Find:

13. $\frac{2}{3} + \frac{7}{8}$	**17.** $\frac{3}{16} + \frac{3}{4} + \frac{5}{12}$	**21.** $\frac{3}{4} + \frac{13}{20} + \frac{4}{5}$
14. $\frac{3}{5} + \frac{3}{10}$	**18.** $\frac{1}{4} + \frac{2}{3}$	**22.** $\frac{2}{7} + \frac{1}{9} + \frac{1}{6}$
15. $\frac{3}{5} + \frac{5}{8} + \frac{1}{2}$	**19.** $\frac{1}{3} + \frac{5}{9}$	**23.** $\frac{5}{12} + \frac{3}{8} + \frac{3}{4}$
16. $\frac{2}{7} + \frac{1}{2} + \frac{3}{14}$	**20.** $\frac{5}{6} + \frac{2}{3} + \frac{1}{4}$	**24.** $\frac{8}{21} + \frac{1}{2} + \frac{2}{3}$

Find $\frac{7}{15} - \frac{5}{12}$

(60 is the LCM of 15 and 12)

$$\frac{7}{15} - \frac{5}{12} = \frac{28 - 25}{60}$$

$$= \frac{\cancel{3}^{1}}{\cancel{60}_{20}}$$

$$= \frac{1}{20}$$

25. $\frac{7}{9} - \frac{5}{12}$ **27.** $\frac{3}{10} - \frac{1}{15}$ **29.** $\frac{2}{5} - \frac{3}{8}$

26. $\frac{1}{4} - \frac{2}{9}$ **28.** $\frac{1}{10} - \frac{1}{20}$ **30.** $\frac{5}{6} - \frac{5}{9}$

Find   a) $1\frac{2}{3} + \frac{1}{6} + 2\frac{1}{2}$ b) $2\frac{1}{4} - \frac{3}{5}$

a) $1\frac{2}{3} + \frac{1}{6} + 2\frac{1}{2} = 3 + \frac{2}{3} + \frac{1}{6} + \frac{1}{2}$ (adding whole numbers)

$$= 3 + \frac{4 + 1 + 3}{6}$$

$$= 3 + \frac{\cancel{8}^{4}}{\cancel{6}_{3}}$$

$$= 3 + 1\frac{1}{3}$$

$$= 4\frac{1}{3}$$

b) $2\frac{1}{4} - \frac{3}{5} = 2 + \frac{5 - 12}{20}$

$$= 1 + \frac{20 + 5 - 12}{20}$$ (changing 1 unit into $\frac{20}{20}$)

$$= 1 + \frac{13}{20}$$

$$= 1\frac{13}{20}$$

Find:

31. $2\frac{3}{5} + 1\frac{1}{8}$ **34.** $2\frac{3}{4} + 1\frac{1}{2} - \frac{1}{3}$ **37.** $2\frac{3}{5} + 1\frac{8}{15}$

32. $4\frac{2}{9} - 3\frac{5}{6}$ **35.** $3\frac{1}{8} - 2\frac{3}{4} + 4\frac{1}{2}$ **38.** $1\frac{3}{8} + 1\frac{1}{4} - 2\frac{1}{2}$

33. $2\frac{1}{5} - 4\frac{1}{8} + 1\frac{7}{10}$ **36.** $1\frac{3}{10} - \frac{9}{20}$ **39.** $2\frac{1}{4} - 1\frac{5}{6} + \frac{2}{3}$

MULTIPLICATION OF FRACTIONS

Remember that $\frac{2}{3} \times \frac{1}{5}$ means $\frac{2}{3}$ of $\frac{1}{5}$.

Now $\frac{1}{3}$ of $\frac{1}{5}$ is $\frac{1}{15}$, so $\frac{2}{3}$ of $\frac{1}{5}$ is $\frac{2}{15}$,

i.e.
$$\frac{2}{3} \times \frac{1}{5} = \frac{2 \times 1}{3 \times 5} = \frac{2}{15}$$

> **To multiply fractions, multiply the numerators together and multiply the denominators together.**

Any mixed numbers must be changed into improper fractions, and factors that are common to the numerator and denominator should be cancelled before multiplication.

EXERCISE 1b

Find $2\frac{1}{2} \times \frac{3}{15}$

$$2\frac{1}{2} \times \frac{3}{15} = \frac{5}{2} \times \frac{3}{15}$$

$$= \frac{\overset{1}{\cancel{5}} \times \overset{1}{\cancel{3}}}{2 \times \cancel{15}_{1}}$$

$$= \frac{1}{2}$$

Find:

1. $\frac{2}{3} \times \frac{5}{6}$ **4.** $\frac{2}{3} \times \frac{1}{4} \times \frac{3}{5}$ **7.** $\frac{2}{5} \times 1\frac{3}{7}$

2. $1\frac{1}{2} \times \frac{8}{9}$ **5.** $1\frac{1}{3} \times \frac{1}{2} \times \frac{5}{7}$ **8.** $\frac{3}{5} \times 10$

3. $4 \times \frac{3}{8}$ **6.** $\frac{3}{4} \times \frac{2}{5}$ **9.** $\frac{2}{5} \times \frac{7}{8} \times \frac{10}{11}$

In questions 10 to 15 find the missing numbers:

10. $\frac{3}{4} \times \frac{}{6} = \frac{1}{4}$ **12.** $\frac{2}{} \times \frac{1}{4} = \frac{1}{6}$ **14.** $\frac{3}{4} \times - = 1$

11. $\frac{1}{3} \times \frac{}{2} = \frac{1}{2}$ **13.** $- \times \frac{2}{3} = 1$ **15.** $- \times \frac{7}{8} = 1$

RECIPROCALS

> If the product of two numbers is 1 then each number is called the reciprocal of the other.

We know that $\frac{1}{3} \times 3 = 1$ so

$\frac{1}{3}$ is the reciprocal of 3　　and　　3 is the reciprocal of $\frac{1}{3}$.

To find the reciprocal of $\frac{3}{4}$ we require the number which, when multiplied by $\frac{3}{4}$, gives 1

Now $\frac{4}{3} \times \frac{3}{4} = 1$

so　$\frac{4}{3}$ is the reciprocal of $\frac{3}{4}$

In all cases the reciprocal of a fraction is obtained by turning the fraction upside down.

Any number can be written as a fraction, e.g. $3 = \frac{3}{1}$, $2.5 = \frac{2.5}{1}$, ...

so　　　　the reciprocal of $\frac{3}{1}$ is $\frac{1}{3}$ or $1 \div 3$,

　　　　the reciprocal of $\frac{2.5}{1}$ is $\frac{1}{2.5}$ or $1 \div 2.5$

> The reciprocal of a number is 1 divided by that number.

DIVISION BY A FRACTION

Consider $\frac{2}{5} \div \frac{3}{7}$.

This can be interpreted as $\frac{2}{5} \times 1 \div \frac{3}{7}$.

Now $1 \div \frac{3}{7}$ is the reciprocal of $\frac{3}{7}$, i.e. $\frac{7}{3}$.

Therefore　　　　$\frac{2}{5} \div \frac{3}{7} = \frac{2}{5} \times \frac{7}{3} = \frac{14}{15}$

i.e.,　to divide by a fraction we multiply by its reciprocal.

EXERCISE 1C　Write down the reciprocals of the following numbers:

1. 4	**4.** 10	**7.** 100
2. $\frac{1}{2}$	**5.** $\frac{1}{8}$	**8.** $\frac{2}{9}$
3. $\frac{2}{5}$	**6.** $\frac{3}{11}$	**9.** $\frac{15}{4}$

Find $3\frac{1}{2} \div \frac{7}{8}$

(Before multiplying or dividing, mixed numbers must be changed into improper fractions.)

$$3\frac{1}{2} \div \frac{7}{8} = \frac{7}{2} \div \frac{7}{8}$$

$$= \frac{\cancel{7}^1}{\cancel{2}_1} \times \frac{\cancel{8}^4}{\cancel{7}_1}$$

$$= 4$$

Find:

10. $\frac{2}{3} \div \frac{1}{2}$ **13.** $5 \div \frac{4}{5}$ **16.** $\frac{3}{7} \div 1\frac{3}{4}$

11. $1\frac{2}{3} \div \frac{5}{6}$ **14.** $\frac{2}{9} \div 1\frac{2}{7}$ **17.** $\frac{5}{9} \div 10$

12. $2\frac{1}{2} \div 4$ **15.** $\frac{1}{2} \div \frac{3}{4}$ **18.** $3 \div \frac{2}{3}$

Find $2\frac{1}{2} + \frac{3}{5} \div 1\frac{1}{2} - \frac{1}{2}\left(\frac{3}{5} + \frac{1}{3}\right)$

(Remember that brackets are worked out first, then multiplication and division and lastly addition and subtraction.)

$$2\frac{1}{2} + \frac{3}{5} \div 1\frac{1}{2} - \frac{1}{2}\left(\frac{3}{5} + \frac{1}{3}\right) = 2\frac{1}{2} + \frac{3}{5} \div 1\frac{1}{2} - \frac{1}{2}\left(\frac{9+5}{15}\right)$$

$$= 2\frac{1}{2} + \frac{3}{5} \div \frac{3}{2} - \frac{1}{2} \times \frac{14}{15}$$

$$= 2\frac{1}{2} + \frac{3}{5} \times \frac{2}{\cancel{3}_1} - \frac{1}{\cancel{2}_1} \times \frac{\cancel{14}^7}{15}$$

$$= 2\frac{1}{2} + \frac{2}{5} - \frac{7}{15}$$

$$= 2 + \frac{15 + 12 - 14}{30}$$

$$= 2 + \frac{13}{30}$$

$$= 2\frac{13}{30}$$

19. $1\frac{2}{3} \times \frac{1}{2} - \frac{2}{5}$ **22.** $5\frac{1}{2} \div 3 + \frac{2}{9}$ **25.** $2\frac{1}{2} \div \frac{7}{9} + 1\frac{1}{3}$

20. $\frac{3}{7} + \frac{1}{4} \div 1\frac{1}{3}$ **23.** $\frac{4}{5} \div \frac{1}{6} + \frac{1}{3} \times 1\frac{1}{2}$ **26.** $\frac{3}{5}\left(1\frac{1}{4} - \frac{2}{3}\right)$

21. $\frac{2}{5} \div \left(\frac{1}{2} + \frac{3}{4}\right)$ **24.** $\frac{9}{11} - \frac{2}{5} \times \frac{3}{4}$ **27.** $3\frac{1}{2} - \frac{2}{3} \times 6$

28. $2\frac{1}{3} + \frac{1}{2}\left(2 \div \frac{4}{5}\right)$

30. $\left(\frac{2}{3} - \frac{1}{2}\right) \div \left(\frac{3}{4} - \frac{1}{3}\right)$

29. $\frac{3}{4}\left(5\frac{1}{3} - 2\frac{1}{5}\right) \div \frac{7}{9}$

31. $\frac{7}{9} - \frac{1}{3}$ of $1\frac{2}{7}$

32. $\frac{1}{2} + \left(\frac{3}{4} \div \frac{1}{6}\right)$ of 3

36. $\frac{2}{3} \times \frac{6}{7} - \frac{5}{8} \div 1\frac{1}{4}$

33. $\dfrac{\frac{1}{3} + \frac{1}{4}}{\frac{5}{6} - \frac{3}{4}}$

37. $\dfrac{\frac{7}{8}}{3\frac{1}{2} - \frac{2}{3}}$

34. $\dfrac{1\frac{1}{5} - \frac{3}{4}}{2\frac{1}{2}}$

38. $\dfrac{3\frac{1}{4}}{2\frac{3}{5}}$

35. $\dfrac{\frac{9}{10}}{\frac{5}{6}}$

39. $\dfrac{\frac{2}{3} \times \frac{3}{4}}{\frac{5}{6} \times \frac{3}{10}}$

DECIMAL FRACTIONS

Long after meeting common fractions, we learn that we can represent fractions of an object by placing a point after the units and continuing to add figures to the right. The first figure after the decimal point is the number of tenths, the next figure is the number of hundredths, and so on.

For example, 0.75 is 7 tenths plus 5 hundredths.

Fractions written this way are called *decimal fractions.*

Usually we refer to common fractions simply as fractions and to decimal fractions simply as decimals.

INTERCHANGING DECIMALS AND FRACTIONS

EXERCISE 1d

Express 0.705 as a fraction

$$0.705 = \frac{7}{10} + \frac{5}{1000} \qquad \text{(this step is usually omitted)}$$

$$= \frac{\cancel{705}^{141}}{\cancel{1000}_{200}}$$

$$= \frac{141}{200}$$

Express the following decimals as fractions:

1.	0.35	**5.**	0.03	**9.**	0.11
2.	0.216	**6.**	0.012	**10.**	2.05
3.	0.204	**7.**	0.005	**11.**	1.104
4.	1.36	**8.**	1.01	**12.**	0.0001

Express $\frac{7}{8}$ as a decimal

$$\frac{7}{8} = 0.875$$

$$\begin{array}{r} 0.875 \\ 8\overline{)7.000} \end{array}$$

Express the following fractions as decimals:

13.	$\frac{3}{20}$	**17.**	$\frac{1}{16}$	**21.**	$\frac{4}{25}$
14.	$\frac{1}{8}$	**18.**	$\frac{27}{50}$	**22.**	$\frac{5}{16}$
15.	$\frac{3}{5}$	**19.**	$1\frac{3}{4}$	**23.**	$2\frac{3}{8}$
16.	$\frac{6}{25}$	**20.**	$\frac{5}{32}$	**24.**	$\frac{1}{500}$

RECURRING DECIMALS

If we try to change $\frac{1}{6}$ to a decimal, i.e.

$$\begin{array}{r} 0.1666\ldots \\ 6\overline{)1.0000\ldots} \end{array}$$

we discover that

a) we cannot write $\frac{1}{6}$ as an exact decimal

b) from the second decimal place, the 6 recurs for as long as we have the patience to continue the division.

Similarly if we convert $\frac{2}{11}$ to a decimal by dividing 2 by 11, we get 0.18181818 ... and we see that

a) $\frac{2}{11}$ cannot be expressed as an exact decimal

b) the pair of figures "18" recurs indefinitely.

Decimals like these are called *recurring decimals*. To save time and space we place a dot over the figure that recurs. In the case of a group of figures recurring we place a dot over the first and last figure in the group.

Therefore   we write 0.166666 . . .   as 0.16̇

and   we write 0.181818 . . .   as 0.1̇8̇

Similarly   we write 0.316316316 . . . as 0.3̇16̇

EXERCISE 1e   Use the dot notation to write the following fractions as decimals:

1. $\frac{1}{3}$ **4.** $\frac{1}{15}$ **7.** $\frac{1}{11}$ **10.** $\frac{1}{14}$

2. $\frac{2}{9}$ **5.** $\frac{1}{7}$ **8.** $\frac{1}{18}$ **11.** $\frac{7}{30}$

3. $\frac{5}{6}$ **6.** $\frac{1}{12}$ **9.** $\frac{5}{12}$ **12.** $\frac{1}{13}$

ADDITION AND SUBTRACTION OF DECIMALS

Decimals are added and subtracted in the same way as whole numbers. It is sensible to write them in a column so that the decimal points are in a vertical line. This ensures that units are added to units, tenths are added to tenths, and so on.

MULTIPLICATION OF DECIMALS

To multiply decimals we can first convert them to fractions.

For example
$$0.05 \times 1.04 = \frac{5}{100} \times \frac{104}{100}$$
(2 d.p.)   (2 d.p.)
$$= \frac{520}{10000}$$
$$= 0.0520$$
(4 d.p.)

From such examples we deduce the following rule.

First ignore the decimal points and multiply the numbers together. Then the sum of the number of decimal places in the original numbers gives the number of decimal places in the answer (including any zeros at the end).

DIVISION BY DECIMALS

To divide a decimal by a whole number, proceed as with whole numbers, adding zeros after the point when necessary. For example, to find $3.14 \div 5$ we have

$$\begin{array}{r} 0.628 \\ 5 \overline{)\ 3.140} \end{array}$$

(Make sure that the decimal points are in a vertical line.)

Therefore $3.14 \div 5 = 0.628$

To divide by a decimal we use the fact that the top and bottom of a fraction can be multiplied by the same number without altering the value of the fraction. Division by a decimal can therefore be converted to division by a whole number.

For example $3.14 \div 0.5 = \dfrac{3.14}{0.5}$

$$= \frac{31.4}{5} \qquad \left(\frac{3.14 \times 10}{0.5 \times 10} \right)$$

$$= 6.28$$

EXERCISE 1f   Calculate, without using a calculator:

1. $1.26 + 3.75$	**4.** $0.04 + 8.76$	**7.** $4.002 + 0.83$
2. $12.4 + 6.7$	**5.** $1.8 + 0.02$	**8.** $0.016 + 1.09$
3. $5.82 + 0.35$	**6.** $25 + 1.36$	**9.** $0.00032 + 0.0017$

10. $5.3 - 2.1$	**13.** $1.3 - 0.09$	**16.** $0.37 - 0.009$
11. $8.2 - 4.9$	**14.** $1.07 - 0.58$	**17.** $2 - 0.17$
12. $0.16 - 0.08$	**15.** $24 - 0.98$	**18.** $0.0127 - 0.0059$

19. 1.2×0.8	**22.** 0.1×0.1	**25.** 42×0.08
20. 0.7×0.06	**23.** 0.5×0.5	**26.** 3.0501×1.1
21. 0.4×0.02	**24.** 1.002×0.36	**27.** 0.0012×0.32

28. $2.8 \div 0.4$	**31.** $0.02 \div 2.5$	**34.** $0.1 \div 0.1$
29. $0.36 \div 1.2$	**32.** $0.018 \div 1.2$	**35.** $0.01 \div 0.5$
30. $1.08 \div 0.4$	**33.** $5.31 \div 0.9$	**36.** $0.0013 \div 1.3$

Evaluate $3.7 + 0.2(1.3 - 0.27)$

$$
\begin{aligned}
3.7 + 0.2(1.3 - 0.27) &= 3.7 + 0.2 \times 1.03 \quad &\text{(bracket)}\\
&= 3.7 + 0.206 \quad &\text{(multiplication)}\\
&= 3.906
\end{aligned}
$$

Evaluate:

37. $2.6 - 1.4 \div 0.7$

38. $200 \times 0.04 - 0.2$

39. $(1.2 - 0.8) \div 0.8$

40. $4.3 \times 2.1 \div 0.07$

41. $3.2(0.6 - 0.09) + 10.25$

42. $2.73 \div 0.9 \times 1.02$

43. $(20 \times 0.06) \div (3.1 - 1.9)$

44. $(2.5 + 1.3) \div (2.06 - 0.16)$

45. $\dfrac{2.4 + 0.98}{1.78 + 0.22}$

46. $\dfrac{0.04 \times 1.02}{3.4 \times 0.06}$

47. $\dfrac{1.2 \times 0.05}{0.07 + 0.08}$

48. $\dfrac{4.02 + 12.09}{0.9 \times 2}$

RELATIVE SIZES

Before we can compare the sizes of a set of numbers we must either change them all into decimals or change them all into fractions with the same denominator. Choose whichever method is easier.

EXERCISE 1g

Place $>$ or $<$ between the two numbers $\frac{3}{5}$, 0.67

(Remember that $>$ means "is greater than"
            and $<$ means "is less than".)

$$\frac{3}{5} = 0.6$$

Comparing 0.6 and 0.67 we see that $0.6 < 0.67$

$\therefore \qquad \frac{3}{5} < 0.67$

Place $>$ or $<$ between each of the following pairs of numbers:

1. $\frac{5}{8}$, 0.63

2. 0.16, $\frac{3}{20}$

3. $\frac{9}{11}$, 0.9

4. $\frac{1}{16}$, 0.07

5. $\frac{2}{7}$, 0.16

6. 0.48, $\frac{4}{9}$

7. 0.33, $\frac{7}{25}$

8. $\frac{3}{11}$, 0.25

9. $\frac{2}{3}$, 0.66

Arrange the following numbers in ascending order (i.e. the smallest first): $\frac{1}{2}$, 0.52, $\frac{11}{20}$, 0.51

Writing them all as fractions with denominator 100 gives

$$\frac{1}{2} = \frac{50}{100}, \ 0.52 = \frac{52}{100}, \ \frac{11}{20} = \frac{55}{100}, \ 0.51 = \frac{51}{100}$$

Therefore the required order is $\frac{1}{2}$, 0.51, 0.52, $\frac{11}{20}$.

In each of the following questions arrange the numbers in ascending order of size:

10. $\frac{2}{3}$, 0.6, $\frac{4}{5}$

11. 0.85, $\frac{4}{5}$, 0.79

12. $\frac{2}{7}$, 0.3, $\frac{1}{5}$

13. 0.75, $\frac{5}{7}$, 0.875, $\frac{7}{9}$

14. 0.16, $\frac{3}{20}$, 0.2, $\frac{6}{25}$

15. $1\frac{1}{5}$, 1.3, 1.24, $1\frac{1}{8}$

INDICES

In the number 2^3, the 3 is called the *index* or *power* and 2^3 means $2 \times 2 \times 2$. Thus, indices are a kind of shorthand notation. All other forms of indices are derived from this.

EXERCISE 1h

Find the value of   a) $2^2 \times 3^3$ b) $(3^3)^2$

a) $2^2 \times 3^3 = 4 \times 27$

          $= 108$

b) $(3^3)^2 = (27)^2$

          $= 729$

Find the value of:

1. 5^2	**6.** $2^4 \times 3^2$	**11.** 3.25×10^2
2. 3^4	**7.** $8^2 \times 5^2$	**12.** 8.01×10^3
3. 2^5	**8.** $6^3 \times 2^2$	**13.** 0.072×10^4
4. 5^3	**9.** $4^4 \times 2^3$	**14.** 1.102×10^3
5. 4^3	**10.** $7^3 \times 9^2$	**15.** 1.1×10^6

Remember that we can multiply one number to a power by the *same* number to another power by adding the powers.

> Where possible write as a single number in index form
> a) $5^2 \times 5^4$ b) $3^2 \times 2^3$
>
> a) $5^2 \times 5^4 = 5^6$
>
> $$(5^2 \times 5^4 = 5 \times 5 \times 5 \times 5 \times 5 \times 5)$$
>
> b) $3^2 \times 2^3$ cannot be written as a single number in index form
>
> $$(3^2 \times 2^3 = 3 \times 3 \times 2 \times 2 \times 2)$$

Write the following, where possible, as a single number in index form:

16. $2^3 \times 2^4$	**19.** $5^1 \times 5^3$	**22.** $4^2 \times 4^7$
17. $3^2 \times 3^5$	**20.** 2×2^4	**23.** $a^2 \times a^3$
18. $5^2 \times 3^5$	**21.** $7^2 \times 7^5$	**24.** $a^2 \times b^3$

We can also divide one number to a power by the *same* number to another power by subtracting the powers.

> Where possible write as a single number in index form
> a) $2^5 \div 2^2$ b) $3^3 \div 5^3$
>
> a) $2^5 \div 2^2 = 2^3$
>
> $$\left(2^5 \div 2^2 = \frac{\cancel{2} \times \cancel{2} \times 2 \times 2 \times 2}{\cancel{2} \times \cancel{2}} = 2^3\right)$$
>
> b) $3^3 \div 5^3$ cannot be simplified.
>
> $$\left(3^3 \div 5^3 = \frac{3 \times 3 \times 3}{5 \times 5 \times 5}\right)$$

Write the following, where possible, as a single number in index form:

25. $2^4 \div 2^2$		**29.** $2^3 \div 3^2$		**33.** $a^3 \div b^2$	
26. $7^3 \div 7^2$		**30.** $3^6 \div 3^2$		**34.** $(2^3)^2$	
27. $5^3 \div 3^2$		**31.** $3^4 \div 3$		**35.** $(3^2)^2$	
28. $4^5 \div 4^2$		**32.** $a^8 \div a^4$		**36.** $(5^3)^2$	

ZERO AND NEGATIVE INDICES

Consider $a^3 \div a^3$.

Subtracting indices gives $a^3 \div a^3 = a^0$

Dividing gives $a^3 \div a^3 = 1$

$$a^0 = 1$$

i.e.   (any number)$^0 = 1$

Now consider $a^3 \div a^5$.

Subtracting indices gives $\qquad a^3 \div a^5 = a^{-2}$

Dividing gives $\quad \dfrac{a^3}{a^5} = \dfrac{\cancel{a} \times \cancel{a} \times \cancel{a}}{\cancel{a} \times \cancel{a} \times \cancel{a} \times a \times a} = \dfrac{1}{a^2}$

Therefore a^{-2} means $\dfrac{1}{a^2}$

A negative sign in front of the index means "the reciprocal of"

i.e. $\qquad\qquad\qquad a^{-b} = \dfrac{1}{a^b}$

EXERCISE 1i

Find the value of 3^{-1}

$$3^{-1} = \frac{1}{3^1} = \frac{1}{3}$$

Find the value of:

1. 2^{-1}	**3.** 5^{-1}	**5.** 8^{-1}	**7.** a^{-1}
2. 10^{-1}	**4.** 7^{-1}	**6.** 4^{-1}	**8.** x^{-1}

Find the value of   a) $\left(\frac{1}{2}\right)^{-1}$ b) $\left(\frac{2}{5}\right)^{-1}$

a) $\left(\frac{1}{2}\right)^{-1} = \left(\frac{2}{1}\right)^{1}$

$= 2$

b) $\left(\frac{2}{5}\right)^{-1} = \left(\frac{5}{2}\right)^{1}$

$= 2\frac{1}{2}$

Find the value of:

9. $\left(\frac{1}{3}\right)^{-1}$ **11.** $\left(\frac{1}{4}\right)^{-1}$ **13.** $\left(\frac{1}{5}\right)^{-1}$ **15.** $\left(\frac{1}{a}\right)^{-1}$

10. $\left(\frac{2}{3}\right)^{-1}$ **12.** $\left(\frac{3}{4}\right)^{-1}$ **14.** $\left(\frac{4}{5}\right)^{-1}$ **16.** $\left(\frac{x}{y}\right)^{-1}$

Find the value of 3^{-2}

$$3^{-2} = \frac{1}{3^2}$$

$$= \frac{1}{9}$$

Find the value of:

17. 2^{-3} **19.** 10^{-3} **21.** 2^{-5} **23.** 10^{-2}

18. 5^{-2} **20.** 6^{-2} **22.** 10^{-4} **24.** 4^{-3}

Find the value of $\left(\frac{1}{3}\right)^{-2}$

$$\left(\frac{1}{3}\right)^{-2} = \left(\frac{3}{1}\right)^{2}$$

$$= 9$$

Find the value of:

25. $\left(\frac{1}{5}\right)^{-3}$ **27.** $\left(\frac{1}{2}\right)^{-5}$ **29.** $\left(\frac{1}{8}\right)^{-3}$ **31.** $\left(\frac{1}{2}\right)^{-3}$

26. $\left(\frac{1}{4}\right)^{-2}$ **28.** $\left(\frac{1}{3}\right)^{-4}$ **30.** $\left(\frac{1}{10}\right)^{-4}$ **32.** $\left(\frac{1}{6}\right)^{-2}$

Find the value of $\left(\frac{2}{5}\right)^{-3}$

$$\left(\frac{2}{5}\right)^{-3} = \left(\frac{5}{2}\right)^{3}$$

$$= \frac{125}{8}$$

$$= 15\frac{5}{8}$$

Find the value of:

33. $\left(\frac{3}{4}\right)^{-2}$ **35.** $\left(\frac{4}{9}\right)^{-2}$ **37.** $\left(\frac{2}{3}\right)^{-4}$ **39.** $\left(\frac{3}{10}\right)^{-4}$

34. $\left(\frac{2}{3}\right)^{-3}$ **36.** $\left(\frac{2}{7}\right)^{-2}$ **38.** $\left(\frac{3}{5}\right)^{-2}$ **40.** $\left(\frac{5}{8}\right)^{-2}$

EXERCISE 1j   Find the value of:

1. $\left(\frac{1}{8}\right)^{-1}$ **6.** $\left(\frac{2}{3}\right)^{0}$ **11.** $\left(\frac{3}{4}\right)^{-3}$ **16.** $\left(\frac{4}{5}\right)^{3}$

2. $\left(\frac{2}{5}\right)^{-2}$ **7.** 5^3 **12.** $\left(\frac{2}{7}\right)^{-1}$ **17.** 12^{-1}

3. 4^{-2} **8.** 9^{-1} **13.** 5^0 **18.** 9^3

4. 8^2 **9.** $\left(\frac{1}{2}\right)^{-4}$ **14.** $\left(\frac{7}{10}\right)^{-3}$ **19.** $\left(\frac{1}{4}\right)^{-3}$

5. $\left(\frac{1}{2}\right)^{0}$ **10.** 6^0 **15.** 2^{-2} **20.** $\left(\frac{3}{7}\right)^{0}$

STANDARD FORM (SCIENTIFIC NOTATION)

Very large or very small numbers are more briefly written in standard form, and it is easier to compare sizes.

Standard form is a number between 1 and 10 multiplied by the appropriate power of 10.

EXERCISE 1k   The following numbers are given in standard form. Write them as ordinary numbers:

1. 3.45×10^2 **4.** 4.7×10^{-3} **7.** 9.02×10^5

2. 1.2×10^3 **5.** 2.8×10^2 **8.** 6.37×10^{-4}

3. 5.01×10^{-2} **6.** 7.3×10^{-1} **9.** 8.72×10^6

> Write the following numbers in standard form
> a) 3840   b) 0.0025
>
> (First, write the given figures as a number between 1 and 10 and then decide what power of 10 to multiply by to bring it back to the correct size.)
>
> $$a) \quad 3840 = 3.84 \times 10^3$$
>
> $$b) \quad 0.0025 = 2.5 \times 10^{-3}$$

Write the following numbers in standard form:

10.	265	**13.**	0.019	**16.**	0.000 85
11.	0.18	**14.**	76 700	**17.**	7000
12.	3020	**15.**	390 000	**18.**	0.004
19.	58 700	**22.**	0.000 007	**25.**	24 000
20.	2600	**23.**	0.8	**26.**	39 000 000
21.	450 000	**24.**	0.000 56	**27.**	0.000 000 000 08

If you have a scientific calculator, it will display very large or very small numbers in scientific notation, but only the power of ten is given; 10 itself does not appear in the display.

Try this: enter 0.000 05, then press $\boxed{x^2}$.

The display will read $\boxed{2.5 \quad -09}$. This means 2.5×10^{-9}

Now try entering 500 000 and then pressing $\boxed{x^2}$.

The display will read $\boxed{2.5 \quad 11}$. This means 2.5×10^{11}

28. Use your calculator to find the value of
a) $(250\,000)^2$ b) $(2\,570\,000)^2$ c) $(0\,0.000\,08)^2$ d) $(0.000\,007)^2$

APPROXIMATIONS (DECIMAL PLACES AND SIGNIFICANT FIGURES)

We have seen that it is sometimes unnecessary and often impossible to give exact values. In the case of measurements this is particularly true. However, we do need to know the degree of accuracy of an answer. For example, if a manufacturer is asked to make screws that are about $12\frac{1}{2}$ cm long, he does not know what is acceptable as being "about $12\frac{1}{2}$ cm long"! But if he is asked to make them 12.5 cm long correct to one decimal place, he knows what tolerances to work to.

DECIMAL PLACES

To correct 0.078 22 to *two* decimal places (d.p.) we look at the *third* decimal place. If it is 5 or larger, we add 1 to the figure in the second decimal place. If it is less than 5, we do not alter the figure in the second decimal place.

In this example the figure in the third decimal place is 8,

so $0.07|822 = 0.08$ correct to 2 d.p.

SIGNIFICANT FIGURES

To determine the 1st, 2nd, 3rd, ... significant figure (s.f.) in a number we can change it into standard form. Then the figure to the left of the decimal point is the 1st s.f., the next figure is the 2nd s.f., ... and so on.

For example $0.07025 = 7.025 \times 10^{-2}$

To correct 0.078 22 to two significant figures we look at the third significant figure; in this case it is 2.

Therefore 0.078 22 = 0.078   correct to 2 s.f.

EXERCISE 1I

Give 0.07 025 correct to   a) 3 d.p.   b) 3 s.f.

a) $0.070|25 = 0.070$ correct to 3 d.p.

(We leave the zero at the end of the corrected answer to indicate that the third decimal place is zero.)

b) $0.0702|5 = 7.02|5 \times 10^{-2}$ (this step can be omitted)

       $= 0.0703$ correct to 3 s.f.

Give each of the following numbers correct to
a) three decimal places   b) three significant figures:

1. 2.7846   **3.** 3.2094   **5.** 0.150 76

2. 0.1572   **4.** 0.073 25   **6.** 0.020 39

7. 0.7801	**9.** 254.1627	**11.** 7.8196
8. 3.2994	**10.** 0.000 925 8	**12.** 0.009 638

> Estimate a rough value for 0.826×38.3 by correcting each number to 1 significant figure. Then use your calculator to give the answer correct to 3 significant figures
>
> $$0.826 \times 38.3 \approx 0.8 \times 40 = 32 \approx 30$$
>
> $0.826 \times 38.3 = 31.6$ correct to 3 s.f.

First estimate a rough value for each of the following calculations, then use your calculator to give the answer correct to three significant figures:

13. 0.035×1.098	**15.** 0.0932×0.48	**17.** 0.295×0.732
14. 258×184	**16.** 18.27×3.82	**18.** 1250×532

19. $86.27 \div 39.8$	**22.** $0.0026 \div 0.0378$
20. $204 \div 942$	**23.** $16.97 \div 3.702$
21. $0.827 \div 0.093$	**24.** $3502 \div 651$

25. $(3.827)^3$	**29.** $\dfrac{2.27 \times 3.84}{5.01}$
26. $\dfrac{3.257}{83.6}$	**30.** $\dfrac{0.016 \times 5.82}{2.31}$
27. $0.532 \times 3.621 \times 36$	**31.** $\dfrac{2.93 \times 0.037\,2}{1.84 \times 0.562}$
28. $(0.32)^3$	

RANGE OF VALUES FOR A CORRECTED NUMBER

Suppose we are told that, correct to the nearest 10, 250 people boarded a particular train. People are counted in whole numbers only. Hence in this case, 245 is the lowest number that gives 250 when corrected to the nearest 10 and 254 is the highest number that can be corrected to 250. We can therefore say that the actual number of people who boarded the train is any whole number from 245 to 254.

Now suppose that we are given a nail and are told that its length is 25 mm correct to the nearest millimetre. The length may be a whole number of millimetres, but may also be any part of a millimetre.

Look at this magnified section of a measuring gauge:

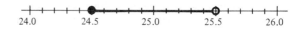

The lowest number that can be rounded up to 25 is 24.5. The highest number that can be rounded down to 25 is not so easy to determine. All we can say is that any number up to, but not including, 25.5 can be rounded down to 25.

The length of the nail is therefore in the range from 24.5 mm up to, but not including, 25.5 mm.

If *l* mm is the length of the nail, we can therefore write

$$24.5 \leqslant l < 25.5$$

To illustrate this on the diagram we use a solid circle to show that 24.5 is included in the range and an open circle to show that 25.5 is not included in the range.

EXERCISE 1m

Illustrate on a number line the range of values of x given by $0.1 < x \leqslant 0.8$

Use a number line like this for questions 1 to 10:

In each case illustrate the range on your number line:

1. $5 \leqslant x \leqslant 10$ **6.** $-5 < x \leqslant 5$

2. $0 < x \leqslant 15$ **7.** $4 \leqslant x \leqslant 12$

3. $-2 \leqslant x \leqslant 6$ **8.** $3 < x \leqslant 18$

4. $5 \leqslant x < 15$ **9.** $-5 < x < 8$

5. $0 < x < 10$ **10.** $-1 < x \leqslant 11$

Use a number line like this for questions 11 to 20. In each case illustrate the range on your number line:

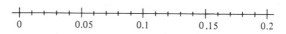

11.	$0 < x < 0.1$	16.	$0.03 < x \leqslant 0.13$
12.	$0.1 \leqslant x < 0.2$	17.	$0.05 \leqslant x \leqslant 0.2$
13.	$0.05 \leqslant x \leqslant 0.15$	18.	$0.07 < x \leqslant 0.14$
14.	$0.08 < x \leqslant 0.16$	19.	$0.10 < x < 0.19$
15.	$0.02 < x < 0.08$	20.	$0.13 \leqslant x < 0.18$

A number is given as 3.15 correct to 2 d.p. Illustrate on a number line the range in which this number lies

(3.15 is between 3.14 and 3.16 so we will use just that part of the number line.)

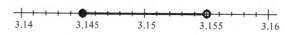

Illustrate on a number line the range of possible values for each of the following corrected numbers:

21.	1.5 correct to 1 d.p.	26.	0.6 correct to 1 d.p.
22.	0.2 correct to 1 d.p.	27.	1.3 correct to 1 d.p.
23.	0.1 correct to 1 d.p.	28.	6.2 correct to 1 d.p.
24.	4.8 correct to 1 d.p.	29.	8.0 correct to 1 d.p.
25.	2.0 correct to 1 d.p.	30.	12.9 correct to 1 d.p.

31.	0.25 correct to 2 d.p.	36.	0.52 correct to 2 d.p.
32.	1.15 correct to 2 d.p.	37.	6.89 correct to 2 d.p.
33.	12.26 correct to 2 d.p.	38.	26.35 correct to 2 d.p.
34.	0.05 correct to 2 d.p.	39.	8.50 correct to 2 d.p.
35.	3.10 correct to 2 d.p.	40.	0.70 correct to 2 d.p.

PROBLEMS

EXERCISE 1n

It is stated that a packet of pins contains 500 pins to the nearest 10. Find the range in which the actual number of pins lies

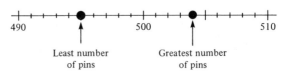

(There must be a whole number of pins in the packet.) If n is the number of pins in the packet then n is a whole number such that $495 \leqslant n \leqslant 504$.

A copper tube is sold as having an internal bore (diameter) of 10 mm to the nearest millimetre. Find the range in which the actual bore lies

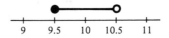

If d mm is the diameter of the tube then $9.5 \leqslant d < 10.5$

In some of the following questions you are asked to find a range of values for a quantity that can only have whole number values. When this is the case, it *must* be clearly stated in your answer:

1. The weight, w kg, of a bag of sand is given as 5.6 kg correct to 1 decimal place. Find the range of values in which the actual weight lies.

2. A shop is said to make a profit of £2500 a month. If this figure is given correct to the nearest £100, find the range in which the actual monthly figure, £x, lies.

3. Correct to one decimal place, the length of a room is given as 2.8 m. Find the range in which the actual length, x m, of the room lies.

4. On a certain model of bicycle, the brake pads have to be 12.5 mm thick to work efficiently. Find the range in which the thickness, x mm, can lie, if the figure given is correct to 1 decimal place.

5. Referring to a football match, a newspaper headline proclaimed "75 000 watch England win". If this figure for the number of spectators is correct to the nearest 1000, find the range in which *x*, the number of people who actually attended the match, lies.

6. One of the component parts of a metal hinge is a pin. In order to work properly this pin must have a diameter of 1.25 mm correct to 2 decimal places. Find the range in which the diameter, *d* mm, must lie.

7. Alan measured the width of a space between two kitchen units in metres correct to 1 decimal place, and wrote down 1.6 m.
a) Find the range within which the width of the space lies.
b) The width of the new cupboard to go into the space is 1.62 m correct to 2 decimal places. Will it fit the space?
c) Comment on Alan's measuring.

8. One hundred people were asked if they ate cornflakes for breakfast that morning. To the nearest 10, 40 people said they did. Of the 100 people interviewed, what is the largest possible number who did *not* eat cornflakes?

9. An advertisement says "For about £50 you can buy this automatic watch." If this price is correct to the nearest £10, what is the most that you would expect to have to pay for such a watch?

10. The length of a car is given as 255 cm. If this figure is correct to the nearest 5 cm, find the range in which the actual length lies.

11. Knitting yarn is sold by weight. It is found that 10 g of double knitting pure wool has a length of 20 m correct to the nearest metre. What is the minimum length of yarn you would expect in a 50 g ball of double knitting pure wool?

12. The weight of a wall tile is given as 40 g to the nearest gram. A DIY store sell these tiles in polythene wrapped packs of ten. Ignoring the weight of the wrapping, find the range in which the weight of one of these packs lies.

13. The length of a side of a square carpet tile is 29.9 cm correct to 1 decimal place. One hundred of these carpet tiles are laid end to end on the floor of a shop. Find the range in which the length of this line of tiles lies. Hence give the difference between the length of the longest possible and the shortest possible line of 100 tiles.

MIXED EXERCISES

EXERCISE 1p   **1.** Find the LCM of   a) 2, 5 and 6   b) 3, 7 and 14.

2. Find the reciprocal of   a) $\frac{3}{4}$ b) $\frac{x}{y}$

3. Calculate   a) $1\frac{1}{3} \div \frac{8}{9}$ b) $1\frac{1}{3} - \frac{8}{9}$

4. Calculate $(3\frac{1}{4} + 2\frac{1}{2}) \times \frac{2}{5}$

5. Find, without using a calculator:
a) $3.27 + 0.09$ b) 3.27×0.09 c) $3.27 \div 0.03$

6. Find the value of   a) 2^4 b) 2^0 c) 2^{-4}

7. Write as a single number in index form
a) $5^6 \div 5^4$ b) $5^{10} \times 5^2$

8. Write the following numbers in standard form
a) 2560   b) 0.000 256

9. The diameter of a ball bearing is given as 1.5 mm correct to 1 decimal place. Find the range in which the diameter lies.

10. To the nearest 10, 70 children brought a packed lunch to school. Find the range in which the actual number of children lies.

EXERCISE 1q   **1.** Find the LCM of   a) 3, 8 and 24   b) 2, 3 and 5

2. Find the reciprocal of   a) $\frac{1}{5}$ b) $1\frac{1}{2}$

3. Calculate   a) $\frac{3}{5} \times 1\frac{1}{4}$ b) $\frac{3}{5} + 1\frac{1}{4}$

4. Calculate $2\frac{1}{2} \times \frac{7}{10} + 1\frac{1}{3}$

5. Find, without using a calculator:
a) $2.5 - 1.05$ b) 2.5×1.05 c) $1.05 \div 2.5$.

6. Find the value of   a) $\left(\frac{1}{2}\right)^2$ b) $\left(\frac{1}{2}\right)^0$ c) $\left(\frac{1}{2}\right)^{-2}$

7. Write in standard form   a) 570 000   b) 0.057

8. A box contains 450 tacks to the nearest 10. Find the range in which the actual number of tacks lies.

9. To fit properly, the diameter of a bicycle tyre has to be 0.75 m correct to 2 decimal places. Find the range in which the diameter can lie.

10. I buy 10 balls of wool, each one of which weighs 50 g to the nearest gram. What is the range in which the weight of the 10 balls of wool lies?

2 EQUATIONS, INEQUALITIES AND FORMULAE

MORE NAMES FOR NUMBERS

The natural numbers, 1, 2, 3, ... and numbers such as $3\frac{2}{5}$, 89.701, ... are not capable of describing every situation. For example, to describe temperatures above and below $0\,°C$ (the freezing point of water) we need *positive numbers* and *negative numbers*.

Positive and negative numbers are together known as *directed numbers*.

Directed numbers are not necessarily whole numbers. For example a depth of 1.2 km below sea level could be described by the directed number -1.2 km.

USING DIRECTED NUMBERS

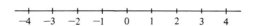

On the number line we can interpret positive numbers as positions to the right of zero and negative numbers as positions to the left of zero.

If a number a is to the left of a number b then $a < b$,

e.g. $2 < 4$ and $-4 < -2$

Conversely if c is to the right of d then $c > d$,

e.g. $3 > 1$ and $-1 > -3$

Adding or subtracting directed numbers can be interpreted as adding or taking away steps to the left or right: e.g. $-(-2)$ could mean "take away two steps to the left". To do this you have to add 2 steps to the right, i.e. $-(-2) = 2$.

Multiplying directed numbers can be interpreted in a similar way, i.e. $(-3)(-2)$ could mean "take away 3 lots of 2 steps to the left". To do this you have to move 6 steps to the right, i.e. $(-3)(-2) = 6$.

Division by a positive number can be interpreted in terms of fractions: e.g. $(-6) \div 2$ means $\frac{1}{2}$ of -6,

i.e. $(-6) \div 2 = -3$

26

Division by a negative number can be converted to division by a positive number by multiplying top and bottom by -1:

e.g. $$(-6) \div (-2) = \frac{-6}{-2} = \frac{(-6) \times (-1)}{(-2) \times (-1)} = \frac{6}{2} = 3$$

For all possible combinations of signs for multiplication and division,

> if the signs are the same, the result is positive;
> if the signs are different, the result is negative.

(Remember that it is possible to divide by any number *except zero*.)

EXERCISE 2a Insert $<$ or $>$ between each of the following pairs of numbers:

1. $3, -1$ **3.** $2, 7$ **5.** $-5, -1$

2. $-4, 6$ **4.** $-2, -7$ **6.** $-5, 3$

Find $32 - 46 + 3$

$$32 - 46 + 3 = 35 - 46$$
$$= -11$$

Find:

7. $8 - 10$ **9.** $-2 + 5$ **11.** $5 - 2 - 7$

8. $-5 - 6$ **10.** $4 - 6 + 3$ **12.** $-3 + 5 - 2$

Find $12 \div (-4)$

$$12 \div (-4) = \frac{12}{-4}$$
$$= -3$$

Find:

13. $(-2) \times (4)$ **16.** $(-6) \times (-2)$ **19.** $(-12) \div 6$

14. $(6) \div (-3)$ **17.** $2 \times (-4)$ **20.** 6×8

15. $8 \div 2$ **18.** $(-8) \div (-4)$ **21.** $(-5) \times (-7)$

SIMPLIFYING ALGEBRAIC EXPRESSIONS

If an algebraic expression contains two terms each with the same letter then we can usually add them.

For example $x + x$ can be written as $2x$

and $2x + 3x$ can be written as $5x$

$$(2x + 3x = x + x + x + x + x)$$

However, if the two terms contain different letters then, as we do not know what numbers the letters stand for, we cannot add them.

For example $x + y$ cannot be simplified, neither can $x - x^2$.

EXERCISE 2b

Simplify where possible   a) $x - (-y)$ b) $x + 2y - 3x$

a) $x - (-y) = x + y$

b) $x + 2y - 3x = 2y - 2x$
(We usually write the positive term first.)

Simplify where possible:

1. $p + q$

2. $a + a$

3. $2s + 3t$

4. $3v + 4v$

5. $5x - 3x$

6. $p - (-q)$

7. $x + 3x - 2y$

8. $3u - (-2u)$

9. $a + 3b - 2a$

10. $c + 2d - (-3c)$

When terms are multiplied together we can write them in a slightly shorter form by omitting the multiplication sign.

For example
$a \times b$ can be written as ab
$p \times p$ can be written as p^2 (using index notation)
$5a \times 2b$ can be written as $10ab$ (multiplication can be done in any order so 5 can be multiplied by 2)

But we cannot simplify ab or $\dfrac{x}{y}$.

EXERCISE 2c

> Simplify   a) $4x \div (-2y)$ b) $2p \times q$
>
> a) $4x \div (-2y) = \dfrac{\overset{2}{\cancel{4}}x}{\underset{1}{\cancel{-2}}y}$ b) $2p \times q = 2pq$
>
> $\qquad\qquad\quad = -\dfrac{2x}{y}$

Simplify where possible:

1.	$x \times y$	**5.**	$u \div v$
2.	$a \times a$	**6.**	$a \div (-b)$
3.	$2s \times 3s$	**7.**	$a \div a$
4.	$4x \times 3x$	**8.**	$6b \div 2c$
9.	$s + t$	**13.**	$a - 3a$
10.	$m - n$	**14.**	$p \times 4p^2$
11.	$m \times (-n)$	**15.**	$8u \div 4w$
12.	$3s - 2t$	**16.**	$2z - 3y$
17.	$3s \times 2t$	**21.**	$8p \div 2q$
18.	$p \times 2p$	**22.**	$(-3s) \times (-2t)$
19.	$q - 5q$	**23.**	$(b) \times (-2b)$
20.	$r - (-4s)$	**24.**	$(-x) \div (-y)$

> Simplify $a(a - b) - (b - a)$
>
> (Remember that $-(b - a)$ means minus b *and* minus $(-a)$)
> $$a(a - b) - (b - a) = a^2 - ab - b - (-a)$$
> $$= a^2 - ab - b + a$$

> Simplify $3(a - b) - 2(a + b)$
>
> $$3(a - b) - 2(a + b) = 3a - 3b - 2a - 2b$$
> $$= a - 5b$$

Simplify:

25. $a - 3(a - b)$ **29.** $y - 2(y - z)$

26. $2a + a(a - 3)$ **30.** $4(x + y) + 2(x + z)$

27. $2(a - b) - (b - a)$ **31.** $3(p + q) - 2(p + r)$

28. $4(a - c) + 2(a - b)$ **32.** $2x - (x + y)$

33. $2(p + q) - 3(p - q)$ **37.** $4(p - q) + 2(q - p)$

34. $a(a + b) - 2(a - b)$ **38.** $w(w + x) - x(w - x)$

35. $x(x - y) + y(y - x)$ **39.** $3(m + n) - 5(m - n)$

36. $4(b - c) - 2(b + c)$ **40.** $5(b - c) - 3(b + c)$

SOLVING EQUATIONS

Consider the statement

$$3x + 6 = 2 - x$$

This is an equation. It means that the left-hand side is *equal* to (i.e. has the same value as) the right-hand side.

We can think of the two sides of an equation as the contents of the two pans on a pair of scales which are exactly balanced. The equality will remain true provided that we always do the same thing to both sides.

Solving an equation means finding the number that the letter stands for so that the two sides *are* equal.

When solving equations like these it is sensible to proceed in the following order:

1.   remove any brackets

2.   collect any like terms on each side

3.   collect the letter terms on one side (choose the side with the greater number to start with remembering that, say, $-2x > -3x$)

4.   collect the number terms on the other side.

EXERCISE 2d

Solve the equation $3x + 6 = 2 - x$

$$3x + 6 = 2 - x$$

Add x to each side $\quad 4x + 6 = 2$

Take 6 from each side $\quad 4x = -4$

Divide each side by 4 $\quad x = -1$

Check: when $x = -1$, $\quad$ LHS $= 3(-1) + 6 = -3 + 6 = 3$

$\quad$ RHS $= 2 - (-1) = 2 + 1 = 3$

$\therefore \quad\quad\quad\quad\quad\quad x = -1$

Solve the following equations:

1. $2p + 5 = 3 - p$

2. $3 - s = 4 - 3s$

3. $x + 5 = 3x - 2$

4. $2a + 3 = 4 - 3a$

5. $3 - 2x = 8x - 7$

6. $3y + 2 = 7 - 2y$

7. $4 - x = 8 - 3x$

8. $a + 7 = 4a - 5$

9. $2x + 5 = 7 - 2x$

10. $3 - 2x = 9 - 5x$

Solve the equation $3 - 2(4 + x) = x - 6$

$$3 - 2(4 + x) = x - 6$$

Expand bracket $\quad 3 - 8 - 2x = x - 6$

Collect like terms $\quad -5 - 2x = x - 6$

Add $2x$ to both sides $\quad -5 = 3x - 6$

Add 6 to both sides $\quad 1 = 3x$

Divide both sides by 3 $\quad \frac{1}{3} = x \quad$ or $\quad x = \frac{1}{3}$

Solve:

11. $3 - 2(x - 2) = 8$

12. $x - 4(x + 3) = 3$

13. $4x = 2 - 3(x + 1)$

14. $2y = 8 - (y - 2)$

15. $3(2 - x) = 4(x - 3)$

16. $5 - 3(x - 4) = 6$

17. $a - 2(a + 3) = 5$

18. $3p - 2 = 4 - 3(p + 2)$

19. $4w = w - (w - 8)$

20. $7(5 - x) = 3(x - 5)$

21. $5(x-2) - 3(x+1) = 0$ **26.** $7(b-4) - 5(b+2) = 0$

22. $3 - 5(2x+1) = 2x$ **27.** $12 - 3(4x-1) = 5$

23. $3(2x+1) - 2(4x-1) = 0$ **28.** $5(3x-1) - 4(3x-2) = 0$

24. $4 - 2(3-2x) = 5$ **29.** $8 - 5(2-x) = 8$

25. $3x - 4(1-3x) = 2x - (x+1)$ **30.** $4x - 2(1-3x) = 3 - (2x-1)$

INEQUALITIES

Consider the statement

$$x > 5$$

This is an *inequality* (as opposed to $x = 5$ which is an equality or equation).

This inequality is true when x stands for any number that is greater than 5. Thus there is a range of numbers that x can stand for and we can illustrate this range on a number line.

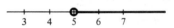

The circle at the left hand end of the range is "open", because 5 is not included in the range.

EXERCISE 2e

Use a number line to illustrate the range of values of x for which $x < -1$

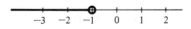

Use a number line to illustrate the range of values of x for which each of the following inequalities is true:

1. $x > 7$ **4.** $x > 0$ **7.** $x < 5$

2. $x < 4$ **5.** $x < -2$ **8.** $x < 0$

3. $x > -2$ **6.** $x > \frac{1}{2}$ **9.** $x < 1.5$

10. State which of the inequalities given in questions 1 to 9 are satisfied by   a) 2   b) −3   c) 0   d) 1.5   e) 0.0005

11. For each of the questions 1 to 9 give a number that satisfies the inequality and is
a) a whole number   b) not a whole number

12. Consider the true inequality $3 > 1$
  a) Add 2 to each side.
  b) Add -2 to each side.
  c) Take 5 from each side.
  d) Take -4 from each side.
  In each case state whether or not the inequality remains true.

13. Repeat question 12 with the inequality $-2 > -3$

14. Repeat question 12 with the inequality $-1 < 4$

15. Try adding and subtracting different numbers on both sides of a true inequality of your own choice.

SOLVING INEQUALITIES

From the last exercise we can see that

> an inequality remains true when the *same* number is added to, or subtracted from, *both* sides.

Now consider the inequality $x - 2 < 3$
Solving this inequality means finding the range of values of x for which it is true.
Adding 2 to each side gives $x < 5$

We have now solved the inequality.

EXERCISE 2f Solve the following inequalities and illustrate your solutions on a number line:

1. $x - 4 < 8$	**4.** $x - 3 > -1$	**7.** $x - 3 < -6$
2. $x + 2 < 4$	**5.** $x + 4 < 2$	**8.** $x + 7 < 0$
3. $x - 2 > 3$	**6.** $x - 5 < -2$	**9.** $x + 2 < -3$

Solve the inequality $4 - x < 3$

$$4 - x < 3$$
Add x to each side $4 < 3 + x$
Take 3 from each side $1 < x$ or $x > 1$

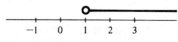

Solve the following inequalities and illustrate your solutions on a number line:

10. $4 - x > 6$ **13.** $5 < x + 5$ **16.** $3 - x > 2$

11. $2 < 3 + x$ **14.** $5 - x < 8$ **17.** $6 < x + 8$

12. $7 - x > 4$ **15.** $2 > 5 + x$ **18.** $2 + x < -3$

19. $2 > x - 3$ **22.** $3 - x < 3$ **25.** $3 > -x$

20. $4 < 5 - x$ **23.** $5 < x - 2$ **26.** $4 - x > -9$

21. $1 < -x$ **24.** $7 > 2 - x$ **27.** $5 - x < -7$

28. Consider the true inequality $12 < 36$
   a) Multiply each side by 2
   b) Divide each side by 4
   c) Multiply each side by 0.5
   d) Divide each side by 6
   e) Multiply each side by -2
   f) Divide each side by -3

   In each case state whether or not the inequality remains true.

29. Repeat question 28 with the true inequality $36 > -12$

30. Repeat question 28 with the true inequality $-18 < -6$

31. Repeat question 28 with a true inequality of your own choice.

32. Can you multiply both sides of an inequality by any one number and be confident that the inequality remains true?

an inequality remains true when both sides are multiplied or divided by the same *positive* number.

Multiplication or division of an inequality by a negative number must be avoided, because it destroys the truth of the inequality.

EXERCISE 2g

> Solve the inequality $2x - 4 > 5$ and illustrate the solution on a number line.
>
> $$2x - 4 > 5$$
> Add 4 to both sides $2x > 9$
> Divide both sides by 2 $x > 4\frac{1}{2}$
>
>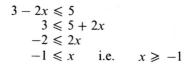

Solve the inequalities and illustrate the solutions on a number line:

1. $3x - 2 < 7$ **5.** $5 + 2x < 6$

2. $1 + 2x > 3$ **6.** $3x + 1 > 5$

3. $4x - 1 > 7$ **7.** $4x - 5 < 4$

4. $3 + 5x < 8$ **8.** $6x + 2 > 11$

> Solve the inequality $3 - 2x \leqslant 5$ and illustrate the solution on a number line. ($\leqslant$ means "less than or equal to")
>
> (As with equations, we collect the letter term on the side with the greater number to start with. In this case we collect on the right, as $-2 < 0$)
>
> $$3 - 2x \leqslant 5$$
> Add $2x$ to each side $3 \leqslant 5 + 2x$
> Take 5 from each side $-2 \leqslant 2x$
> Divide each side by 2 $-1 \leqslant x$ i.e. $x \geqslant -1$
>
>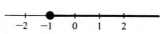
>
> (A solid circle is used for the end of the range because -1 *is* included.)

Solve the inequalities and illustrate the solution on a number line:

9. $3 \leqslant 5 - 2x$ **12.** $4 \geqslant 9 - 5x$

10. $5 \geqslant 2x - 3$ **13.** $10 < 3 - 7x$

11. $4 - 3x \leqslant 10$ **14.** $8 - 3x \geqslant 2$

15. $x - 1 > 2 - 2x$ **18.** $2x + 1 \leqslant 7 - 4x$

16. $2x + 1 \geqslant 5 - x$ **19.** $1 - x > 2x - 2$

17. $3x + 2 \leqslant 5x + 2$ **20.** $2x - 5 > 3x - 2$

Find, where possible, the range of values of x which satisfy both of the inequalities   a) $x \geqslant 2$ and $x > -1$ b) $x \leqslant 2$ and $x > -1$ c) $x \geqslant 2$ and $x < -1$

a)

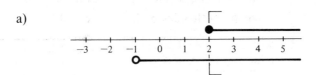

(Illustrating the ranges on a number line, we can see that both inequalities are satisfied for values on the number line where the ranges overlap.)

$\therefore$ $x \geqslant 2$ and $x > -1$ are both satisfied for $x \geqslant 2$

b)

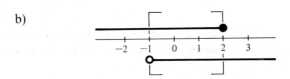

$x \leqslant 2$ and $x > -1$ are both satisfied for $-1 < x \leqslant 2$

c)

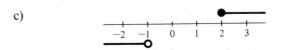

There are no values of x for which $x \geqslant 2$ and $x < -1$ (The lines do not overlap.)

Find, where possible, the range of values of x for which the two inequalities are both true:

21. a) $x > 2$ and $x > 3$

b) $x \geqslant 2$ and $x \leqslant 3$

c) $x < 2$ and $x > 3$

22. a) $x \geqslant 0$ and $x \leqslant 1$

b) $x \leqslant 0$ and $x \leqslant 1$

c) $x < 0$ and $x > 1$

23. a) $x \leqslant 4$ and $x > -2$

b) $x \geqslant 4$ and $x < -2$

c) $x \leqslant 4$ and $x < -2$

24. a) $x < -1$ and $x > -3$

b) $x < -1$ and $x < -3$

c) $x > -1$ and $x < -3$

Solve each of the following pairs of inequalities and then find the range of values of x which satisfy both of them:

25. $x - 4 < 8$ and $x + 3 > 2$

26. $3 + x \leqslant 2$ and $4 - x \leqslant 1$

27. $x - 3 \leqslant 4$ and $x + 5 \geqslant 3$

28. $2x + 1 > 3$ and $3x - 4 < 2$

29. $5x - 6 > 4$ and $3x - 2 < 7$

30. $3 - x > 1$ and $2 + x > 1$

31. $1 - 2x \leqslant 3$ and $3 + 4x < 11$

32. $0 > 1 - 2x$ and $2x - 5 \leqslant 1$

Find the values of x for which $x - 2 < 2x + 1 < 3$

($x - 2 < 2x + 1 < 3$ represents two inequalities,
i.e. $x - 2 < 2x + 1 < 3$)

$$x - 2 < 2x + 1 \qquad\qquad 2x + 1 < 3$$
$$-2 < x + 1 \qquad\qquad\quad 2x < 2$$
$$-3 < x \ \text{ i.e. } \ x > -3 \qquad\quad x < 1$$

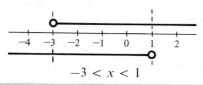

$$-3 < x < 1$$

Find the range of values for which the following inequalities are true:

33. $x + 4 > 2x - 1 > 3$

34. $x - 3 \leqslant 2x \leqslant 4$

35. $3x + 1 < x + 4 < 2$

36. $2 - x < 3x + 2 < 8$

37. $2 - 3x \leqslant 4 - x \leqslant 3$

38. $x - 3 < 2x + 1 < 5$

39. $2x < x - 3 < 4$

40. $4x - 1 < x - 4 < 2$

41. $4 - 3x < 2x - 5 < 1$

42. $x < 3x - 1 < x + 1$

CONSTRUCTING FORMULAE

Electricity bills are presented every quarter. They are made up of a fixed standing charge plus the cost of the number of units used in the quarter.

By using letters for the unknown quantities, we can construct a formula for a quarterly electricity bill.

If £C is the total bill, £R is the standing charge, the cost of one unit is x pence and N units are used, then

$$\text{the cost of the units is } Nx \text{ pence or } £\frac{Nx}{100}$$

$$\text{therefore} \qquad C = R + \frac{Nx}{100}$$

Notice that the units are the same throughout, i.e. the cost of the units was converted from pence to pounds so that we added pounds to pounds, not pounds to pence.

EXERCISE 2h

A number p is equal to the sum of a number q and twice a number r. Write down a formula for p in terms of q and r.

$$p = q + 2r$$

In questions 1 to 12 write down a formula connecting the given letters:

1. A number a is equal to the sum of two numbers b and c.

2. A number m is equal to twice the sum of two numbers n and p.

3. A number z is equal to the product of two numbers x and y.

4. A number a is equal to twice the product of two numbers b and c.

5. A number v is equal to the square of a number n.

6. A number d is equal to the difference of two numbers e and f, where e is greater than f.

7. A number x is equal to half a number y.

8. A number a is equal to a number b divided by twice a number c.

9. A number k is equal to the sum of twice a number u and three times a number v.

10. A number x is equal to twice a number y, minus a number z.

11. A number n is equal to the sum of a number p and its square.

12. A number v is equal to the sum of a number u and the product of the numbers a and t.

13. Cloth is sold at £p per metre. The cost of N metres is £R. Find a formula for R in terms of N and p.

14. Oranges are sold at x pence each. The cost of n oranges is y pence. Find a formula for y in terms of n and x.

15. A ship moving with constant speed takes x minutes to cover one nautical mile. It takes X minutes to cover y nautical miles. Find a formula for X in terms of x and y.

16. A shop sells two brands of baked beans. It has N tins of baked beans altogether; y of them are one brand and z of them are the other brand. Find a formula for N in terms of y and z.

17. A rectangle is x centimetres wide and y centimetres long and its perimeter is P metres. Find a formula for P in terms of x and y.

18. Fertiliser is applied at the rate of a grams per square metre. It takes b kilograms to cover a field of area c square metres. Find a formula for b in terms of a and c.

19. A money box contains l one-penny coins and m two-pence coins. The total value of money in the money box is n pence. Find a formula for n in terms of l and m.

20. A bag of coins contains x ten-pence coins and y twenty-pence coins. The total value of the coins is £R. Find a formula for R in terms of x and y.

SUBSTITUTING NUMBERS INTO FORMULAE

EXERCISE 2i

Given that $s = ut - \frac{1}{2}gt^2$ find s when $u = 8$, $t = 6$ and $g = -10$

$$s = ut - \frac{1}{2}gt^2$$

When $u = 8$, $t = 6$ and $g = -10$

$$s = (8)(6) - (\tfrac{1}{2})(-10)(6)^2$$

$$= 48 - (-5)(36)$$

$$= 48 - (-180)$$

$$= 48 + 180$$

$$= 228$$

(Notice that we have put each number in brackets; this is particularly important in the case of negative numbers.)

1. Given that $p = r - nt$, find p when $r = 40$, $n = 8$ and $t = 4$

2. Given that $v = \dfrac{u - t}{3}$, find v when $u = 4$ and $t = -2$

3. Given that $z = \dfrac{1}{x} + \dfrac{1}{y}$, find z when $x = 2$ and $y = 4$

4. Given that $a = bc + d$, find a when $b = 3$, $c = 2$ and $d = -4$

5. If $x = y^2$, find x when $y = 5$

6. If $C = rt$, find C when $r = -3$ and $t = -10$

7. If $x = rt - v$, find x when $r = 2$, $t = 10$ and $v = -4$

8. If $p = x + x^2$, find p when $x = 2$

9. Given that $s = \frac{1}{2}(a + b + c)$, find s when $a = 6$, $b = 9$ and $c = 5$

10. Given that $v = ab + bc + cd$, find v when $a = 2$, $b = \frac{1}{2}$, $c = 4$ and $d = -2$

11. If $p = r(2t - s)$, find p when $r = \frac{1}{2}$, $t = 3$ and $s = -2$

12. If $a = (b + c)^2$, find a when $b = 8$ and $c = -5$

13. If $r = \dfrac{2}{s+t}$, find r when $s = \frac{1}{2}$ and $t = \frac{1}{4}$

14. If $n = \dfrac{r}{p-q}$, find n when $r = 4$, $p = 3$ and $q = -5$

15. Given that $a = bc - \frac{1}{2}dc$, find a when $b = 3$, $c = -4$ and $d = 7$

16. Given that $V = \frac{1}{2}(X - Y)^2$, find V when $X = 3$ and $Y = -5$

17. Given that $P = 2Q + 5RT$, find P when $Q = 8$, $R = -2$ and $T = -\frac{1}{2}$

18. Given that $a = (b - c)(c - d)$, find a when $b = 2$, $c = 4$ and $d = 7$

CHANGING THE SUBJECT OF A FORMULA

Consider the formula
$$A = B + C$$

Taking C from each side gives
$$A - C = B$$
i.e.
$$B = A - C$$

We have changed the subject of this formula from A to B. Changing the subject of a formula is very similar to solving equations.

EXERCISE 2j Make the letter in brackets the subject of the formula:

1. $p = s + r$ (s) **6.** $k = l + m$ (m)

2. $x = 3 + y$ (y) **7.** $u = v - 5$ (v)

3. $a = b - c$ (b) **8.** $z = x + y$ (y)

4. $X = Y - Z$ (Y) **9.** $N = P - Q$ (P)

5. $r = s + 2t$ (s) **10.** $v = u + 10t$ (u)

11. $x = 2y$ (y) **16.** $a = 3b$ (b)

12. $v = \frac{1}{2}t$ (t) **17.** $X = \frac{1}{10}N$ (N)

13. $a = bc$ (b) **18.** $v = ut$ (u)

14. $t = \dfrac{u}{3}$ (u) **19.** $z = \dfrac{w}{100}$ (w)

15. $l = \dfrac{m}{k}$ (m) **20.** $n = \dfrac{p}{q}$ (p)

Make t the subject of the formula $v = u + 2t$

$$v = u + 2t$$

Take u from each side $\quad v - u = 2t$

$\therefore \qquad\qquad 2t = v - u$

Divide each side by 2 $\qquad t = \dfrac{(v - u)}{2}$

Make the letter in the bracket the subject of the formula:

21. $p = 2s + r$ $\quad (s)$ **25.** $x = 2w - y$ $\quad (w)$

22. $v = u - 3t$ $\quad (t)$ **26.** $l = k + 4t$ $\quad (t)$

23. $a = b - 4c$ $\quad (c)$ **27.** $w = x - 6y$ $\quad (y)$

24. $V = 2v + 3u$ $\quad (v)$ **28.** $N = It - 2s$ $\quad (s)$

29. $x = \dfrac{3y}{4}$ $\quad (y)$ **33.** $V = \dfrac{2R}{I}$ $\quad (R)$

30. $u = v + 5t$ $\quad (t)$ **34.** $p = 2r - w$ $\quad (r)$

31. $A = P + \frac{1}{10}I$ $\quad (I)$ **35.** $a = b + \frac{1}{2}c$ $\quad (c)$

32. $z = x - \dfrac{y}{3}$ $\quad (y)$ **36.** $p = q - \dfrac{r}{5}$ $\quad (r)$

37. Make u the subject of the formula $v = u + at$
Find u when $v = 80$, $a = -10$ and $t = 6$

38. Make B the subject of the formula $A = \dfrac{C}{100} + B$
Find B when $A = 20$ and $C = 250$

39. Make C the subject of the formula $P = \dfrac{C}{N}$
Find C when $N = 20$ and $P = \frac{1}{2}$

40. Make x the subject of the formula $z = \frac{1}{2}x - 3t$
Find x when $z = 4$ and $t = -3$

41. A number a is equal to the sum of a number b and twice a number c
   a) Find a formula for a in terms of b and c
   b) Find a when $b = 8$ and $c = -2$
   c) Make b the subject of the formula.

42. A number x is equal to the product of a number z and twice a number y
   a) Find a formula for x in terms of z and y
   b) Find x when $z = 3$ and $y = 2$
   c) Make y the subject of the formula.

43. A number d is equal to the square of a number e plus twice a number f
   a) Find a formula for d in terms of e and f
   b) Make f the subject of the formula
   c) Find f when $d = 10$ and $e = 3$

44. A retailer sells bags of crisps at x pence each. The cost of n dozen bags of crisps is £R
   a) Find a formula for R in terms of x and n
   b) Find R when $x = 10$ and $n = 4$

DIMENSIONS OF FORMULAE

We have encountered various formulae for finding lengths, areas and volumes. The letters that appear in these formulae represent numbers of length, area and volume units. A formula may also contain numbers and symbols representing numbers such as π.

An expression that contains only one symbol representing a number of length units is *one-dimensional*. Its value must be a number of length units (e.g. cm, km, feet etc).

An expression containing the product of two length unit symbols gives an area. It is *two-dimensional* and its value is the number of (length)^2 units (e.g. m^2, cm^2).

If there are three length-unit symbols or a length-unit symbol and an area-unit symbol multiplied together then that expression is *three-dimensional*. The result is a volume measured in (length)^3 units (e.g. mm^3, m^3).

A number that does not represent a number of length or area units does not alter the dimensions of an expression, e.g. if x is a number of length units then

$$3x \text{ is one-dimensional} \quad \text{(length)}$$
$$4x^2 \text{ is two-dimensional} \quad \text{(area)}$$
$$\pi x^3 \text{ is three-dimensional} \quad \text{(volume)}$$

$$(\pi \text{ is a number})$$

Checking dimensions and units helps to identify whether a given quantity represents length, area or volume. For example, if a sentence contains "$z\,\text{cm}^3$" then z must represent a number of volume units.

It also allows us to recognize incorrect formulae. Suppose, for example, that we are told that the formula for the volume, V cubic units, of a certain container is $V = 3\pi xy$, where x and y are numbers of length units.

V cubic units is a volume and therefore three dimensional.

But $3\pi xy$ is only two dimensional.

So this formula cannot be correct.

EXERCISE 2k

1. State whether each of the following quantities is a length, an area or a volume.
   a) 10 cm
   b) 21 cm^3
   c) 85 cm^2
   d) 4 m^3
   e) 630 mm
   f) 93 km^2

2. State whether each of the following quantities should be measured in length, area or volume units.
   a) Diameter of a circle
   b) Amount of air in a room
   c) Space inside a sphere
   d) Perimeter
   e) Region inside a square
   f) Surface of a sphere

3. a, b, and c represent numbers of centimetres. Give a suitable unit (e.g. cm^2) for the subject of each of the following formulae.
   a) $G = a + b$
   b) $H = 4ab$
   c) $K = 4\pi a^2$
   d) $P = abc$
   e) $Q = \pi c$
   f) $R = \frac{4}{3}\pi a^3$

In questions 4 and 5, a, b and c represent numbers of length units, A, B and C represent numbers of area units and V represents a number of volume units.

4. State whether each of the letters P to Y used in the following formulae represents numbers of length or area or volume units.

a) $P = \pi bc$ d) $S = \pi a^2 b$ g) $W = 2a + 3b$

b) $Q = A + B$ e) $T = 4bA$ h) $X = a^2 + b^2$

c) $R = \dfrac{ab}{c}$ f) $U = \dfrac{2A}{c}$ i) $Y = \dfrac{V}{a}$

5. Some of the following formulae are wrongly constructed. State which formulae are incorrect and justify your statement.

a) $B = ac$ c) $C = a^2 + b^3$ e) $V = ab + c$

b) $C = \pi a^2$ d) $V = 6a^2 b$ f) $A = a(b + c)$

6. Peter had to find the area of a circle whose diameter was 18 cm. He wrote down

$$Area = 2\pi r \,\text{cm}^2$$
$$= 2 \times \pi \times 9 \,\text{cm}^2$$
$$= 56.5 \,\text{cm}^2$$

Louise knew nothing about circle formulae but was able to tell Peter that he was wrong. How did she know?

7. Jane copied down a formula that she was supposed to use to find the volume of a solid but she found that she could not read the index number. All she had was $V = \pi x^? y$. She knew that x and y were numbers of length units.

What is the index number?

MIXED EXERCISES

EXERCISE 2I

1. Evaluate   a) $8 \times (-\frac{1}{2})$ b) $5 + 2 - 8$ c) $8 \div (-4)$

2. Simplify   a) $x + 3x$ b) $2b - (-4b)$ c) $(x)(x)(-3x)$

3. Simplify   a) $2a - (a - b)$ b) $3(a + b) - 2(a - b)$

4. Solve   a) $x - 3 = 2 - 3x$ b) $3(x - 2) - 8 = 0$

5. Solve   a) $2x > 4$ b) $2(x - 3) \leqslant 6$ c) $x - 1 < 2 - x < 4$

6. Make r the subject of the formula   a) $v = 4r + u$ b) $s = \dfrac{5r}{p}$

7. Find P, given that $P = \dfrac{100I}{RT}$, when

a) $I = 3$, $R = 4$ and $T = 2$ b) $I = 6$, $R = 5$, $T = 3$

EXERCISE 2m   **1.** Evaluate   a) $3 - (-10)$ b) $(-\frac{1}{2}) \times (-4)$ c) $(-20) \div (-5)$

2. Simplify   a) $4a - 3b + 6a$ b) $x + x^2 + 3x$ c) $3a \times 4b$

3. Simplify   a) $x - 3(x - y)$ b) $2(x + 3y) - 4(2x + y)$

4. Solve   a) $4 - 3a = 5 - 2a$ b) $2(x + 4) = 3(5 - 2x)$

5. Solve   a) $2x > 3 - x$ b) $3 - x < 4$ c) $2x < 3 - x < 4$

6. Make d the subject of the formula   a) $C = \pi d$ b) $a = 7d - s$

7. Find u, given that $u = v - gt$, when
a) $v = 16$, $g = -10$ and $t = 4$ b) $v = -6$, $g = 10$ and $t = 8$

3 SEQUENCES

SEQUENCES

A sequence is a set of numbers arranged in a particular order. The simplest sequence is the sequence of natural numbers, i.e. counting numbers

$$1, 2, 3, 4, 5, 6, \ldots$$

Other simple sequences are

$$1, 3, 5, 7, 9, \ldots \qquad \text{(odd numbers)}$$
$$2, 3, 5, 7, 11, \ldots \qquad \text{(prime numbers)}$$
$$-1, -2, -3, -4, \ldots \quad \text{(negative integers)}$$

To continue a sequence we need to recognise a pattern. Consider, for example,

$$2, 5, 8, 11, \ldots$$

Each number in this sequence is three units greater than the preceding number. Hence the sequence continues 14, 17, 20, ...

The pattern is not always obvious. If you cannot immediately see a pattern, try looking for sums, products, multiples, differences, two simple sequences put together by alternating their terms, etc.

If not enough terms are given to define a sequence uniquely, there are sometimes two or more possible rules for continuing it. For example, the rule for the sequence 1, 2, 4, ... might be that each term is doubled to give the next but equally well we might add a number that increases by 1 each time: first add 1 to give 2, then 2 to give 4, then 3 to give 7.

EXERCISE 3a For each sequence in questions 1 to 8 give the next two terms and the rule for continuing the sequence.

1. 1, 4, 9, 16, ...

2. 3, 6, 9, 12, ...

3. 7, 14, 21, 28, ...

4. −6, −2, 2, 6, ...

5. 15, 11, 7, 3, ...

6. 2, 1, $\frac{1}{2}$, $\frac{1}{4}$, ...

7. 2, 6, 12, 20, ...

8. 1, 2, 6, 24, 120, ...

Fill the gaps in the following sequences.

9. 3, 9, __, 21, 27, __, ...

10. 3, __, 48, 192, 768, ...

In each question from 11 to 18, the first two terms of a sequence, and the rule for continuing that sequence, are given. Give the next three terms.

11. Add 2 to the previous term;   6, 8, . . .

12. Divide the previous term by 2;   4, 2, . . .

13. Add a number that increases by 2 each time;   2, 4, . . .

14. Multiply the previous term by 3;   2, 6, . . .

15. Add the previous two terms;   2, 4, . . .

16. Divide the previous term by 10;   1, 0.1, . . .

17. Subtract 2 from the previous term;   3, 1, . . .

18. Multiply the previous term by -2;   2, -4, . . .

In each question from 19 to 22, give two possible rules for continuing the sequence.

19. 1, 3, 9, . . .

20. 1, 3, 7, . . .

21. 0, 1, 4, . . .

22. 3, 6, 12, . . .

In each question from 23 to 28, give the next two terms of the sequence.

23. $\frac{1}{2}, \frac{1}{3}, \frac{1}{4}, \ldots$

24. 2, 3, 5, 7, 11, . . .

25. 2, 3, 6, 4, 18, 5, . . .

26. 100, 99, 95, 86, 70, . . .

27. 1, 3, 2, 9, 3, 27, 4, . . .

28. 1, 1, 2, 4, 3, 9, 4, . . .

29. This is a sequence of pairs of numbers

$$(1, 2), \ (2, 5), \ (3, 10), \ (4, 17), \ \ldots$$

a) Find the next pair in the sequence.
b) Find the 10th pair in the sequence.

30. Triangular numbers can be represented by dots arranged in triangles.

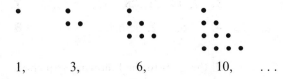

1,   3,   6,   10,   . . .

Write down the next four numbers in this sequence.

31.

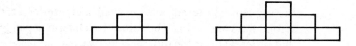

This is the beginning of a sequence of tile patterns.
Write down the first five terms of the sequence given by the number of tiles in each pattern.

32.

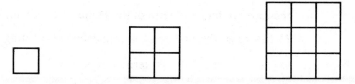

This is a sequence of square patterns.

a) Draw the next two patterns in the sequence.

b) A sequence is formed from the number of squares of all sizes in each pattern. Its first two terms are 1, 5, . . .
Write down the next three terms of the number sequence.

33.   1, 1, 2, 3, 5, 8, . . .

This is an example of a Fibonacci sequence; the first two terms are given and the next is formed by adding the two terms before it, and so on.

a) Give the next three terms of the given sequence.

b) Form a new Fibonacci sequence starting with 2, 5, . . .

34.

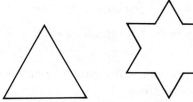

This sequence of 'snowflake' patterns is based on equilateral triangles. The second shape is made by drawing an equilateral triangle on the middle third of each side of the first shape. The third shape is obtained from the second in the same way, and so on for subsequent patterns.

Write down the first six terms of the number sequence where each term is the number of sides of the corresponding 'snowflake'.

THE *n*TH TERM OF A SEQUENCE

In a sequence the terms occur in a particular order, i.e. there is a first term, a second term and so on. The value of each term depends upon its position in that order. (This is the difference between a sequence of numbers and a set of numbers which can be in any order.)

The letter *n* is used for a natural number so we can refer to the *n*th term of a sequence in the same way as we refer to the 4th term or the 8th term.

If we are given a formula such as

$$\text{nth term} = n(n + 1)$$

then we can find any term of the sequence by giving *n* a numerical value.

In this case, the first term is found by substituting 1 for *n* in $n(n+1)$,

i.e.   1st term $= 1 \times 2 = 2$

Similarly,   2nd term $= 2 \times 3 = 6$ (substituting 2)

3rd term $= 3 \times 4 = 12$

. . . .

10th term $= 10 \times 11 = 110$

and so on.

EXERCISE 3b

The *n*th term of a sequence is given by the formula

$$\text{nth term} = (n - 1)^2$$

Give the first two terms and the eighth term of the sequence.

$n = 1$ 1st term $= (1 - 1)^2 = 0$

$n = 2$ 2nd term $= (2 - 1)^2 = 1^2 = 1$

$n = 8$ 8th term $= (8 - 1)^2 = 7^2 = 49$

Write down the first four terms and the seventh term of the sequence for which the *n*th term is given.

1. $2n + 1$

2. $2n - 1$

3. 2^n

4. n^2

5. $(n - 1)(n + 1)$

6. $n + 4$

7. $3 + 2n$

8. $\dfrac{1}{n}$

FINDING AN EXPRESSION FOR THE *n*TH TERM

When the pattern in a sequence is known, an expression for the *n*th term can often be found.

Consider the sequence   2, 4, 6, 8, . . .

We need to find the relationship between each term and the number, *n*, that gives its position. It is helpful to start by making a table of values of *n* and the corresponding terms in the sequence.

n	1	2	3	4	. . .
*n*th term	2	4	6	8	. . .

The pattern here is that each term is twice its position number, so the 10th term will be 2×10 and the *n*th term is $2n$.
Now we can check that this does give the correct sequence, i.e.

$n = 1$, 1st term $= 2 \times 1 = 2$, if $n = 2$, 2nd term $= 2 \times 2 = 4$, and so on.

EXERCISE 3c   Find, in terms of *n*, an expression for the *n*th term of each of the following sequences.

1. 3, 6, 9, 12, . . .   **4.** 0, 1, 2, 3, . . .

2. $-1, -2, -3, -4, \ldots$ **5.** 4, 8, 12, 16, . . .

3. 2, 3, 4, 5, . . .   **6.** 2, 4, 8, 16, . . .

Find a formula for the *n*th term of the sequence 1, 4, 7, 10, . . .

n	1	2	3	4
*n*th term	1	4	7	10

(The terms increase by 3 each time, so multiples of 3 are involved.)

$3n$	3	6	9	12

(These are each 2 more than the given terms, so take 2 from $3n$.)

$$n\text{th term} = 3n - 2$$

(Check that this fits all the given terms.)

$$n = 1 \quad \text{1st term} = 3 \times 1 - 2 = 3 - 2 = 1$$
$$n = 2 \quad \text{2nd term} = 3 \times 2 - 2 = 6 - 2 = 4$$
$$n = 3 \quad \text{3rd term} = 3 \times 3 - 2 = 9 - 2 = 7$$
$$n = 4 \quad \text{4th term} = 3 \times 4 - 2 = 12 - 2 = 10$$

Find an expression for the nth term of each of the following sequences.

7. 7, 9, 11, 13, ...

8. 0, 3, 6, 9, 12, ...

9. $\frac{1}{3}, \frac{1}{4}, \frac{1}{5}, \frac{1}{6}, \ldots$

10. $1 \times 3, 2 \times 4, 3 \times 5, \ldots$

11. 1, 8, 27, 64, ...

12. 3, 2, 1, 0, -1, ...

13. Find an expression for the nth term of each sequence in Exercise 3a, questions 11 to 14.

14.

S ⌐
  └ 1 m   1 m

One of the races at a school sports day is set out with bean bags placed at 1 metre intervals along the track.

A competitor starts at S, runs to the first bag, picks it up and returns it to S. Then she runs to the second bag, picks it up and returns it to S, and so on.

How far has a competitor run when she has returned

a) 1 bean bag   b) 4 bean bags   c) n bean bags?

4 MATRICES

Mrs Smith and Mrs Jones go shopping for oranges, lemons and grapefruit. Mrs Smith buys six oranges, two lemons and three grapefruit while Mrs Jones buys five oranges, one lemon and two grapefruit. We can arrange this information in a table:

	Oranges	Lemons	Grapefruit
Mrs Smith	6	2	3
Mrs Jones	5	1	2

We can write this information more briefly in the form

$$\begin{pmatrix} 6 & 2 & 3 \\ 5 & 1 & 2 \end{pmatrix}$$

All the numerical information is given, though we have missed out the descriptions of what the various rows and columns mean.

A *rectangular array* of numbers is called a *matrix* (plural *matrices*). It is held together by curved brackets. Each item is called an *entry*. A matrix can have any number of *rows* and *columns*; rows go across and columns go down. If we had the shopping lists of three people buying four different sorts of fruit we would have a matrix with three rows and four columns.

e.g.
$$\begin{pmatrix} 4 & 3 & 1 & 2 \\ 1 & 2 & 0 & 4 \\ 9 & 0 & 1 & 1 \end{pmatrix}$$

We use the number of rows and the number of columns to describe the size of a matrix. This matrix is a 3×4 matrix (number of rows first, then the number of columns).

EXERCISE 4a

a) Give the size of the matrix $\begin{pmatrix} 1 & 4 & 3 & 2 \\ 7 & 0 & 6 & 5 \end{pmatrix}$

b) What is the entry in the second row, third column?

a) The matrix is a 2 × 4 matrix

b) The entry in the second row, third column is 6

Give the sizes of the matrices in questions 1 to 6:

1. $\begin{pmatrix} 2 & 4 \\ 1 & 2 \end{pmatrix}$ **3.** $\begin{pmatrix} 1 \\ 2 \end{pmatrix}$ **5.** $\begin{pmatrix} 1 & 2 & 3 \end{pmatrix}$

2. $\begin{pmatrix} 2 & 3 & 4 \\ 1 & 1 & 4 \end{pmatrix}$ **4.** $\begin{pmatrix} 2 \end{pmatrix}$ **6.** $\begin{pmatrix} 2 & 3 \\ 4 & 5 \\ 6 & 7 \end{pmatrix}$

7. In the matrix $\begin{pmatrix} 1 & 2 & 3 \\ 4 & 5 & 6 \\ 7 & 8 & 9 \end{pmatrix}$ give the entry in

a) the second row, third column
b) the third row, second column
c) the first row, second column
d) the third row, first column

8. Write down the second row and the third column of the matrix

$$\begin{pmatrix} 5 & 2 & 4 \\ 3 & 1 & 7 \\ 9 & 6 & 2 \end{pmatrix}$$

a) Give the entry in the second row of the third column.
b) Give the entry in the third row of the second column.
c) Give the entry in the first row of the third column.

9. Write down the matrix with three rows and three columns in which the first row is a row of zeros, the second row is a row of ones and the third row is a row of twos.

10. Write down the 3×2 matrix in which the first column is a column of threes and the second column is a column of ones.

ADDITION OF MATRICES

Mrs Smith's and Mrs Jones' first shopping lists were represented by the matrix $\begin{pmatrix} 6 & 2 & 3 \\ 5 & 1 & 2 \end{pmatrix}$

The next week, Mrs Smith buys three oranges, no lemons and one grapefruit and Mrs Jones buys four oranges, two lemons and one grapefruit. We can arrange this in a table:

	Oranges	Lemons	Grapefruit
Mrs Smith	3	0	1
Mrs Jones	4	2	1

and represent it by the matrix $\begin{pmatrix} 3 & 0 & 1 \\ 4 & 2 & 1 \end{pmatrix}$

We see that over the two weeks Mrs Smith buys nine oranges, two lemons and four grapefruit and Mrs Jones buys nine oranges, three lemons and three grapefruit. This information can be written briefly in the form

$$\begin{pmatrix} 6 & 2 & 3 \\ 5 & 1 & 2 \end{pmatrix} + \begin{pmatrix} 3 & 0 & 1 \\ 4 & 2 & 1 \end{pmatrix} = \begin{pmatrix} 9 & 2 & 4 \\ 9 & 3 & 3 \end{pmatrix}$$

Each of these matrices is a 2×3 matrix.

We have added the entries in corresponding positions, e.g. in the first row, third column, $3 + 1 = 4$

In the third week, Mrs Smith buys two oranges and one lemon and Mrs Jones buys four oranges and one lemon.

The matrix showing this information is $\begin{pmatrix} 2 & 1 \\ 4 & 1 \end{pmatrix}$

This is a 2×2 matrix.

We do not know how many grapefruit were bought in the third week. This makes it impossible to add the matrices for the last two weeks.

i.e. $\begin{pmatrix} 3 & 0 & 1 \\ 4 & 2 & 1 \end{pmatrix} + \begin{pmatrix} 2 & 1 \\ 4 & 1 \end{pmatrix}$ cannot be worked out.

We can add matrices if they are the same size, but not if they are of different sizes.

EXERCISE 4b

Find, where possible

a) $\begin{pmatrix} 6 & 3 \\ 1 & 2 \end{pmatrix} + \begin{pmatrix} 4 & 1 \\ 3 & 2 \end{pmatrix}$

b) $\begin{pmatrix} 1 \\ 2 \end{pmatrix} + \begin{pmatrix} 4 & 1 \\ 0 & 3 \end{pmatrix}$

a) $\begin{pmatrix} 6 & 3 \\ 1 & 2 \end{pmatrix} + \begin{pmatrix} 4 & 1 \\ 3 & 2 \end{pmatrix} = \begin{pmatrix} 10 & 4 \\ 4 & 4 \end{pmatrix}$

b) $\begin{pmatrix} 1 \\ 2 \end{pmatrix} + \begin{pmatrix} 4 & 1 \\ 0 & 3 \end{pmatrix}$ not possible

Find, where possible:

1. $\begin{pmatrix} 2 \\ 3 \end{pmatrix} + \begin{pmatrix} 10 \\ 12 \end{pmatrix}$

2. $\begin{pmatrix} 9 & 2 \\ 4 & 1 \end{pmatrix} + \begin{pmatrix} 6 & 2 \\ 3 & 0 \end{pmatrix}$

3. $\begin{pmatrix} 1 \\ 2 \\ 3 \end{pmatrix} + \begin{pmatrix} 3 & 6 \\ 4 & 7 \\ 5 & 8 \end{pmatrix}$

4. $(6 \quad 1) + (3 \quad 4)$

5. $\begin{pmatrix} 4 & 1 & 2 \\ 3 & 5 & 6 \end{pmatrix} + \begin{pmatrix} 7 & 1 & 0 \\ 3 & 2 & 1 \end{pmatrix}$

6. $\begin{pmatrix} 7 & 8 \\ 9 & 1 \end{pmatrix} + \begin{pmatrix} 4 & 3 \\ 2 & 4 \end{pmatrix}$

7. $(2 \quad 1 \quad 4) + (3 \quad 2 \quad 1)$

8. $\begin{pmatrix} 1 \\ 2 \end{pmatrix} + (3 \quad 4)$

9. $\begin{pmatrix} 1 & 2 \\ 3 & 4 \\ 5 & 6 \end{pmatrix} + \begin{pmatrix} 5 & 6 \\ 4 & 3 \\ 2 & 1 \end{pmatrix}$

10. $(6 \quad 3) + (4 \quad 5)$

Negative numbers can be used.

Find $\begin{pmatrix} 4 & -3 \\ 4 & -1 \end{pmatrix} + \begin{pmatrix} -3 & 2 \\ 1 & 4 \end{pmatrix}$

$$\begin{pmatrix} 4 & -3 \\ 4 & -1 \end{pmatrix} + \begin{pmatrix} -3 & 2 \\ 1 & 4 \end{pmatrix} = \begin{pmatrix} 1 & -1 \\ 5 & 3 \end{pmatrix}$$

Find, where possible:

11. $\begin{pmatrix} 2 & 4 \\ -1 & 4 \end{pmatrix} + \begin{pmatrix} -1 & 4 \\ -3 & 3 \end{pmatrix}$

17. $\begin{pmatrix} 3 & 6 \\ 2 & -5 \end{pmatrix} + \begin{pmatrix} -1 & 4 \\ 3 & 2 \end{pmatrix}$

12. $\begin{pmatrix} 1 \\ -4 \end{pmatrix} + \begin{pmatrix} -3 \\ 6 \end{pmatrix}$

18. $\begin{pmatrix} 1 & -2 \\ -2 & 1 \end{pmatrix} + \begin{pmatrix} 4 & -3 \\ 5 & -1 \end{pmatrix}$

13. $\begin{pmatrix} 4 & 3 \\ -2 & 1 \end{pmatrix} + \begin{pmatrix} -6 & 4 \\ -3 & 2 \end{pmatrix}$

19. $\begin{pmatrix} 3 & 2 \\ -1 & -4 \end{pmatrix} + \begin{pmatrix} -3 & 6 \\ 9 & 2 \end{pmatrix}$

14. $\begin{pmatrix} 1 & 2 \end{pmatrix} + \begin{pmatrix} 3 & 4 \end{pmatrix}$

20. $\begin{pmatrix} 4 \\ -2 \end{pmatrix} + \begin{pmatrix} 4 & -2 \end{pmatrix}$

15. $\begin{pmatrix} -2 \\ -3 \\ 4 \end{pmatrix} + \begin{pmatrix} -1 \\ 0 \\ 2 \end{pmatrix}$

21. $\begin{pmatrix} 3 & 2 & 0 \\ 1 & 4 & 3 \end{pmatrix} + \begin{pmatrix} 2 & 4 \\ -3 & 4 \\ 1 & 2 \end{pmatrix}$

16. $\begin{pmatrix} 1 & -6 & 1 \end{pmatrix} + \begin{pmatrix} -4 & 1 \end{pmatrix}$

22. $\begin{pmatrix} 1 & 4 & -3 \end{pmatrix} + \begin{pmatrix} 0 & 2 & 0 \end{pmatrix}$

23. $\begin{pmatrix} 3 & -1 & 1 \\ 5 & 6 & -7 \end{pmatrix} + \begin{pmatrix} -1 & 4 & 3 \\ 0 & -6 & -5 \end{pmatrix}$

MULTIPLES OF MATRICES

If Mrs Smith and Mrs Jones each have the same shopping list for three weeks running we can see that

$$\begin{pmatrix} 3 & 1 & 4 \\ 1 & 2 & 1 \end{pmatrix} + \begin{pmatrix} 3 & 1 & 4 \\ 1 & 2 & 1 \end{pmatrix} + \begin{pmatrix} 3 & 1 & 4 \\ 1 & 2 & 1 \end{pmatrix} = 3\begin{pmatrix} 3 & 1 & 4 \\ 1 & 2 & 1 \end{pmatrix}$$

$$= \begin{pmatrix} 9 & 3 & 12 \\ 3 & 6 & 3 \end{pmatrix}$$

In the same way $5\begin{pmatrix} 1 & 4 \\ 3 & -2 \end{pmatrix} = \begin{pmatrix} 5 & 20 \\ 15 & -10 \end{pmatrix}$

When we multiply a matrix by five, we multiply *every* entry by five.

SUBTRACTION OF MATRICES

We can subtract matrices if they are the same size.

Two weeks' shopping   First week's shopping   Second week's shopping

$$\begin{pmatrix} 7 & 9 & 2 \\ 6 & 4 & 1 \end{pmatrix} \quad - \quad \begin{pmatrix} 3 & 2 & 1 \\ 3 & 3 & 1 \end{pmatrix} \quad = \quad \begin{pmatrix} 4 & 7 & 1 \\ 3 & 1 & 0 \end{pmatrix}$$

EXERCISE 4c

Find, where possible

a) $4\begin{pmatrix} 4 & 1 \\ 3 & -2 \end{pmatrix}$ b) $\begin{pmatrix} 4 & 2 \\ 1 & 1 \end{pmatrix} - \begin{pmatrix} 3 & 4 \\ 0 & 1 \end{pmatrix}$

a) $4\begin{pmatrix} 4 & 1 \\ 3 & -2 \end{pmatrix} = \begin{pmatrix} 16 & 4 \\ 12 & -8 \end{pmatrix}$

b) $\begin{pmatrix} 4 & 2 \\ 1 & 1 \end{pmatrix} - \begin{pmatrix} 3 & 4 \\ 0 & 1 \end{pmatrix} = \begin{pmatrix} 1 & -2 \\ 1 & 0 \end{pmatrix}$

Find, where possible:

1.
$$3\begin{pmatrix} 1 \\ 2 \\ 4 \end{pmatrix}$$

4.
$$6\begin{pmatrix} 1 & 4 \\ 3 & -2 \end{pmatrix}$$

2.
$$2\begin{pmatrix} 1 & 4 & 0 \\ 2 & -1 & 3 \end{pmatrix}$$

5.
$$6\begin{pmatrix} -1 & -5 \\ 1 & 2 \end{pmatrix}$$

3.
$$\frac{1}{2}\begin{pmatrix} 2 & 4 \\ 1 & 6 \\ 3 & 8 \end{pmatrix}$$

6.
$$\frac{2}{3}\begin{pmatrix} 6 & 0 \\ 1 & 2 \\ 3 & 5 \end{pmatrix}$$

7.
$$\begin{pmatrix} 3 & 2 \\ 1 & 4 \end{pmatrix} - \begin{pmatrix} 1 & 4 \\ 0 & 1 \end{pmatrix}$$

10.
$$\begin{pmatrix} 4 & 5 & 3 \\ 1 & -2 & 1 \end{pmatrix} - \begin{pmatrix} 2 & 1 & 1 \\ 4 & 1 & 2 \end{pmatrix}$$

8.
$$\begin{pmatrix} 1 \\ 3 \end{pmatrix} - \begin{pmatrix} 2 & 3 \\ 4 & 1 \end{pmatrix}$$

11.
$$\begin{pmatrix} 1 & 6 & 2 \end{pmatrix} - \begin{pmatrix} 4 & 3 \end{pmatrix}$$

9.
$$\begin{pmatrix} 1 \\ 2 \\ 3 \end{pmatrix} - \begin{pmatrix} 4 \\ -1 \\ 3 \end{pmatrix}$$

12.
$$\begin{pmatrix} 1 & 2 & 3 \\ 4 & 5 & 6 \\ 7 & 8 & 9 \end{pmatrix} - \begin{pmatrix} 4 & 3 & 1 \\ -5 & 0 & 2 \\ 6 & -3 & 4 \end{pmatrix}$$

MIXED QUESTIONS

EXERCISE 4d Find, where possible:

1.
$$\begin{pmatrix} 1 & 4 \\ 3 & 2 \end{pmatrix} + \begin{pmatrix} -2 & 4 \\ 3 & -1 \end{pmatrix}$$

3.
$$\begin{pmatrix} 1 & 2 & 4 \end{pmatrix} - \begin{pmatrix} 3 & 2 \end{pmatrix}$$

2.
$$\begin{pmatrix} 1 & 4 \\ 3 & 2 \end{pmatrix} - \begin{pmatrix} -2 & 4 \\ 3 & -1 \end{pmatrix}$$

4.
$$\begin{pmatrix} 3 & 2 \\ 4 & -1 \end{pmatrix} + \begin{pmatrix} 4 & -3 \\ 1 & 0 \end{pmatrix}$$

5.
$$\frac{1}{3}\begin{pmatrix} 4 \\ 5 \\ -1 \end{pmatrix}$$

9.
$$\begin{pmatrix} 1 \\ 2 \\ 3 \end{pmatrix} + \begin{pmatrix} 4 \\ 5 \\ -1 \end{pmatrix} + \begin{pmatrix} 3 \\ 2 \\ 1 \end{pmatrix}$$

6.
$$\begin{pmatrix} 1 \\ 2 \\ 3 \end{pmatrix} - \begin{pmatrix} 4 \\ 5 \\ 6 \end{pmatrix}$$

10.
$$4\begin{pmatrix} 6 & 2 & -1 \\ 4 & 3 & 4 \end{pmatrix}$$

7.
$$2\begin{pmatrix} 1 & 4 \\ 3 & 2 \end{pmatrix}$$

11.
$$\begin{pmatrix} 6 & 2 & 1 \\ 4 & 3 & 4 \end{pmatrix} + \begin{pmatrix} -2 & 4 \\ & 3 & -1 \end{pmatrix}$$

8.
$$\frac{1}{2}\begin{pmatrix} -2 & 4 \\ 3 & -1 \end{pmatrix}$$

12.
$$\begin{pmatrix} 6 & 2 & -1 \\ 4 & 3 & 4 \end{pmatrix} - \begin{pmatrix} -2 & 4 & 1 \\ 3 & -1 & 0 \end{pmatrix}$$

USE OF LETTERS

In Book 1, we saw that we could use a single small letter in heavy type to denote a vector,

e.g.
$$\mathbf{a} = \begin{pmatrix} 1 \\ 2 \end{pmatrix}$$

In the same way we can represent a matrix by giving it a capital letter in heavy type:

e.g.
$$\mathbf{A} = \begin{pmatrix} 1 & 2 \\ 4 & 1 \end{pmatrix} \quad \text{and} \quad \mathbf{B} = \begin{pmatrix} 4 & 1 \\ 3 & 0 \end{pmatrix}$$

then
$$\mathbf{A} + \mathbf{B} = \begin{pmatrix} 1 & 2 \\ 4 & 1 \end{pmatrix} + \begin{pmatrix} 4 & 1 \\ 3 & 0 \end{pmatrix} = \begin{pmatrix} 5 & 3 \\ 7 & 1 \end{pmatrix}$$

and
$$2\mathbf{A} = 2\begin{pmatrix} 1 & 2 \\ 4 & 1 \end{pmatrix} = \begin{pmatrix} 2 & 4 \\ 8 & 2 \end{pmatrix}$$

We cannot *write* **A**, so we write $\underline{A}$.

EXERCISE 4e The questions in this exercise refer to the following matrices:

$$A = \begin{pmatrix} 4 & 3 & 1 \\ 1 & 2 & 3 \end{pmatrix} \qquad B = \begin{pmatrix} 4 \\ 1 \end{pmatrix} \qquad C = \begin{pmatrix} 2 & 3 \\ 1 & -2 \end{pmatrix}$$

$$D = \begin{pmatrix} 6 & 2 \\ 1 & 4 \end{pmatrix} \qquad E = \begin{pmatrix} 6 & -1 & 2 \end{pmatrix} \qquad F = \begin{pmatrix} 3 & 2 \end{pmatrix}$$

$$G = \begin{pmatrix} 5 & 1 & 3 \\ 6 & -1 & 4 \end{pmatrix}$$

Give the size of **A**.

$$\textbf{A} \text{ is a } 2 \times 3 \text{ matrix}$$

---

1. Give the sizes of the matrices **B** to **G**.

Find, where possible a) **C** − **D**   b) **B** + **F**.

a)
$$\textbf{C} - \textbf{D} = \begin{pmatrix} 2 & 3 \\ 1 & -2 \end{pmatrix} - \begin{pmatrix} 6 & 2 \\ 1 & 4 \end{pmatrix}$$

$$= \begin{pmatrix} -4 & 1 \\ 0 & -6 \end{pmatrix}$$

b)
$$\textbf{B} + \textbf{F} = \begin{pmatrix} 4 \\ 1 \end{pmatrix} + \begin{pmatrix} 3 & 2 \end{pmatrix} \qquad \text{not possible}$$

Find, where possible:

2. A + G	**6.** E + F	**10.** 6B
3. B + C	**7.** $\frac{1}{2}$F	**11.** $\frac{3}{4}$D
4. D − C	**8.** G + B	**12.** C + D + G
5. 3A	**9.** G − A	**13.** F − E

MATRIX MULTIPLICATION

We can use the idea of shopping lists to find an operation that we call matrix multiplication.

Mrs Smith buys six oranges and one lemon and Mrs Jones buys four oranges and two lemons.

The matrix showing this information is $\begin{pmatrix} 6 & 1 \\ 4 & 2 \end{pmatrix}$

The oranges cost 12p each and the lemons 8p and this information can be written as $\begin{pmatrix} 12 \\ 8 \end{pmatrix}$

We can work out that Mrs Smith spends 80p and Mrs Jones 64p. This can be written in a table

	Amount spent in pence
Mrs Smith	80
Mrs Jones	64

or as a matrix $\begin{pmatrix} 80 \\ 64 \end{pmatrix}$

We can say $\begin{pmatrix} 6 & 1 \\ 4 & 2 \end{pmatrix} \begin{pmatrix} 12 \\ 8 \end{pmatrix} = \begin{pmatrix} 80 \\ 64 \end{pmatrix}$

This is a sort of multiplication, though different from any multiplication we have done so far.

It is called *matrix multiplication* or *row–column multiplication*.

Notice how the numbers have been paired:

$\begin{pmatrix} 6 & 1 \end{pmatrix} \begin{pmatrix} 12 \\ 8 \end{pmatrix} = \begin{pmatrix} 80 \end{pmatrix}$ $6 \times 12 + 1 \times 8 = 80$

EXERCISE 4f

Find $\begin{pmatrix} 4 & 3 \\ 6 & 1 \end{pmatrix} \begin{pmatrix} 3 \\ 2 \end{pmatrix}$

$\begin{pmatrix} 4 & 3 \\ 6 & 1 \end{pmatrix} \begin{pmatrix} 3 \\ 2 \end{pmatrix} = \begin{pmatrix} 18 \\ 20 \end{pmatrix}$ $4 \times 3 + 3 \times 2 = 18$
$6 \times 3 + 1 \times 2 = 20$

Find the following products:

1. $\begin{pmatrix} 1 & 4 \\ 3 & 2 \end{pmatrix} \begin{pmatrix} 5 \\ 6 \end{pmatrix}$

6. $\begin{pmatrix} 6 & 2 \\ 1 & 4 \end{pmatrix} \begin{pmatrix} 2 \\ 3 \end{pmatrix}$

2. $\begin{pmatrix} 6 & 2 \\ 3 & 5 \end{pmatrix} \begin{pmatrix} 2 \\ 1 \end{pmatrix}$

7. $\begin{pmatrix} 5 & 4 \\ 3 & 1 \end{pmatrix} \begin{pmatrix} 2 \\ 4 \end{pmatrix}$

3. $\begin{pmatrix} 3 & 2 \\ 4 & 3 \end{pmatrix} \begin{pmatrix} 1 \\ 1 \end{pmatrix}$

8. $\begin{pmatrix} 6 & 4 \\ 1 & 2 \end{pmatrix} \begin{pmatrix} 5 \\ 7 \end{pmatrix}$

4. $\begin{pmatrix} 1 & 3 \\ 0 & 1 \end{pmatrix} \begin{pmatrix} 3 \\ 2 \end{pmatrix}$

9. $\begin{pmatrix} 11 & 12 \\ 10 & 9 \end{pmatrix} \begin{pmatrix} 4 \\ 1 \end{pmatrix}$

5. $\begin{pmatrix} 5 & 2 \\ 1 & 2 \end{pmatrix} \begin{pmatrix} 1 \\ 2 \end{pmatrix}$

10. $\begin{pmatrix} 3 & 4 \\ 1 & 2 \end{pmatrix} \begin{pmatrix} 6 \\ 2 \end{pmatrix}$

Mrs Smith bought six oranges and one lemon and Mrs Jones bought four oranges and two lemons. Oranges cost 12p each and lemons 8p each.

This gives the matrix multiplication $\begin{pmatrix} 6 & 1 \\ 4 & 2 \end{pmatrix} \begin{pmatrix} 12 \\ 8 \end{pmatrix} = \begin{pmatrix} 80 \\ 64 \end{pmatrix}$

If they chose to do their shopping at a second shop where the oranges cost 11p each and the lemons 9p, the matrix multiplication would be

$\begin{pmatrix} 6 & 1 \\ 4 & 2 \end{pmatrix} \begin{pmatrix} 11 \\ 9 \end{pmatrix} = \begin{pmatrix} 75 \\ 62 \end{pmatrix}$

The table for the bills is as follows:

	First shop	Second shop
Mrs Smith	80	75
Mrs Jones	64	62

This gives the 2×2 matrix $\begin{pmatrix} 80 & 75 \\ 64 & 62 \end{pmatrix}$

We can combine the previous two matrix multiplications into one:

First Second
shop shop

Mrs Smith

Mrs Jones
$$\begin{pmatrix} 6 & 1 \\ 4 & 2 \end{pmatrix} \times \begin{pmatrix} 12 & 11 \\ 8 & 9 \end{pmatrix} = \begin{pmatrix} 80 & 75 \\ 64 & 62 \end{pmatrix}$$

Notice that we pair the entries in the *first row* of the first matrix with the entries in the *first column* of the second matrix to give the entry in the *first row, first column* of the final matrix.

$$\begin{pmatrix} 6 & 1 \\ & \end{pmatrix}\begin{pmatrix} 12 \\ 8 \end{pmatrix} = \begin{pmatrix} 80 \\ & \end{pmatrix} \qquad 6 \times 12 + 1 \times 8 = 80$$

Then the entries in the *first row* of the first matrix paired with the entries in the *second column* of the second matrix give the entry in the *first row, second column* of the final matrix.

$$\begin{pmatrix} 6 & 1 \\ & \end{pmatrix}\begin{pmatrix} 11 \\ 9 \end{pmatrix} = \begin{pmatrix} 75 \\ & \end{pmatrix} \qquad 6 \times 11 + 1 \times 9 = 75$$

Then we use the second row in the same way.

EXERCISE 4g

Find the missing entries in the matrix multiplication

$$\begin{pmatrix} 2 & 4 \\ 3 & 2 \end{pmatrix}\begin{pmatrix} 4 & 1 \\ 3 & 2 \end{pmatrix} = \begin{pmatrix} 20 & \\ & 7 \end{pmatrix}$$

$$\begin{pmatrix} 2 & 4 \\ 3 & 2 \end{pmatrix}\begin{pmatrix} 4 & 1 \\ 3 & 2 \end{pmatrix} = \begin{pmatrix} 20 & 10 \\ 18 & 7 \end{pmatrix}$$

(1st row, 2nd column
$2 \times 1 + 4 \times 2 = 10$)
(2nd row, 1st column
$3 \times 4 + 2 \times 3 = 18$)

Find the missing entries in the following matrix multiplications:

1. $\begin{pmatrix} 1 & 4 \\ 3 & 1 \end{pmatrix}\begin{pmatrix} 3 & 2 \\ 1 & 4 \end{pmatrix} = \begin{pmatrix} & 18 \\ & 10 \end{pmatrix}$

2. $\begin{pmatrix} 3 & 2 \\ 4 & 6 \end{pmatrix}\begin{pmatrix} 4 & 1 \\ 1 & 3 \end{pmatrix} = \begin{pmatrix} & 9 \\ 22 & \end{pmatrix}$

3. $\begin{pmatrix} 9 & 1 \\ 0 & 2 \end{pmatrix}\begin{pmatrix} 1 & 4 \\ 3 & 1 \end{pmatrix} = \begin{pmatrix} 12 & \\ 6 & \end{pmatrix}$

4. $\begin{pmatrix} 5 & 6 \\ 7 & 9 \end{pmatrix}\begin{pmatrix} 1 & 2 \\ 3 & 1 \end{pmatrix} = \begin{pmatrix} & \\ 34 & 23 \end{pmatrix}$

<u>5.</u> $\begin{pmatrix} 1 & 2 \\ 3 & 1 \end{pmatrix}\begin{pmatrix} 5 & 6 \\ 7 & 9 \end{pmatrix} = \begin{pmatrix} & 24 \\ & 27 \end{pmatrix}$

<u>6.</u> $\begin{pmatrix} 5 & 1 \\ 0 & 6 \end{pmatrix}\begin{pmatrix} 4 & 3 \\ 2 & 1 \end{pmatrix} = \begin{pmatrix} 22 & \\ & 6 \end{pmatrix}$

<u>7.</u> $\begin{pmatrix} 3 & 9 \\ 2 & 1 \end{pmatrix}\begin{pmatrix} 1 & 1 \\ 1 & 1 \end{pmatrix} = \begin{pmatrix} 12 & \\ 3 & \end{pmatrix}$

<u>8.</u> $\begin{pmatrix} 5 & 3 \\ 1 & 1 \end{pmatrix}\begin{pmatrix} 1 & 2 \\ 4 & 3 \end{pmatrix} = \begin{pmatrix} & \\ & 5 \end{pmatrix}$

Find the following products:

9. $\begin{pmatrix} 4 & 6 \\ 1 & 3 \end{pmatrix}\begin{pmatrix} 1 & 4 \\ 3 & 6 \end{pmatrix}$

<u>12.</u> $\begin{pmatrix} 1 & 4 \\ 3 & 6 \end{pmatrix}\begin{pmatrix} 4 & 6 \\ 1 & 3 \end{pmatrix}$

10. $\begin{pmatrix} 4 & 8 \\ 1 & 1 \end{pmatrix}\begin{pmatrix} 5 & 6 \\ 3 & 1 \end{pmatrix}$

<u>13.</u> $\begin{pmatrix} 6 & 4 \\ 5 & 1 \end{pmatrix}\begin{pmatrix} 2 & 6 \\ 8 & 1 \end{pmatrix}$

11. $\begin{pmatrix} 5 & 1 \\ 3 & 2 \end{pmatrix}\begin{pmatrix} 3 & 2 \\ 1 & 4 \end{pmatrix}$

<u>14.</u> $\begin{pmatrix} 3 & 2 \\ 1 & 4 \end{pmatrix}\begin{pmatrix} 5 & 1 \\ 3 & 2 \end{pmatrix}$

Sometimes the entries are negative.

Find $\begin{pmatrix} 1 & -2 \\ 2 & 1 \end{pmatrix}\begin{pmatrix} 2 & 4 \\ -3 & 1 \end{pmatrix}$

$\begin{pmatrix} 1 & -2 \\ 2 & 1 \end{pmatrix}\begin{pmatrix} 2 & 4 \\ -3 & 1 \end{pmatrix} = \begin{pmatrix} 8 & 2 \\ 1 & 9 \end{pmatrix}$

(1st row, 2nd column)
$1 \times 4 + (-2) \times 1$
$= 4 - 2 = 2$

Find the following products:

15. $\begin{pmatrix} 3 & 2 \\ 1 & 4 \end{pmatrix} \begin{pmatrix} -2 & 4 \\ 3 & 1 \end{pmatrix}$

18. $\begin{pmatrix} 5 & -2 \\ 4 & 3 \end{pmatrix} \begin{pmatrix} -4 & -3 \\ 2 & 1 \end{pmatrix}$

16. $\begin{pmatrix} 3 & 4 \\ -1 & 2 \end{pmatrix} \begin{pmatrix} 1 & 4 \\ 3 & 2 \end{pmatrix}$

19. $\begin{pmatrix} 4 & 3 \\ -1 & 4 \end{pmatrix} \begin{pmatrix} 3 & 2 \\ 3 & 1 \end{pmatrix}$

17. $\begin{pmatrix} 5 & 2 \\ 3 & 4 \end{pmatrix} \begin{pmatrix} -1 & -2 \\ 4 & 3 \end{pmatrix}$

20. $\begin{pmatrix} 4 & -3 \\ 2 & -1 \end{pmatrix} \begin{pmatrix} -1 & -2 \\ 4 & -3 \end{pmatrix}$

ORDER OF MULTIPLICATION

We can see from the answers to questions 4 and 5 and others in the previous exercise, that the order in which we write the matrices *does* matter when we multiply. For example,

if $\qquad A = \begin{pmatrix} 3 & 1 \\ 2 & 4 \end{pmatrix}$ and $B = \begin{pmatrix} 5 & 6 \\ 1 & 1 \end{pmatrix}$

then $\qquad AB = \begin{pmatrix} 16 & 19 \\ 14 & 16 \end{pmatrix}$ but $BA = \begin{pmatrix} 27 & 29 \\ 5 & 5 \end{pmatrix}$.

That is, $\qquad\qquad\qquad AB \neq BA$.

Occasionally, with a few special matrices, the order does not affect the answer.

EXERCISE 4h The questions in this exercise refer to the following matrices:

$$A = \begin{pmatrix} 4 & 3 \\ 2 & 1 \end{pmatrix} \qquad B = \begin{pmatrix} 2 & 1 \\ 4 & 3 \end{pmatrix} \qquad C = \begin{pmatrix} 7 & 8 \\ 1 & 1 \end{pmatrix} \qquad D = \begin{pmatrix} 2 & 0 \\ 0 & 2 \end{pmatrix}$$

Find:

1. AB	**4.** CA	**7.** BC	**10.** DB
2. BA	**5.** AD	**8.** CB	**11.** CD
3. AC	**6.** DA	**9.** BD	**12.** DC

13. In some cases in questions 1 to 12 the order mattered and in some cases it did not. What is special about the matrices for which the order did not matter?

14. Make up a set of matrices similar to the ones used in this exercise and see whether the same results occur when you multiply them together in pairs.

MULTIPLICATION OF MATRICES OF DIFFERENT SIZES

We can multiply matrices of sizes other than 2 × 2. If Mrs Smith and Mrs Jones are buying oranges, lemons and grapefruit then we might have the information in the following tables:

Shopping lists	Oranges	Lemons	Grapefruit
Mrs Smith	4	0	1
Mrs Jones	2	3	2

Costs in pence	First shop	Second shop
Oranges	12	11
Lemons	8	9
Grapefruit	15	12

Then we would have the matrix multiplication

$$\begin{pmatrix} 4 & 0 & 1 \\ 2 & 3 & 2 \end{pmatrix} \begin{pmatrix} 12 & 11 \\ 8 & 9 \\ 15 & 12 \end{pmatrix}$$

Again we pair entries in a row in the first matrix with the entries in a column in the second matrix:

$$\begin{pmatrix} 4 & 0 & 1 \end{pmatrix} \begin{pmatrix} 12 \\ 8 \\ 15 \end{pmatrix} = \begin{pmatrix} 63 \end{pmatrix} \qquad 4 \times 12 + 0 \times 8 + 1 \times 15 = 63$$

This gives us Mrs Smith's bill at the first shop; she would spend 63p.

$$\begin{pmatrix} 4 & 0 & 1 \end{pmatrix} \begin{pmatrix} 11 \\ 9 \\ 12 \end{pmatrix} = \begin{pmatrix} 56 \end{pmatrix} \qquad 4 \times 11 + 0 \times 9 + 1 \times 12 = 56$$

Mrs Smith would spend 56p at the second shop.

The complete multiplication is as follows:

$$\begin{pmatrix} 4 & 0 & 1 \\ 2 & 3 & 2 \end{pmatrix} \begin{pmatrix} 12 & 11 \\ 8 & 9 \\ 15 & 12 \end{pmatrix} = \begin{pmatrix} 63 & 56 \\ 78 & 73 \end{pmatrix}$$

For example, pairing the entries in the *second row* of the first matrix with the entries in the *first column* of the second matrix gives the entry in the *second row*, *first column* of the resulting matrix,

i.e. $2 \times 12 + 3 \times 8 + 2 \times 15 = 78$

EXERCISE 4i

Find   a) $\begin{pmatrix} 3 & 1 & 4 \\ 4 & 0 & 3 \end{pmatrix} \begin{pmatrix} 1 \\ 2 \\ 3 \end{pmatrix}$ b) $\begin{pmatrix} 1 & 2 \end{pmatrix} \begin{pmatrix} 1 & 3 \\ 4 & 2 \end{pmatrix}$

a) $\begin{pmatrix} 3 & 1 & 4 \\ 4 & 0 & 3 \end{pmatrix} \begin{pmatrix} 1 \\ 2 \\ 3 \end{pmatrix} = \begin{pmatrix} 17 \\ 13 \end{pmatrix}$

(first row, first column)
$3 \times 1 + 1 \times 2 + 4 \times 3 = 17$

(second row, first column)
$4 \times 1 + 0 \times 2 + 3 \times 3 = 13$

b) $\begin{pmatrix} 1 & 2 \end{pmatrix} \begin{pmatrix} 1 & 3 \\ 4 & 2 \end{pmatrix} = \begin{pmatrix} 9 & 7 \end{pmatrix}$

(first row, first column)
$1 \times 1 + 2 \times 4 = 9$

(first row, second column)
$1 \times 3 + 2 \times 2 = 7$

Find:

1. $\begin{pmatrix} 1 & 4 \\ 3 & 1 \end{pmatrix} \begin{pmatrix} 3 \\ 1 \end{pmatrix}$

3. $\begin{pmatrix} 1 & 2 \end{pmatrix} \begin{pmatrix} 4 \\ 3 \end{pmatrix}$

2. $\begin{pmatrix} 1 & 4 & 1 \\ 3 & 2 & 4 \end{pmatrix} \begin{pmatrix} 6 \\ 1 \\ 3 \end{pmatrix}$

4. $\begin{pmatrix} 3 & 2 & 1 \\ 3 & 4 & 8 \end{pmatrix} \begin{pmatrix} 2 & 1 \\ 4 & 3 \\ 6 & 1 \end{pmatrix}$

5. $\begin{pmatrix} 4 & 2 \\ 3 & 1 \end{pmatrix} \begin{pmatrix} 4 & 6 & 3 \\ 8 & 1 & 2 \end{pmatrix}$

8. $\begin{pmatrix} 4 & 1 & 2 \\ 3 & 8 & 1 \\ 5 & 6 & 2 \end{pmatrix} \begin{pmatrix} 1 & 2 \\ 4 & 3 \\ 1 & 0 \end{pmatrix}$

6. $\begin{pmatrix} 1 & 4 \\ 2 & 5 \\ 3 & 6 \end{pmatrix} \begin{pmatrix} 4 \\ 5 \end{pmatrix}$

9. $\begin{pmatrix} 7 & 3 \end{pmatrix} \begin{pmatrix} 1 & 4 & 3 \\ 2 & 1 & 2 \end{pmatrix}$

7. $\begin{pmatrix} 1 & 4 & 1 \\ 2 & 3 & 0 \end{pmatrix} \begin{pmatrix} 4 & 1 & 2 \\ 3 & 8 & 1 \\ 5 & 6 & 2 \end{pmatrix}$

10. $\begin{pmatrix} 1 & 2 & 3 \end{pmatrix} \begin{pmatrix} 4 \\ 1 \\ 3 \end{pmatrix}$

COMPATIBILITY FOR MULTIPLICATION

It was possible to find all the matrix products in the last exercise but, just as there are pairs of matrices which cannot be added together, there are pairs of matrices that we cannot multiply together.

Consider $\begin{pmatrix} 1 & 2 \\ 3 & 4 \end{pmatrix} \begin{pmatrix} 1 & 4 \\ 2 & 5 \\ 3 & 6 \end{pmatrix}$

If we try to multiply the entries in the first row of the first matrix with entries in the first column of the second matrix,

i.e. $\begin{pmatrix} 1 & 2 \end{pmatrix} \begin{pmatrix} 1 \\ 2 \\ 3 \end{pmatrix}$

we find that the entries do not fit together in pairs.

We need to make sure that the number of entries in the row we use is the same as the number of entries in the column that goes with it. The easiest way to check this is to write the size below each matrix:

$$\begin{pmatrix} 3 & 1 & 4 \\ 4 & 0 & 3 \end{pmatrix} \begin{pmatrix} 1 \\ 2 \\ 3 \end{pmatrix} = \begin{pmatrix} 17 \\ 3 \end{pmatrix}$$

$$2 \times \mathbf{3} \qquad \mathbf{3} \times 1 \qquad 2 \times 1$$

There are three columns in the first matrix and three rows in the second, so these two matrices are compatible for multiplication. The two outer numbers give us the size of the resulting matrix.

EXERCISE 4j

Under each matrix write its size and find the product

$$\begin{pmatrix} 1 & 2 \end{pmatrix} \begin{pmatrix} 1 & 2 & 3 \\ 3 & 4 & 6 \end{pmatrix}$$

Give the size of this product.

$$\begin{pmatrix} 1 & 2 \end{pmatrix} \begin{pmatrix} 1 & 2 & 3 \\ 3 & 4 & 6 \end{pmatrix} = \begin{pmatrix} 7 & 10 & 15 \end{pmatrix}$$

$$1 \times \mathbf{2} \qquad \mathbf{2} \times 3 \qquad\qquad 1 \times 3$$

$1 \times 1 + 2 \times 3 = 7$
$1 \times 2 + 2 \times 4 = 10$
$1 \times 3 + 2 \times 6 = 15$

In questions 1 to 8, under each matrix write its size and give the size of the product. Find the product. (The pairs of matrices in these questions are all compatible for multiplication.)

1. $\begin{pmatrix} 3 & 2 \\ 4 & 1 \end{pmatrix} \begin{pmatrix} 1 \\ 2 \end{pmatrix}$

2. $\begin{pmatrix} 1 & 4 & 5 \\ 3 & 2 & 1 \end{pmatrix} \begin{pmatrix} 1 \\ 4 \\ 1 \end{pmatrix}$

3. $\begin{pmatrix} 1 & 2 \end{pmatrix} \begin{pmatrix} 4 \\ 3 \end{pmatrix}$

4. $\begin{pmatrix} 3 & 2 & 1 \\ 3 & 4 & 8 \end{pmatrix} \begin{pmatrix} 2 & 1 \\ 4 & 3 \\ 6 & 1 \end{pmatrix}$

5. $\begin{pmatrix} 3 & 2 \\ 4 & 5 \end{pmatrix} \begin{pmatrix} 1 & 2 \\ 4 & 7 \end{pmatrix}$

6. $\begin{pmatrix} 1 \\ 2 \end{pmatrix} \begin{pmatrix} 3 & 4 \end{pmatrix}$

7. $\begin{pmatrix} 9 & 4 \end{pmatrix} \begin{pmatrix} 1 & 4 \\ 3 & 0 \end{pmatrix}$

8. $\begin{pmatrix} 1 \\ 2 \\ 3 \end{pmatrix} \begin{pmatrix} 4 & 5 & 6 \end{pmatrix}$

Find, if possible, $\begin{pmatrix} 4 & 3 \\ 1 & 2 \end{pmatrix} \begin{pmatrix} 3 & 1 & 4 \end{pmatrix}$

$$\begin{pmatrix} 4 & 3 \\ 1 & 2 \end{pmatrix} \begin{pmatrix} 3 & 1 & 4 \end{pmatrix}$$

$$2 \times 2 \qquad 1 \times 3$$

It is not possible.

Find, where possible, the following products. Some of the matrices are not compatible for multiplication; in these cases, write "not possible":

9. $\begin{pmatrix} 3 & 4 \\ 1 & 2 \end{pmatrix} \begin{pmatrix} 4 \\ 1 \end{pmatrix}$

14. $\begin{pmatrix} 1 & 4 \\ 0 & 6 \end{pmatrix} \begin{pmatrix} 3 & 2 \end{pmatrix}$

10. $\begin{pmatrix} 2 & 4 \\ 1 & 0 \end{pmatrix} \begin{pmatrix} 2 & 3 \\ 1 & 4 \\ 1 & 5 \end{pmatrix}$

15. $\begin{pmatrix} 2 & 5 & 4 \end{pmatrix} \begin{pmatrix} 3 \\ 4 \\ 1 \end{pmatrix}$

11. $\begin{pmatrix} 3 & 2 \\ 4 & 5 \end{pmatrix} \begin{pmatrix} 1 & 2 \\ 4 & 7 \end{pmatrix}$

16. $\begin{pmatrix} 1 \\ 2 \end{pmatrix} \begin{pmatrix} 3 & 2 \\ 4 & 1 \end{pmatrix}$

12. $\begin{pmatrix} 7 & 2 & 3 \\ 4 & 1 & 0 \end{pmatrix} \begin{pmatrix} 1 & 2 \\ 4 & 5 \end{pmatrix}$

17. $\begin{pmatrix} 3 & 2 \end{pmatrix} \begin{pmatrix} 1 & 4 \\ 0 & 6 \end{pmatrix}$

13. $\begin{pmatrix} 1 & 2 \\ 4 & 5 \end{pmatrix} \begin{pmatrix} 7 & 2 & 3 \\ 4 & 1 & 0 \end{pmatrix}$

18. $\begin{pmatrix} 3 \\ 4 \\ 1 \end{pmatrix} \begin{pmatrix} 2 & 4 & 5 \end{pmatrix}$

MATRICES CONTAINING NEGATIVE NUMBERS

EXERCISE 4k

Find $\begin{pmatrix} 1 & -2 \\ -4 & 3 \end{pmatrix} \begin{pmatrix} 4 \\ -2 \end{pmatrix}$

$$\begin{pmatrix} 1 & -2 \\ -4 & 3 \end{pmatrix} \begin{pmatrix} 4 \\ -2 \end{pmatrix} = \begin{pmatrix} 8 \\ -22 \end{pmatrix} \qquad \begin{matrix} 1 \times 4 + (-2) \times (-2) = 8 \\ (-4) \times 4 + 3 \times (-2) = -22 \end{matrix}$$

$\qquad 2 \times 2 \qquad 2 \times 1 \qquad 2 \times 1$

Find the following matrix products:

1. $\begin{pmatrix} 4 & -2 \\ 2 & -3 \end{pmatrix} \begin{pmatrix} 1 \\ -1 \end{pmatrix}$

2. $\begin{pmatrix} 4 & -3 & 0 \\ 1 & 4 & -3 \end{pmatrix} \begin{pmatrix} 1 \\ -2 \\ 4 \end{pmatrix}$

3. $\begin{pmatrix} 2 & -1 & 4 \end{pmatrix} \begin{pmatrix} 1 & 2 \\ 0 & -2 \\ -1 & -3 \end{pmatrix}$ **7.** $\begin{pmatrix} 6 & 1 & -3 \end{pmatrix} \begin{pmatrix} -4 \\ 1 \\ 1 \end{pmatrix}$

4. $\begin{pmatrix} 1 & -2 & 3 \\ -4 & 1 & 4 \\ 5 & 2 & 6 \end{pmatrix} \begin{pmatrix} -2 \\ -3 \\ -1 \end{pmatrix}$ **8.** $\begin{pmatrix} -4 \\ 1 \\ 1 \end{pmatrix} \begin{pmatrix} 6 & 1 & -3 \end{pmatrix}$

5. $\begin{pmatrix} -4 & 2 \\ 2 & 5 \end{pmatrix} \begin{pmatrix} -3 & 4 \\ -2 & -5 \end{pmatrix}$ **9.** $\begin{pmatrix} 2 & 3 \\ -2 & -3 \end{pmatrix} \begin{pmatrix} 5 & 6 & -2 \\ -1 & 2 & 1 \end{pmatrix}$

6. $\begin{pmatrix} -6 & -2 \end{pmatrix} \begin{pmatrix} 5 & -1 \\ 4 & -3 \end{pmatrix}$ **10.** $\begin{pmatrix} 5 & -1 \\ 6 & 2 \\ -2 & 1 \end{pmatrix} \begin{pmatrix} 2 & 3 \\ -2 & -3 \end{pmatrix}$

EXERCISE 4I   The questions in this exercise refer to the following matrices:

$$A = \begin{pmatrix} 4 & 3 \\ 1 & 2 \end{pmatrix} \quad B = \begin{pmatrix} 1 & -2 \\ 3 & 0 \end{pmatrix} \quad C = \begin{pmatrix} 6 & 2 \\ -1 & 0 \end{pmatrix} \quad D = \begin{pmatrix} 1 \\ 2 \end{pmatrix}$$

$$E = \begin{pmatrix} 3 & 4 \end{pmatrix} \quad F = \begin{pmatrix} 6 \\ 1 \\ -4 \end{pmatrix} \quad G = \begin{pmatrix} 1 & 2 & 3 \\ 4 & -5 & 1 \\ 3 & 2 & 4 \end{pmatrix} \quad H = \begin{pmatrix} 3 \end{pmatrix}$$

Find, where possible, the following products:

1. AB	**6.** AE	**11.** EC
2. BA	**7.** AF	**12.** EF
3. AD	**8.** GA	**13.** EH
4. AG	**9.** AH	**14.** HE
5. DE	**10.** ED	**15.** HD

16. Find the other possible products of pairs of matrices chosen from **A** to **G**.

ADDITION, SUBTRACTION AND MULTIPLICATION

EXERCISE 4m The questions in this exercise refer to the following matrices:

$$A = \begin{pmatrix} 1 & 2 \\ -4 & 3 \end{pmatrix} \qquad B = \begin{pmatrix} 6 & 2 \\ 1 & 0 \end{pmatrix} \qquad C = \begin{pmatrix} 1 & -1 & 3 \\ 2 & 2 & 0 \end{pmatrix}$$

$$D = \begin{pmatrix} 1 & 2 \\ 3 & 4 \\ 0 & 1 \end{pmatrix} \qquad E = \begin{pmatrix} 1 & -2 \end{pmatrix} \qquad F = \begin{pmatrix} 3 & 1 \end{pmatrix}$$

Find, where possible:

1. **AB**
2. **A + B**
3. **A − B**
4. **B − A**

5. **A + C**
6. **EF**
7. **E − F**
8. **CD**

9. **DC**
10. **C − D**
11. **BE**
12. **EB**

MULTIPLYING MORE THAN TWO MATRICES TOGETHER

Consider $\quad A = \begin{pmatrix} 1 & 4 \\ -1 & 0 \end{pmatrix} \quad B = \begin{pmatrix} 6 & 2 \\ 0 & 3 \end{pmatrix} \quad$ and $\quad C = \begin{pmatrix} 1 & 2 \\ 3 & 4 \end{pmatrix}.$

If we wish to find **ABC** we do this in two steps.
We can think of **ABC** either as **(AB)C** or as **A(BC)**

either $\quad AB = \begin{pmatrix} 1 & 4 \\ -1 & 0 \end{pmatrix}\begin{pmatrix} 6 & 2 \\ 0 & 3 \end{pmatrix} = \begin{pmatrix} 6 & 14 \\ -6 & -2 \end{pmatrix}$

so $\quad ABC = \begin{pmatrix} 6 & 14 \\ -6 & -2 \end{pmatrix}\begin{pmatrix} 1 & 2 \\ 3 & 4 \end{pmatrix} = \begin{pmatrix} 48 & 68 \\ -12 & -20 \end{pmatrix}$

or $\quad BC = \begin{pmatrix} 6 & 2 \\ 0 & 3 \end{pmatrix}\begin{pmatrix} 1 & 2 \\ 3 & 4 \end{pmatrix} = \begin{pmatrix} 12 & 20 \\ 9 & 12 \end{pmatrix}$

so $\quad ABC = \begin{pmatrix} 1 & 4 \\ -1 & 0 \end{pmatrix}\begin{pmatrix} 12 & 20 \\ 9 & 12 \end{pmatrix} = \begin{pmatrix} 48 & 68 \\ -12 & -20 \end{pmatrix}$

We must pay careful attention to the order of the matrices. For example, if we found **AC** and then multiplied by **B**, the result would *not* be **ABC**.

POWERS OF MATRICES

A^2 means $A \times A$, A^3 means $A \times A \times A$.

EXERCISE 4n   The questions in this exercise refer to the following matrices:

$$A = \begin{pmatrix} 4 & -1 \\ 3 & 2 \end{pmatrix} \qquad B = \begin{pmatrix} 2 & 3 \\ -4 & 2 \end{pmatrix} \qquad C = \begin{pmatrix} 4 & 0 \\ 3 & -1 \end{pmatrix}$$

Find the following products. Make use of the answers and working from one question to help you with another:

1. AB	**4.** CBA	**7.** ABA	**10.** AB^2
2. ABC	**5.** CAB	**8.** A^2B	**11.** B^3
3. A^2	**6.** A^3	**9.** ACA	**12.** CBC

Where possible, check your answers by finding them in two different ways. For instance, in question 2, **ABC** can be found by thinking of it as **(AB)C** or as **A(BC)**.

MIXED EXERCISES

EXERCISE 4p   The questions in this exercise refer to the following matrices:

$$A = \begin{pmatrix} 5 & 2 \\ 1 & 4 \end{pmatrix} \qquad B = \begin{pmatrix} 3 & -1 \\ 4 & -3 \end{pmatrix} \qquad C = \begin{pmatrix} 1 \\ 4 \end{pmatrix}$$

1.   What are the sizes of **A** and **C**?

2.   Are **A** and **B** compatible for multiplication?

3.   Are **A** and **C** compatible for multiplication?

4.   Find **AB** if it is possible.

5.   Find A^2 and C^2 if it is possible.

6.   Find **A** − **C** if it is possible.

7.   Find **3B**.

8.   Find **2A** + **B** if it is possible.

9.   What is the entry in the second row of the first column of **C**?

10.   Which is possible to find, **BC** or **CB**?

EXERCISE 4q

The questions in this exercise refer to the following matrices:

$$P = \begin{pmatrix} 2 & 1 & -1 \\ 4 & 3 & 1 \end{pmatrix} \qquad Q = \begin{pmatrix} 3 & 2 \\ -1 & 4 \end{pmatrix} \qquad R = \begin{pmatrix} 3 \\ 4 \end{pmatrix}$$

1. Find $2P$.

2. Find $2P + Q$ if it is possible.

3. Find $Q + 2R$ if it is possible.

4. What are the sizes of P and Q?

5. Are P and R compatible for multiplication?

6. What is the entry in the first row of the first column of R?

7. What is the entry in the second row of the third column of P?

8. Which is it possible to find, PQ or QP?

9. Find QR if it is possible.

10. Find Q^2 and P^2 if it is possible.

5 PERCENTAGES

MEANING OF PERCENTAGES

We first met percentages in Book 2A, Chapter 4, and we begin this chapter by reminding ourselves of the work we did then:

$$65\% \text{ means } 65 \text{ out of } 100$$

Therefore, as a fraction $65\% = \frac{65}{100} = \frac{13}{20}$

and as a decimal $65\% = \frac{65}{100} = 0.65$

Similarly the fraction $\frac{7}{20} = \frac{7}{20} \times 100\% = 35\%$

and $\frac{7}{20} = \frac{35}{100} = 0.35$

EXERCISE 5a Complete the following table:

	Fraction	Percentage	Decimal
1.	$\frac{3}{5}$		
2.		40%	
3.			0.55
4.	$\frac{17}{20}$		
5.		54%	
6.			0.24
7.			0.92
8.		84%	
9.	$\frac{37}{40}$		
10.	$\frac{2}{3}$		

> If 20% of the children in a class have blue eyes, what percentage do not?
>
> (All the children either have, or do not have, blue eyes.)
> The percentage with blue eyes is 20% and the percentage who do not have blue eyes is
>
> $$100\% - 20\% = 80\%$$

11. Seventy-six per cent of the passengers on an aeroplane are British. What percentage of them are not?

12. Deductions from a man's wage were: income tax 21%, pension scheme 8%, other deductions 7%. What percentage did he keep?

13. In the third year at Stanley School, 35% take woodwork only, 25% take metalwork only and 20% take both subjects. What percentage of the third year pupils study neither?

> Express 45 cm as a percentage of 2 m
>
> (First express both lengths in the same unit. We will express 2 m in cm.)
>
> $$2\,m = 2 \times 100\,cm = 200\,cm$$
>
> Then 45 cm as a percentage of 2 m is
>
> $$\frac{45}{200} \times 100\% = 22\tfrac{1}{2}\%$$

Express the first quantity as a percentage of the second:

14. 20 cm, 50 cm **17.** 15 cm, 3 m **20.** 54 mm, 20 cm

15. $3\,cm^2$, $12\,cm^2$ **18.** $210\,mm^2$, $84\,cm^2$ **21.** 800 g, 2 kg

16. £1.53, £4.50 **19.** $200\,mm^3$, $10\,cm^3$ **22.** $4500\,cm^3$, 2 litres

> Find 24% of 7.5 m
>
> $$24\% \text{ of } 7.5\,m = \frac{24}{100} \times 7.5\,m$$
>
> $$= 1.8\,m$$

Find the value of:

23. 30% of 250

24. 5% of £18.40

25. $12\frac{1}{2}$% of 4.88 cm

26. 12% of 4.5 km

27. 84% of 225 g

28. $33\frac{1}{3}$% of 126 m²

A pupil measures the length of a line as 8.2 cm. If its actual length is 8 cm find the percentage error.

$$\text{Error} = 0.2\,\text{cm}$$

$$\text{Percentage error} = \frac{\text{error}}{\text{actual value}} \times 100$$

$$= \frac{0.2}{8} \times 100$$

$$= 2.5$$

Therefore the error is 2.5%

Find the percentage error for each of the following values:

29. Measured length 12.3 m, actual length 12 m.

30. Measured area 147 cm², actual area 150 cm².

31. Measured volume 456.75 cm³, actual volume 450 cm³.

32. Measured weight 975 g, actual weight 1000 g.

33. Estimated cost £25.60, actual cost £25.

In Treetown 1680 out of a workforce of 12 000 are unemployed. What percentage is this?

$$\text{Percentage unemployed} = \frac{1680}{12\,000} \times 100\% = 14\%$$

---

34. Paul scored 27 out of a possible 45 in a science test. What percentage was this?

35. In a form of 30 pupils, 21 of them are boys. What percentage are girls?

36. A department store employs 250 female staff. If 26 of them have long hair, what percentage of the female staff do not have long hair?

If 72% of the 25 pupils in a class are good at mathematics how many are not good at it?

If 72% are good at mathematics
$(100 - 72)\% = 28\%$ are not good at it

$$28\% \text{ of class} = \frac{28}{100} \times 25 = 7$$

Therefore 7 are not good at mathematics.

---

37. There are 1460 pupils in my school and 35% are non-swimmers. How many pupils can swim?

38. All the students at a college live either in hall or in lodgings. If 62% of the 2650 students live in lodgings, how many live in hall?

39. Sixty-two per cent of the audience of 1650 at a concert were females. How many males attended?

Increase 250 by 42%

The new value is 142% of the old value.

i.e. $$\text{new value} = \frac{142}{100} \times 250$$
$$= 355$$

---

40. Increase 300 by 27%   **42.** Increase 240 by 45%

41. Increase 44 by 12%   **43.** Increase 23.4 by 35%

44. Decrease 200 by 14%   **46.** Decrease 350 by 16%

45. Decrease 84 by 23%   **47.** Decrease 16.4 by 65%

48. John is 8% heavier now than he was two years ago. If he weighed 55 kg then, how much does he weigh now?

49. The price of a car that cost £9000 last year was increased by 7.5% on 1st January this year. What is its present price?

50. A school employs 120 teachers. Next year they must reduce this number by 15%. How many teachers will there be next year?

PERCENTAGE INCREASE AND PERCENTAGE DECREASE

Percentage increase or decrease arises in many different areas of life today. We read that certain workers are to receive an increase in their wages of 8%; that value added tax (VAT) may be increased from 15% to 18%; that the basic rate of income tax should be reduced from 30% to 27%; or that all the items in a sale are offered at a discount of 20%.

Changes are expressed in percentage terms, because this makes it easier to calculate the actual change in a particular case and to compare one change with another.

If a wage of £100 per week is increased by 8% then the new wage is

$$\frac{108}{100} \times £100 = £108$$

If an article costs £55 plus VAT at 15%, then the full cost is

$$\frac{115}{100} \times £55 = £63.25$$

If a woman earns £550 and has to pay tax on it at the rate of 30% she actually receives

$$\frac{70}{100} \times £550 = £385$$

If a discount of 25% is offered in a sale, a piece of furniture, originally marked at £760, will cost

$$\frac{75}{100} \times £760 = £570$$

Retailers buy in goods which are usually sold at an increased price. The increase is often called "the mark-up" and is normally given as a percentage of the *buying in* price. Occasionally goods are sold at a decreased price, i.e. they are marked down. The mark-down is also given as a percentage of the *buying in* price.

If a store makes a mark-up of 50% on an article it buys for £100, its mark-up is

$$\frac{50}{100} \times £100 = £50$$

and the selling price is

$$\frac{150}{100} \times £100 = £150$$

EXERCISE 5b

A second-hand car dealer bought a car for £3500 and sold it for £4340. Find his percentage mark-up.

$$\text{Mark-up} = \text{selling price} - \text{buying-in price}$$
$$= £4340 - £3500$$
$$= £840$$

$$\% \text{ mark-up} = \frac{\text{mark-up}}{\text{buying-in price}} \times 100$$

$$= \frac{£840}{£3500} \times 100$$

Therefore the mark-up is 24%.

Find the percentage mark-up:

1. Buying-in price £12, mark-up £3

2. Buying-in price £28, mark-up £8.40

3. Buying-in price £16, mark-up £4

4. Buying-in price £55, mark-up £5.50

A retailer bought a leather chair for £375 and sold it for £285. Find his percentage mark-down.

$$\text{Mark-down} = \text{buying-in price} - \text{selling price}$$
$$= £375 - £285$$
$$= £90$$

$$\% \text{ mark-down} = \frac{\text{mark-down}}{\text{buying-in price}} \times 100$$

$$= \frac{\overset{6}{£90}}{\underset{25}{£375}} \times \overset{4}{100}$$

Therefore the mark-down is 24%.

Find the percentage mark-down:

5. Buying-in price £20, mark-down £4

6. Buying-in price £125, mark-down £25

7. Buying-in price £64, mark-down £9.60

8. Buying-in price £160, mark-down £38.40

An article costing £30 is sold at a gain of 25%. Find the selling price

Method 1 $\text{Gain} = \dfrac{\overset{1}{\cancel{25}}}{\underset{4}{\cancel{100}}} \times £30$

$= £7.50$

Selling price $= £30 + £7.50$

Therefore the selling price is £37.50

Method 2 $\text{Selling price} = \dfrac{\overset{5}{\cancel{125}}}{\underset{4}{\cancel{100}}} \times £30$

Therefore the selling price is £37.50

Find the selling price:

9. Cost £50, gain 12%   **12.** Cost £36, loss 50%

10. Cost £64, gain $12\frac{1}{2}$%   **13.** Cost £75, loss 64%

11. Cost £29, gain 110%   **14.** Cost £128, loss $37\frac{1}{2}$%

Find the weekly cash increase for each of the following employees:

15. Ian Dickenson earning £120 p.w. receives a rise of 10%

16. Nairn Williams earning £180 p.w. receives a rise of 12%

17. Sylvia Smith earning £225 p.w. receives a rise of 8%

18. Lyn Wyman earning £270 p.w. receives a rise of 9%

19. Joe Bright earning £300 p.w. receives a rise of 7%

Which is the better cash pay rise, and by how much?

20. a) 12% on a weekly pay of £100, or
b) 8% on a weekly pay of £250

21. a) 7% on a weekly pay of £90, or
b) $3\frac{1}{2}$% on a weekly pay of £200

22. a) 4% on a weekly pay of £300, or
b) 3% on a weekly pay of £400

Find the purchase price of the given object:

23. An electric fire marked £30 plus VAT at 15%

24. A food mixer marked £64 plus VAT at 18%

25. A calculator marked £8.40 plus VAT at $17\frac{1}{2}$%

26. A car tyre costing £38 plus VAT at 15%

27. A bathroom suite costing £600 plus VAT at 17%

28. A ticket for a concert costing £8 plus VAT at 15%

INCOME TAX AND SALE REDUCTIONS

EXERCISE 5c

Peter Smith has a taxable income of £8000 per annum. How much tax will he pay if the basic rate is 30%?

$$\text{Tax due} = \frac{30}{100} \times £8000$$
$$= £2400$$

Assuming that the basic rate of income tax is 30%, find the yearly tax on the following taxable incomes:

1. £5000

2. £8000

3. £6500

4. £12 500

5. £6450

6. £8260

Find the yearly income tax due on a taxable income of:

7. £10 000 if the basic tax rate is 33%

8. £8000 if the basic tax rate is 28%

9. £16 000 if the basic tax rate is 25%

10. £24 000 if the basic tax rate is 32%

Edgar Brooks earns £10 000 each year. If his tax free allowances amount to £4840, how much tax will he pay when the basic rate is 30%?

$$\text{Taxable income} = £10\,000 - £4840$$

$$= £5160$$

$$\text{Tax due} = \frac{30}{100} \times £5160$$

$$= £1548$$

Use the following details to find the income tax due in each case:

	Name	Gross Income	Allowances	Basic Tax Rate
11.	Miss Deats	£8000	£2000	30%
12.	Mr Evans	£10 000	£3000	30%
13.	Mrs Khan	£15 000	£3200	25%
14.	Mr Amos	£9000	£2600	33%
15.	Miss Eyles	£20 000	£4750	28%

During the January sales, a department store offers a discount of 10% off marked prices. What is the purchase price of   a) a dinner service marked £84.50
b) a pair of jeans marked £16.30?

Method 1

a) Discount of 10% on a marked price of £84.50 is

$$\frac{10}{100} \times £84.50 = £8.45$$

∴   Purchase price $= £84.50 - £8.45$

$$= £76.05$$

b) Discount of 10% on a marked price of £16.30 is

$$\frac{10}{100} \times £16.30 = £1.63$$

∴   Purchase price $= £16.30 - £1.63$

$$= £14.67$$

Method 2

a) If the discount is 10%, the cash price is 90% of the marked selling price

i.e. purchase price of dinner service is

$$\frac{90}{100} \times £84.50 = £76.05$$

b) Similarly, purchase price of jeans is

$$\frac{90}{100} \times £16.30 = £14.67$$

In a sale, a shop offers a discount of 20%. What would be the cash price for each of the following articles?

16.   A dress marked £35

17.   A lawn mower marked £115

18.   A pair of shoes marked £32

19.   A set of garden tools priced £72.50

20.   Light fittings marked £42 each

In a sale, a department store offers a discount of 50% on the following articles. Find their sale price:

21.   A pair of curtains marked £76.50

22.   A leather football marked £32.30

23.   A boy's jacket marked £28.60

24.   A girl's coat marked £64.50

25.   In order to clear a large quantity of woollen goods a shopkeeper puts them on sale at a discount of $33\frac{1}{3}\%$. Find the cash price of
a) a jumper marked £18.30   b) a skirt marked £22.20

FINDING THE ORIGINAL QUANTITY

Sometimes we are given an increased or decreased quantity and we want to find the original quantity. For example, if the cost of a chair including VAT at 15% is £172.50, we might wish to find the price of the chair before the tax was added.

EXERCISE 5d

An article is sold for £252. If this gives a mark-up of 5% find the buying-in price.

If there is a mark-up of 5% and the buying-in price is £x

then   Selling price $= \dfrac{105}{100}$ of the buying-in price

i.e. $252 = \dfrac{105}{100} \times x$

$100 \times 252 = \cancel{100} \times \dfrac{105}{\cancel{100}} \times x$

∴ $252 \times \dfrac{100}{105} = x$

i.e. $x = 240$

Therefore the buying-in price is £240

In questions 1 to 20, selling price is abbreviated to SP.

Find the buying-in price:

1. SP £98, mark-up 40%   **6.** SP £40, gain 25%

2. SP £64, mark-up 60%   **7.** SP £920, gain 15%

3. SP £28, mark-up 75%   **8.** SP £1008, gain 12%

4. SP £12, mark-up 100%   **9.** SP £888, gain 11%

5. SP £4.50, mark-up $66\frac{2}{3}$%   **10.** SP £21.50, gain $7\frac{1}{2}$%

A book is sold for £6.30 at a loss of 30%. Find the buying-in price.

If there is a loss of 30% and the buying-in price is x pence

then   Selling price $= \dfrac{70}{100}$ of the buying-in price

i.e. $630 = \dfrac{70}{100} x$

∴ $630 \times \dfrac{100}{70} = x$

i.e. $x = 900$

The buying-in price of the book is 900 p, or £9

Find the buying-in price:

11. SP £30, mark-down 25%   **16.** SP £45, mark-down 10%

12. SP £56, mark-down 30%   **17.** SP £120, mark-down 25%

13. SP £70, loss 65%   **18.** SP £8.50, loss 50%

14. SP £12, loss $33\frac{1}{3}$%   **19.** SP £64, loss 60%

15. SP £8.16, loss 40%   **20.** SP £1200, loss 40%

After a pay rise of 5%, Peter's weekly pay is £126. How much did he earn before the rise?

If Peter's original pay was £x and he receives a 5% pay rise his new pay will be 105% of his original pay.
But his new pay is £126,

$$\therefore \qquad 126 = \frac{105}{100} \times x$$

i.e. $$x = \frac{126 \times 100}{105}$$

$$= 120$$

Peter's original weekly pay was £120

The following table shows the weekly wage of a number of employees after percentage increases as shown. Find the original weekly wage of each employee:

	Name	% increase in pay	Weekly wage after increase
21.	George Black	10%	£132
22.	Anne Reed	8%	£135
23.	John Rowlands	15%	£299
24.	Beryl Lewis	7%	£196.88
25.	Enid Jones	4%	£95.68

The purchase price of a watch is £70.50. If this includes VAT at $17\frac{1}{2}\%$, find the price before VAT was added.

If the cost of an article is £C,
the cost of including VAT at $17\frac{1}{2}\%$ will be $\frac{117.5}{100} \times £C$

i.e. $1.175 \times C = 70.5$

i.e. $C = 70.5 \div 1.175$

$= 60$

The price before VAT was added was £60

26. The purchase price of a hairdryer is £13.80. If this includes VAT at 15%, find the price before VAT was added.

27. I paid £763.75 for a dining table and four chairs. If the price includes VAT at $17\frac{1}{2}\%$ find the price before VAT was added.

28. John's income last week was £112 after income tax had been deducted at 30%. Calculate his pay before the tax was deducted.

29. Water increases in volume by 4% when it is frozen. How much water is required to make $884\,cm^3$ of ice?

30. The stretched length of an elastic string is 31 cm. If this is 24% more than its unstretched length, find its unstretched length.

PROBLEMS INVOLVING PERCENTAGE INCREASE AND DECREASE

Remember that a percentage increase or decrease is always calculated as a percentage of the original quantity.

EXERCISE 5e   **1.** A house is bought for £68 000 and sold at a profit of 14%. Find the selling price.

2. Carpets that had been bought for £18.50 per square metre were sold at a loss of 26%. Find the selling price per square metre.

3. Potatoes bought at £4.50 per 50 kg bag are sold at 12 p per kg. Find the percentage profit.

4. A shopkeeper buys 80 articles for £200 and sells them for £3.50 each. Find his percentage profit.

5. An art dealer sold a picture for £1980 thus making a profit of 65%. What did she pay for it?

6. My present average weekly grocery bill is £40.50, which is 8% more than the same goods cost me, on average, each week last year. What was my average weekly grocery bill last year?

7. When water freezes, its volume increases by 4%. What volume of water is required to make 221 cm^3 of ice?

8. What is Fred Procter's gross weekly wage, if, after paying deductions of 35%, he is left with £111.80?

9. Engine modifications were made to a particular model of car. As a result the number of kilometres it travels on one litre of petrol increases by 8%. If the new petrol consumption is 16.2 km/l what was the previous value?

10. Between two elections, the size of the electorate in a constituency fell by 16%. For the second election, 37 191 people were entitled to vote. How many could have voted at the first election?

11. When the rate of VAT is 15% an LP record costs £6.90. What will it cost if the rate of VAT is
a) increased to 20%   b) decreased to 10%?

12. The purchase price of a diamond ring increases by £20 when the rate of VAT is increased from 12% to 20%. Find the original purchase price of the ring.

INTEREST

Many institutions such as banks and building societies offer savings accounts. If you put money into such an account, the bank uses your money for other purposes and pays you for that use. The amount that the bank pays for the use of your money is called *interest*.

The time may come when you wish to use someone else's money to buy an expensive item such as a car or even a house. You will normally have to pay for the use of borrowed money, i.e. you will have to repay more than you borrow. The difference between what you borrow and what you repay is called *interest*.

Interest on money borrowed (or lent) is usually a percentage of the sum borrowed (or lent). This percentage is often given as a charge per year (per annum, or p.a.) and it is then called the *interest rate*. For example, if £100 is put into a building society account with an interest rate of 8% p.a., then after one year, the society pays

$$8\% \text{ of } £100, \quad \text{i.e.} \quad £8$$

EXERCISE 5f

1. Pam is given £500 for her 18th birthday. She puts the money in a savings account with an interest rate of 12% p.a. How much is added to her account after one year?

2. Ann Peters is given a loan of £650 from the bank which she agrees to repay after one year. How much does she have to repay if the interest rate is $12\frac{1}{2}\%$ p.a.?

Find the interest payable after one year on each of the following sums of money invested (i.e. put in a savings account) at the given interest rate.

3. £352 at 7.5% p.a.

5. £2600 at 8.3% p.a.

4. £10 000 at 9.25% p.a.

6. £5840 at 6.4% p.a.

7. What annual rate of interest is necessary to give interest of £238 after one year on an investment of £2800?

8. What is the amount of a loan that costs £45 when repaid after one year when the interest rate is 9% p.a.?

9. Find the amount borrowed if £287.50 has to be repaid after one year when the interest rate is 15% p.a.

COMPOUND PERCENTAGE PROBLEMS

There are many occasions when a percentage increase or decrease happens more than once. Suppose that a house is bought for £20 000 and increases in value (appreciates) by 10% of its value each year.

After one year, its value will be 110% of its initial value, i.e.

$$\frac{110}{100} \times £20\,000 = £22\,000$$

The next year it will increase by 10% of the £22 000 it was worth at the beginning of the year,
i.e. its value after two years will be

$$\frac{110}{100} \times £22\,000 = £24\,200$$

While some things increase in value year after year, many things decrease in value (depreciate) each year. Should you buy a car or a motorcycle it will probably depreciate in value more quickly than anything else you buy.

If you invest money in a Building Society or Post Office Savings Account and do not spend the interest, your money will increase by larger amounts each year.

This kind of interest is called *compound interest*.

EXERCISE 5g In this exercise give all answers that are not exact correct to the nearest penny.

Find the compound interest on £260 invested for 2 years at 8% p.a.

Simple interest for first year at $8\% = \frac{8}{100} \times £260$

$$= £\frac{2080}{100}$$

$$= £20.80$$

New principal = original principal + interest

$$= £260 + £20.80$$

$$= £280.80$$

Simple interest for second year at 8%

$$= \frac{8}{100} \times £280.80$$

$$= £22.464$$

$$= £22.46 \quad \text{(correct to the nearest penny)}$$

∴ total of interest for the two years is

$$£20.80 + £22.46 = £43.26$$

i.e. compound interest on £260 for 2 years at 8% is £43.26

Find the compound interest on:

1. £200 for 2 years at 10% p.a.

2. £300 for 2 years at 12% p.a.

3. £400 for 3 years at 8% p.a.

4. £650 for 3 years at 9% p.a.

5. £520 for 2 years at 13% p.a.

6. £690 for 2 years at 14% p.a.

7. £624 for 3 years at 12% p.a.

8. A house is bought for £40 000 and appreciates at 10% each year. What will it be worth in 2 years' time?

9. A postage stamp increases in value by 15% each year. If it is bought for £50, what will it be worth in 3 years time?

10. A motorcycle bought for £1500 depreciates in value by 10% each year. Find its value after 3 years.

11. A motor car bought for £20 000 depreciates in any one year by 20% of its value at the beginning of that year. Find its value after 2 years.

6 STRAIGHT LINE GRAPHS

STRAIGHT LINES

Straight lines can lie anywhere and at any angle to the x-axis.

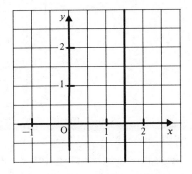

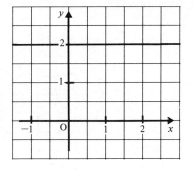

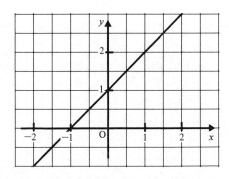

The lines go on for ever in both directions. We draw just part of the line.

93

LINES PARALLEL TO THE AXES

If we consider any point on the line given in the diagram we find that, no matter what the y coordinate is, the x coordinate is always 2.

We say that the equation of this line is $x = 2$

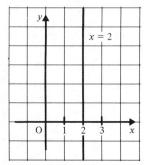

Notice that y is not mentioned because there is no restriction on the value it can take.

In the same way, any point on the line given in this diagram has a y coordinate of 3, while the x coordinate can take any value. The equation of this line is $y = 3$

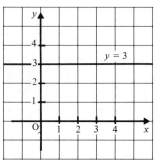

EXERCISE 6a   Give the equations of the lines in questions 1 to 4:

1.

2.

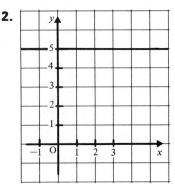

3. 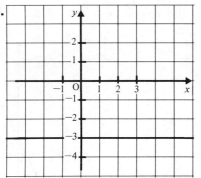   **4.**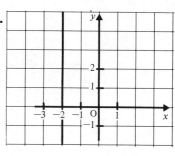

Draw similar diagrams to show the lines with the following equations:

5. $x = 1$ **7.** $y = 4$

6. $x = -4$ **8.** $y = -3$

SLANT LINES

If we take any point on the line shown, say $(2, 4)$ or $(3\frac{1}{2}, 5\frac{1}{2})$ or $(-1, 1)$ or $(0, 2)$, we find that the y coordinate is always 2 units more than the x coordinate.

The equation of the line is therefore $y = x + 2$

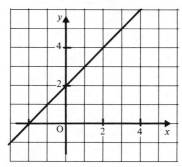

Conversely, if we are given the equation of a line we can draw the line by finding points on it. Two points are enough to draw a straight line but a third is useful as a check. If the three points do not lie in a straight line, at least one point is incorrect. Check all three points.

Suppose that we want to draw the line whose equation is $y = 2x + 1$. Think of this as an instruction for finding the y coordinate to go with a chosen x coordinate.

If $x = 1,$ $y = 2 + 1 = 3$ so $(1, 3)$ is a point on the line.
If $x = 3,$ $y = 2 \times 3 + 1 = 7$ so $(3, 7)$ is on the line.
If $x = -3, y = 2 \times (-3) + 1 = -5$ so $(-3, -5)$ is on the line.

It is simpler to list this information in a table.

x	-3	1	3
y	-5	3	7

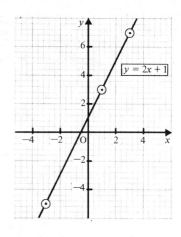

EXERCISE 6b   In all questions in this exercise use 1 cm to 1 unit.

In each of questions 1 to 4, copy and complete the table. On graph paper, draw x and y axes for the ranges indicated in the brackets. Mark the points and, if they are in a straight line, draw that line:

1. $y = x + 4$ $(-2 \leqslant x \leqslant 4, 0 \leqslant y \leqslant 8)$

x	-2	0	4
y			

2. $y = 2x + 1$ $(-2 \leqslant x \leqslant 3, -4 \leqslant y \leqslant 8)$

x	-2	0	3
y			

3. $y = 4 - x$ $(-3 \leqslant x \leqslant 3, 0 \leqslant y \leqslant 8)$

x	-3	0	3
y			

4. $y = 2 - 3x$ $(-1 \leqslant x \leqslant 3, -7 \leqslant y \leqslant 5)$

x	-1	0	3
y			

In each of the questions 5 to 8, make a table, choosing your own values of x within the given range. (Choose one low value, one high value and one in between, such as zero.) Draw x and y axes for the ranges of values indicated. Draw the line.

5. $y = x - 3$ $(-2 \leqslant x \leqslant 5, -5 \leqslant y \leqslant 2)$

6. $y = \frac{1}{2}x + 4$ $(-2 \leqslant x \leqslant 4, 0 \leqslant y \leqslant 6)$

7. $y = 3 - 2x$ $(-2 \leqslant x \leqslant 4, -5 \leqslant y \leqslant 7)$

8. $y = 3x - 4$ $(0 \leqslant x \leqslant 6, -5 \leqslant y \leqslant 14)$

Make a table which can be used for drawing the graph of
$y = 6 + 2x$, taking values of x in the range $-3 \leqslant x \leqslant 3$.
Decide from the table what range of values of y is suitable.

$y = 6 + 2x$

x	-3	0	3
y	0	6	12

The range for y must be $0 \leqslant y \leqslant 12$

For each of the questions 9 to 12, make a table, choosing your own
values of x within the given range. Decide on a suitable range of
values of y after completing the table.

Draw x and y axes using the given range for the x-axis and the range
you have chosen for the y-axis. Draw the line:

9. $y = 3 - x$ $(-3 \leqslant x \leqslant 3)$

10. $y = \frac{1}{2}x - 1$ $(-3 \leqslant x \leqslant 3)$

11. $y = 2x + 2$ $(-2 \leqslant x \leqslant 4)$

12. $y = 5 - 3x$ $(-1 \leqslant x \leqslant 3)$

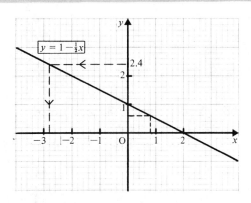

You are given the graph of the line with equation
$y = 1 - \frac{1}{2}x$. From the graph, find
a) the value of y, if x is 0.8
b) the value of x, if y is 2.4

a) From the graph, if $x = 0.8$, $y = 0.6$

b) From the graph, if $y = 2.4$, $x = -2.8$

13. You are given the graph of the line with equation $y = x + 1$
From the graph, find
a) y if $x = \frac{1}{2}$ b) x if $y = 1.4$ c) x if $y = -0.6$

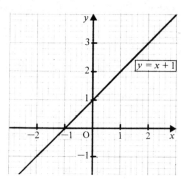

14. You are given the graph of the line with equation $y = 2x - 1$
From the graph, find
a) y if $x = \frac{1}{2}$ b) x if $y = -2.6$ c) y if $x = -1.2$

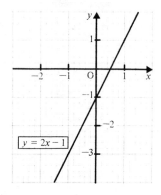

15. You are given the graph of the line with equation $y = -x - 1$
From the graph, find
a) y if $x = 1.6$ b) x if $y = 0.8$ c) x if $y = -2.2$

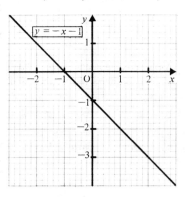

16. You are given the graph of the line with equation $y = 3 - 3x$
From the graph, find
a) y if $x = -0.2$ b) x if $y = 1.2$ c) x if $y = -0.6$

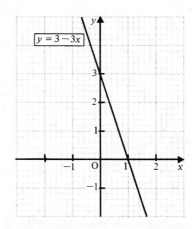

17. Using the graph you drew for question 5, find
a) y if $x = \frac{1}{2}$ b) x if $y = 1.4$ c) x if $y = -0.6$

18. Using the graph you drew for question 6, find
a) y if $x = 1.6$ b) x if $y = 4.6$ c) x if $y = -1.6$

19. Using the graph you drew for question 7, find
a) y if $x = 2.2$ b) x if $y = 0.2$ c) x if $y = -4$

20. Using the graph you drew for question 8, find
a) y if $x = 4.2$ b) x if $y = 4.4$ c) x if $y = 5$

POINTS ON A LINE

EXERCISE 6c

Do the points $(3, 6)$ and $(-2, -10)$ lie on the line whose equation is $y = 3x - 4$?

$$y = 3x - 4$$

When $x = 3$ $y = 3 \times 3 - 4 = 5$

$\therefore$ $(3, 6)$ does not lie on the line.

When $x = 2$ $y = 3 \times (-2) - 4 = -10$

$\therefore$ $(2, -10)$ lies on the line.

Find whether the given points lie on the line whose equation is given:

1. $y = 2x - 1$; $(5, 9)$, $(1, 2)$ **4.** $y = \frac{1}{2}x + 4$; $(4, 5)$, $(3, 5\frac{1}{2})$

2. $y = 3 + 3x$; $(2, 9)$, $(-4, -9)$ **5.** $y = 5 + 4x$; $(\frac{1}{2}, 7)$, $(-2, -3)$

3. $y = 6 - 2x$; $(3, 1)$, $(4, -1)$ **6.** $y = 6 - \frac{1}{2}x$; $(-2, 5)$, $(-2, 7)$

COMPARING SLOPES

EXERCISE 6d   **1.** Draw, on the same pair of axes, the graphs of the lines
a) $y = 2x - 3$ b) $y = 2x$ c) $y = 2x - 2$
for $-3 \leqslant x \leqslant 3$ and $-7 \leqslant y \leqslant 9$ using 1 cm to 1 unit.
Label each line with its equation.
What can you say about the lines?
What do the equations have in common?

2. Draw a pair of axes using $-2 \leqslant x \leqslant 2$ and $-7 \leqslant y \leqslant 9$. On these axes draw the three lines with the following equations
a) $y = -3x - 1$ b) $y = -3x$ c) $y = -3x + 3$
What can you say about the lines?
What do the equations have in common?

3. Draw a pair of axes using $-4 \leqslant x \leqslant 4$, $-7 \leqslant y \leqslant 6$.
On these axes draw the three lines with the following equations
a) $y = \frac{1}{2}x - 4$ b) $y = \frac{1}{2}x + 1$ c) $y = \frac{1}{2}x + 3$
Comment on the three lines and their equations.

4. Draw x and y axes, marking values from -5 to 5 on each axis.
Draw the three lines with the equations
a) $y = x$ b) $y = x - 3$ c) $y = x + 2$
What do you notice?

5. Draw x and y axes, using $-3 \leqslant x \leqslant 3$, $-3 \leqslant y \leqslant 5$. Draw the three lines with the following equations
a) $y = 2x + 2$ b) $y = -2x + 1$ c) $y = 2x$
Which two lines are parallel?

6. Draw x and y axes, marking values from -5 to 5 on each axis.
Draw the lines with the equations
a) $y = 4 - x$ b) $y = -x$ c) $y = -3 - x$
What do you notice?

GRADIENTS

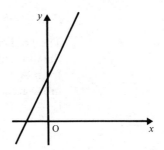

Different lines have different gradients or slopes. Some lines point steeply up to the right; they have large *positive* gradients.

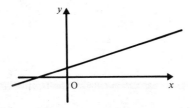

Some have a shallow slope up to the right; they have small positive gradients.

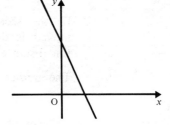

Some slope the other way; they have *negative* gradients.

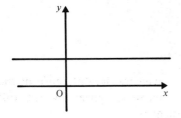

Some have zero gradient and are parallel to the *x*-axis.

Some are parallel to the *y*-axis.

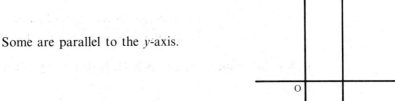

We can see from the questions in the last exercise that the *coefficient* of *x* (2 in question 1, -3 in question 2 and $\frac{1}{2}$ in question 3) has something to do with gradient.

CALCULATING THE GRADIENT OF A LINE

First Method

The gradient is found by comparing the amount you move up as you go from A to B, with the amount you move across. In this case the gradient is $\frac{4}{3}$.

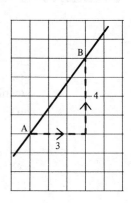

In the second case the distance moved is downwards so we take this distance as negative.

The gradient is $\frac{-3}{5}$, i.e. $-\frac{3}{5}$.

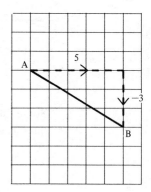

Second Method

The distance moved up is given by the difference in the y coordinates and the distance moved across is given by the difference in the x coordinates, so the gradient is

$$\frac{\text{the difference in the } y \text{ coordinates}}{\text{the difference in the } x \text{ coordinates}}$$

If A is the point $(3, 5)$ and B is $(2, 7)$ then the gradient is

$$\frac{5-7}{3-2} = \frac{-2}{1} = -2$$

Notice that the coordinates of B are taken from the coordinates of A for both x and y. We may change the order as long as we change it for both x and y, i.e. the gradient is also $\frac{7-5}{2-3} = -2$

EXERCISE 6e

Find the gradients of the lines joining the points
a) $(4, 2)$ and $(6, 7)$ b) $(2, 3)$ and $(4, -3)$

a)

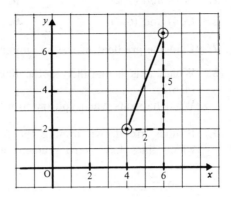

Either: from the diagram the gradient is $\frac{5}{2}$

Or: gradient $= \dfrac{\text{difference in the } y \text{ coordinates}}{\text{difference in the } x \text{ coordinates}}$

$$= \frac{7-2}{6-4} = \frac{5}{2}$$

b)

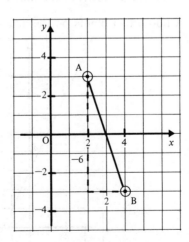

Either: from the diagram the gradient is $\dfrac{-6}{2} = -3$

Or: gradient $= \dfrac{-3-3}{4-2} = \dfrac{-6}{2} = -3$

(Notice that we could also say

$$\text{gradient} = \frac{3-(-3)}{2-4} = \frac{6}{-2} = -3)$$

Find the gradients of the lines joining the following pairs of points:

1. $(5, 1)$ and $(7, 9)$ **4.** $(-2, 4)$ and $(2, 1)$

2. $(3, 6)$ and $(5, 2)$ **5.** $(1, 2)$ and $(6, -7)$

3. $(3, 4)$ and $(6, 7)$ **6.** $(-3, 4)$ and $(-6, 2)$

7. Find the gradient of the line joining the points $(4, 3)$ and $(7, 3)$.

8. Which axis is parallel to the line joining the points $(4, 3)$ and $(4, 6)$?
What happens when you try to work out the gradient?

9. If lines are drawn joining the following pairs of points, state which lines have zero gradient and which are parallel to the y-axis:

a) $(0, 4)$ and $(0, -2)$ c) $(-6, 0)$ and $(-2, 0)$

b) $(3, 0)$ and $(-10, 0)$ d) $(0, 6)$ and $(0, 12)$

10. If $(2, 1)$ is a point on a line and its gradient is 3, draw the line and find the coordinates of two other points on it.

EXERCISE 6f

Find the gradient of the line $y = -2x + 3$

Choose two points on the line, e.g. $(-2, 7)$ and $(2, -1)$.

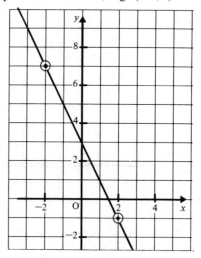

$$\text{Gradient} = \frac{-1-7}{2-(-2)} = \frac{-8}{4}$$

$$= -2$$

(This is the same as the coefficient of x in the equation of the line.)

Choose two points on each line and hence find the gradient of the line. In each case compare your answer with the coefficient of x:

1. $y = 2x + 3$ **3.** $y = 2x - 1$

2. $y = x + 4$ **4.** $y = 3 - 2x$

5. State the gradient of the line $y = 4x + 1$, without calculation if possible.

6. Give the gradients of the lines with equations
   a) $y = 4x + 4$ c) $y = x - 3$
   b) $y = 2 - 3x$ d) $y = \frac{1}{2}x + 1$

7. Sketch a line with a gradient of   a) 3   b) -1

8. Sketch a line with a gradient of   a) 4   b) $\frac{1}{2}$

9. Sketch a line with a gradient of   a) -2 b) 1

THE INTERCEPT ON THE y-AXIS

Consider the line with equation $y = x + 3$.
When $x = 0$, $y = 3$, so the line crosses the y-axis where $y = 3$

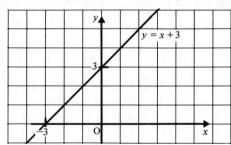

If the equation is $y = 2x - 3$ then, when $x = 0$, $y = -3$, i.e. the line cuts the y-axis where $y = -3$

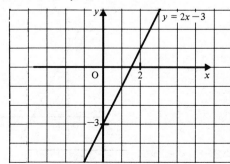

As long as the equation is of the form $y = mx + c$, i.e. like $y = 2x + 3$ or $y = -5x + 1$ then

> the number term c tells us where the line cuts the y-axis
> and m, the coefficient of x, tells us the gradient.

If the equation is $y = 3x$ then the number term is 0, i.e. the intercept is 0. Therefore, the line goes through the origin.

EXERCISE 6g

Give the gradient and the intercept on the y-axis of the line with equation $y = x + 5$. Sketch the line

The gradient is 1

The intercept on the y-axis is 5

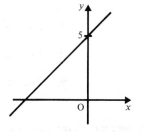

Give the gradients and the intercepts on the y-axis of the lines with the following equations. *Sketch* each line:

1. $y = 2x + 4$ **3.** $y = 3x - 4$

2. $y = 5x + 3$ **4.** $y = x - 6$

Give the gradient and the intercept on the y-axis of the line with equation $y = 4 - 2x$. Sketch the line

Rewrite the equation as $y = -2x + 4$

The gradient is -2. The intercept on the y-axis is 4

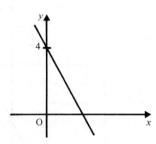

Give the gradients and the intercepts on the *y*-axis of the lines with the following equations. *Sketch* each line:

5. $y = 3 - 2x$

6. $y = -4x + 2$

7. $y = 2 + 5x$

8. $y = \frac{1}{2}x - 1$

9. $y = -\frac{1}{3}x + 4$

10. $y = 3x - 7$

11. $y = 7 - 3x$

12. $y = \frac{1}{3}x + 7$

13. $y = 9 - 0.4x$

14. $y = 4 + 5x$

Give the gradient and the intercept on the *y*-axis of the line with equation $2y = 3x - 1$. Sketch the line

$$2y = 3x - 1$$

Divide both sides by 2 $y = \frac{3}{2}x - \frac{1}{2}$

The gradient is $\frac{3}{2}$

The intercept on the *y*-axis is $-\frac{1}{2}$

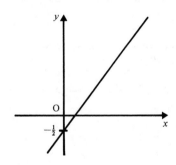

Give the gradients and the intercepts on the *y*-axis of the lines with the following equations. *Sketch* each line:

15. $2y = 4x + 5$

16. $3y = x - 6$

17. $5y = 5 + 2x$

18. $4y = 8 - 3x$

A line has a gradient of -2 and the intercept on the *y*-axis is 4. Find the equation of the line

The equation is $y = -2x + 4$

Write down the equations of the lines with the given gradients and intercepts on the y-axis (y intercepts):

	Gradient	y intercept
19.	2	7
20.	3	1
21.	1	3
22.	2	−5

	Gradient	y intercept
23.	$\frac{1}{2}$	6
24.	−2	1
25.	1	−2
26.	$-\frac{1}{2}$	4

PARALLEL LINES

EXERCISE 6h

Which of the lines with the following equations are parallel:
$y = 2x + 3$, $y = 4 - 2x$, $y = 4 + 2x$, $2y = x + 1$, $y = x + 3$?

The gradients of the lines are 2, −2, 2, $\frac{1}{2}$ and 1, so the first and third lines are parallel.

In questions 1 to 4, state which of the lines with the given equations are parallel.

1. $y = 3x + 1$, $y = \frac{1}{3}x - 4$, $y = x + 1$, $y = 4 - 3x$, $y = 5 + 3x$, $y = 3x - 4$

2. $y = 2 - x$, $y = x + 2$, $y = 4 - x$, $2y = 3 - 2x$, $y = -x + 1$, $y = -x$

3. $3y = x$, $y = \frac{1}{3}x + 2$, $y = \frac{1}{3} + x$, $y = \frac{1}{3} + \frac{1}{3}x$, $y = \frac{1}{3}x - 4$

4. $y = \frac{1}{2}x + 2$, $y = 2 - \frac{1}{2}x$, $y = -x - 4$, $y = \frac{1}{2}x - 1$, $2y = 3 - x$

5. What is the gradient of the line with equation $y = 2x + 1$? Give the equation of the line that is parallel to the first line and which cuts the y-axis at the point $(0, 3)$.

6. What is the gradient of the line with equation $y = 6 - 3x$? If a parallel line goes through the point $(0, 1)$, what is its equation?

7. Give the equation of the line through the origin that is parallel to the line with equation $y = 4x + 2$

8. Give the equations of any three lines that are parallel to the line with equation $y = 4 - x$

9. Give the equations of the lines through the point $(0, 4)$ that are parallel to the lines with equations
a) $y = 4x + 1$ b) $y = 6 - 3x$ c) $y = \frac{1}{2}x + 1$

10. Give the equations of the lines, parallel to the line with equation $y = \frac{1}{3}x + 1$, that pass through the points
a) $(0, 6)$ b) $(0, 0)$ c) $(0, -3)$

11. Give the equations of the lines with gradient 2 which pass through the points
a) $(0, 2)$ b) $(0, 10)$ c) $(0, -4)$

12. Which two of the lines with the following equations are parallel?
$y = 3 + 2x$, $y = 3 - 2x$, $y = 2x - 3$

13. Find the gradients and the intercepts on the y-axis of the lines with equations $y = 4 - 3x$ and $y = 4x - 3$. Give the equation of line which is parallel to the first line and cuts the y-axis in the same point as the second line.

14. A line of gradient -4 passes through the origin.
a) Give its equation
b) Give the equation of the line that is parallel to the first line and which passes through the point $(0, -7)$

DIFFERENT FORMS OF THE EQUATION OF A STRAIGHT LINE

The terms in the equation of a straight line can only be x terms, y terms or number terms.

An equation containing terms like x^2, y^2, $\frac{1}{x}$, $\frac{1}{x^2}$ is not the equation of a straight line.

Sometimes the equation of a straight line is not given exactly in the form $y = mx + c$. It could be $2x + y = 6$ or $\frac{x}{4} + \frac{y}{2} = 1$

An easy way to draw a line when its equation is in one of these forms is to start by finding the point where it cuts each axis.

EXERCISE 6i

Draw on graph paper the line with equation

$$3x - 4y = 12 \qquad \text{(use } 0 \leqslant x \leqslant 5\text{)}$$

Find the gradient of the line.

$$3x - 4y = 12$$

When $x = 0$, $-4y = 12$, i.e. $y = -3$

When $y = 0$, $3x = 12$, i.e. $x = 4$

(Draw the line using these two points only. Choose a point on the line, e.g. $(2, -1\frac{1}{2})$, to check.)

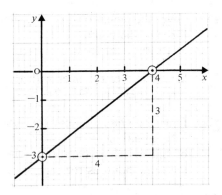

Check: when $x = 2$, $y = -1\frac{1}{2}$,

$3x - 4y = 6 - 4 \times (-1\frac{1}{2}) = 6 + 6 = 12$, which is correct.

From the graph, the gradient $= \frac{3}{4}$

Draw on graph paper the lines with the following equations. Use 1 cm to 1 unit. Find the gradient of each line:

1. $3x + 5y = 15$ $\quad 0 \leqslant x \leqslant 6$ $\quad$ **5.** $2x + y = 5$ $\quad -2 \leqslant x \leqslant 4$

2. $2x + 6y = 12$ $\quad 0 \leqslant x \leqslant 6$ $\quad$ **6.** $x + 3y = 5$ $\quad 0 \leqslant x \leqslant 5$

3. $x - 4y = 8$ $\quad 0 \leqslant x \leqslant 8$ $\quad$ **7.** $x - 3y = 6$ $\quad 0 \leqslant x \leqslant 6$

4. $x + y = 6$ $\quad 0 \leqslant x \leqslant 7$ $\quad$ **8.** $2x - y = 3$ $\quad -2 \leqslant x \leqslant 2$

9. On the same axes $(-6 \leqslant x \leqslant 6$ and $-6 \leqslant y \leqslant 6)$ draw the lines $x + y = 1$, $x + y = 4$, $x + y = 6$, $x + y = -4$. Find their gradients.

EXERCISE 6j

Draw on graph paper the line with equation

$$\frac{x}{3} + \frac{y}{2} = 1 \qquad (-1 \leqslant x \leqslant 4)$$

Find its gradient.

When $x = 0$, $\frac{y}{2} = 1$, so $y = 2$

When $y = 0$, $\frac{x}{3} = 1$, so $x = 3$

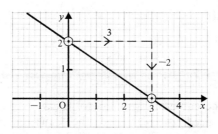

Check: from the graph, when $x = 1\frac{1}{2}$, $y = 1$

Therefore $\frac{x}{3} + \frac{y}{2} = \frac{1\frac{1}{2}}{3} + \frac{1}{2} = \frac{1}{2} + \frac{1}{2} = 1$ which is correct.

The gradient is $-\frac{2}{3}$

Draw on graph paper the lines whose equations are given below. Find the gradient of each line:

1. $\frac{x}{4} + \frac{y}{3} = 1$

4. $\frac{x}{3} + \frac{y}{6} = 1$

2. $\frac{x}{5} + \frac{y}{3} = 1$

5. $\frac{x}{1} - \frac{y}{2} = 1$

3. $\frac{x}{4} - \frac{y}{2} = 1$

6. $\frac{y}{3} - \frac{x}{4} = 1$

7. Without drawing a diagram, state where the lines with the following equations cut the axes

a) $\frac{x}{2} + \frac{y}{4} = 1$ b) $\frac{x}{12} - \frac{y}{9} = 1$

8. Form the equations of the lines which cut the axes at
a) $(0,5)$ and $(6,0)$ b) $(0,-3)$ and $(4,0)$.

9. Sketch the line with equation $\dfrac{x}{6}+\dfrac{y}{2}=1$ and find its gradient.

GETTING INFORMATION FROM THE EQUATION OF A LINE

From the last exercise, we can see that if the equation of a line is in
the form $\dfrac{x}{a}+\dfrac{y}{b}=1$ (i.e. like questions 1 to 6),

then the line cuts the x-axis at $x=a$
and the y-axis at $y=b$.

Then if we sketch the line we can work out the gradient.

If the equation is in the form $ax+by=c$, like those in Exercise 6i,
we need to rearrange the equation so that it is in the form
$y=mx+c$. Then the gradient and the intercept on the y-axis can be
seen.

EXERCISE 6k

Find the gradient and the intercept on the y-axis of the
line with equation $2x+3y=6$

Take $2x$ from both sides $3y=6-2x$

Divide both sides by 3 $y=2-\dfrac{2}{3}x$

i.e. $y=-\dfrac{2}{3}x+2$

The gradient is $-\dfrac{2}{3}$ and the intercept on the y-axis is 2

Find the gradient and the intercept on the y-axis of each of the
following lines:

1. $3x+5y=15$

2. $2x+6y=12$

3. $x-4y=8$

4. $x-3y=6$

5. $y-3x=6$

6. $x+3y=6$

Find the gradient and the intercept on the y-axis of the line with equation $\dfrac{x}{2} + \dfrac{y}{5} = 1$

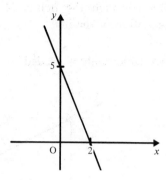

Either

This line cuts the axes at $(2, 0)$ and $(0, 5)$, hence the gradient is $-\dfrac{5}{2}$ and the intercept on the y-axis is 5

Or

$$\frac{x}{2} + \frac{y}{5} = 1$$

Multiply both sides by 5

$$\frac{5x}{2} + y = 5$$

Take $\dfrac{5x}{2}$ from both sides

$$y = 5 - \frac{5x}{2}$$

i.e. $y = -\dfrac{5}{2}x + 5$

Then the gradient is $-\dfrac{5}{2}$ and the intercept is 5

Find the gradient and the intercept on the y-axis of each of the following lines:

7. $\dfrac{x}{4} + \dfrac{y}{3} = 1$ **10.** $\dfrac{x}{2} + \dfrac{y}{6} = 1$ **<u>13.</u>** $y = 4x + 2$

8. $\dfrac{x}{5} + \dfrac{y}{3} = 1$ **11.** $\dfrac{x}{3} + \dfrac{y}{4} = 1$ **<u>14.</u>** $x + y = 4$

9. $\dfrac{x}{4} - \dfrac{y}{2} = 1$ **12.** $\dfrac{x}{3} - \dfrac{y}{4} = 1$ **<u>15.</u>** $\dfrac{x}{2} + \dfrac{y}{4} = 1$

16. $2x + 5y = 15$ **18.** $2y = 4x + 5$ **<u>20.</u>** $x + y = -3$

17. $y = 5 - \dfrac{1}{2}x$ **19.** $\dfrac{x}{2} - \dfrac{y}{4} = 1$ **<u>21.</u>** $3x + 4y = 12$

THE EQUATION OF A LINE THROUGH TWO GIVEN POINTS

EXERCISE 6I

Find the gradient and the intercept on the y-axis of the line through the points $(4,2)$ and $(0,4)$. Hence give the equation of the line.

(A sketch only is needed.)

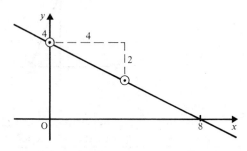

$$\text{Gradient} = \frac{2-4}{4-0} = \frac{-2}{4} = -\frac{1}{2}$$

$$\text{Intercept} = 4$$

∴ the equation is $y = -\frac{1}{2}x + 4$

(Multiplying by 2 and rearranging allows this equation to be written in the form $2y = 8 - x$ or $2y + x = 8$.)

Find the gradient and the intercept on the y-axis of the line through the given points. Hence give the equation of the line:

1. $(0,4)$ and $(3,0)$

2. $(0,7)$ and $(2,3)$

3. $(5,4)$ and $(0,1)$

4. $(0,2)$ and $(3,-2)$

5. $(0,-4)$ and $(2,3)$

6. $(-6,-3)$ and $(0,-1)$

7. $(6,2)$ and $(0,1)$

8. $(5,1)$ and $(0,-3)$

9. $(0,-4)$ and $(3,1)$

10. $(-5,0)$ and $(0,-5)$

11. $(0,12)$ and $(-6,0)$

12. $(-6,1)$ and $(0,6)$

13. A, B and C are the points $(0,4)$, $(5,6)$ and $(3,-1)$. Find the equations of lines AB and AC.

Find the gradient and the equation of the line through $(6, 2)$ and $(4, 8)$

The gradient is $\dfrac{2-8}{6-4} = \dfrac{-6}{2} = -3$

(We do not know the intercept on the y-axis.)

Let the equation be $y = -3x + c$.

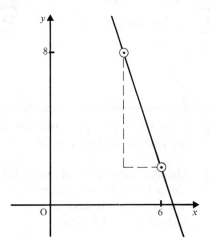

When $x = 6$, $y = 2$ so $2 = -18 + c$

$\therefore$ $c = 20$

$\therefore$ the equation is $y = -3x + 20$

(Check: when $x = 4$, the equation gives $y = -12 + 20$, i.e. $y = 8$, which is correct.)

In questions 14 to 35, find the gradient and the equation of the line through the given pair of points:

14. $(4, 1)$ and $(7, 10)$ **17.** $(-2, 3)$ and $(-1, 5)$

15. $(1, 4)$ and $(2, 1)$ **18.** $(4, -1)$ and $(3, -6)$

16. $(3, 7)$ and $(5, 12)$ **19.** $(5, -2)$ and $(-4, 7)$

20. $(-6, 7)$ and $(1, 0)$ **22.** $(-9, -3)$ and $(6, 0)$

21. $(4, -3)$ and $(2, -7)$ **23.** $(3, 2)$ and $(1, 7)$

Questions 24 to 35 can be done by using either the method of questions 1 to 6 or the method and ideas of Exercise 6j:

24. $(4,0)$ and $(0,5)$ **26.** $(3,0)$ and $(0,-2)$

25. $(3,0)$ and $(0,2)$ **27.** $(2,0)$ and $(0,6)$

28. $(4,2)$ and $(6,8)$ **32.** $(0,5)$ and $(-2,0)$

29. $(0,4)$ and $(3,1)$ **33.** $(-6,3)$ and $(5,5)$

30. $(2,1)$ and $(0,-6)$ **34.** $(-1,-2)$ and $(-5,-6)$

31. $(-1,4)$ and $(5,-2)$ **35.** $(3,2)$ and $(7,1)$

PROBLEMS

EXERCISE 6m

1. Find the equation of the line through the point $(6,2)$, which is parallel to the line $y = 3x - 1$.

2. On the same pair of axes, for $-8 \leqslant x \leqslant 8$ and $-12 \leqslant y \leqslant 4$, using 1 cm to 1 unit, draw the four lines $y = 2x - 4$, $y = 2x + 6$, $x + 2y + 10 = 0$ and $2y + x = 0$. What type of quadrilateral is formed by the four lines?

3. On the same pair of axes, for $-4 \leqslant x \leqslant 10$ and $-6 \leqslant y \leqslant 10$ using 1 cm to 1 unit, draw the four lines $y = 3x + 4$, $y = 3x - 6$, $y = -3x$ and $y = 10 - 3x$. Name the type of quadrilateral formed by the lines.

4. Find the point where the lines $y = 2x + 2$ and $y = 4 - 2x$ meet.

5. Find the equation of the line with gradient -2 that passes through the midpoint of the line joining the points $(3,2)$ and $(7,4)$. (The midpoint can be found either from a drawing on squared paper or from a rough sketch.)

6. Find the equation of the line that is parallel to the line joining $(-1,4)$ and $(3,2)$ and which passes through the point $(4,0)$.

7. On the same pair of axes, for $-2 \leqslant x \leqslant 5$ and $-5 \leqslant y \leqslant 2$, using 1 cm to 1 unit, draw the four lines $4y = x + 1$, $4y = x - 16$, $4x + y - 13 = 0$ and $4x + y + 4 = 0$. Name the type of quadrilateral formed by the lines.

MIXED EXERCISES

EXERCISE 6n

1. What is the gradient of the line with equation $y = 2x + 1$?

2. At what point does the line with equation $y = 4 - 2x$ cut the y-axis?

3. At what point does the line with equation $\dfrac{x}{4} - \dfrac{y}{3} = 1$ cut the x-axis?

4. On the line with equation $y = 6 - 2x$, if $x = -3$, what is y?

5. What is the equation of the line whose gradient is 5 and which passes through the origin?

6. At what point does the line with equation $2x + 3y = 24$ cut the x-axis?

7. Does the point $(2, 4)$ lie on the line with equation $y = 6x - 8$?

8. Find the gradient of the line joining the points $(6, 4)$ and $(1, 1)$.

EXERCISE 6p

1. What is the gradient of the line with equation $y = 4 - 3x$?

2. Does the point $(6, -1)$ lie on the line $3x + 11y = 8$?

3. What is the equation of the line that passes through the origin and has a gradient of -4?

4. At what point does the line with equation $y = 4 - x$ cut the y-axis?

5. At what points does the line with equation $x + y = 6$ cut the two axes?

6. Find the gradient of the line through the points $(3, 1)$ and $(5, -2)$.

7. Give the equation of the line that is parallel to the line with equation $y = 4 + \frac{1}{2}x$ and which passes through the origin.

8. At what points does the line $\dfrac{x}{2} + \dfrac{y}{3} = 1$ cut the two axes?

7 SIMULTANEOUS EQUATIONS

EQUATIONS WITH TWO UNKNOWN QUANTITIES

Up to now the equations we have solved have had only one unknown quantity, but there can be more.

Looking at the equation $2x + y = 8$, we can see that there are many possible values which will fit, for instance $x = 2$ and $y = 4$, or $x = 1$ and $y = 6$. We could also have $x = -1$ and $y = 10$ or even $x = 1.681$ and $y = 4.638$. Indeed, there is an infinite set of pairs of solutions.

If however we are *also* told that $x + y = 5$, then we shall find that not every pair of numbers which satisfies the first equation also satisfies the second. While $x = 2$, $y = 4$ satisfies the first equation, it does not satisfy $x + y = 5$. On the other hand $x = 3$, $y = 2$ satisfies both equations.

These two equations together form a pair of *simultaneous* equations. "Simultaneous" means that the two equations are both satisfied by the same values of x and y; that is, when x has the same value in both equations then y has the same value in both.

There are several different methods for solving simultaneous equations. We start with the simplest.

ELIMINATION METHOD

Whenever we meet a new type of equation, we try to reorganise it so that it is similar to equations we have already met.

Previous equations have had only one unknown quantity, so we try to *eliminate* one of the two unknowns.

Consider the pair of equations

$$2x + y = 8 \qquad [1]$$
$$x + y = 5 \qquad [2]$$

In this case, if we try subtracting the second equation from the first we find that the y term disappears but the x term does not:

i.e. [1] − [2] gives $\qquad\qquad x = 3$

Then, substituting 3 for x in equation [2], we see that $3 + y = 5$ so $y = 2$

118

We can check that $x = 3$ and $y = 2$ also satisfy equation [1].

Notice that it is essential to number the equations and to say that you are subtracting them.

Sometimes it is easier to subtract the first equation from the second rather than the second equation from the first. (In this case we would write equation [1] again, underneath equation [2].)
Sometimes we can eliminate x rather than y.

EXERCISE 7a

Solve the equations $\begin{aligned} x + y &= 5 \\ 3x + y &= 7 \end{aligned}$

$$x + y = 5 \qquad [1]$$
$$3x + y = 7 \qquad [2]$$
$$x + y = 5 \qquad [1]$$

[2] − [1] gives $2x = 2$

$$x = 1$$

(To find y choose the simpler equation, i.e. the first.)

Substituting in [1] gives $1 + y = 5$

$$y = 4$$

(Check in the equation *not* used for finding y.)

Check in [2]   Left-hand side $= 3 + 4$

$$= 7 = \text{right-hand side}$$

Therefore the solution is $x = 1$, $y = 4$

Solve the following pairs of equations:

1. $\quad x + y = 5$
$\quad 4x + y = 14$

2. $\quad 5x + y = 14$
$\quad 3x + y = 10$

3. $\quad 2a + b = 11$
$\quad 4a + b = 17$

4. $\quad 2x + 3y = 23$
$\quad x + 3y = 22$

5. $\quad 5x + 2y = 14$
$\quad 7x + 2y = 22$

6. $\quad x + 2y = 12$
$\quad x + y = 7$

7. $\quad 4p + 3q = -5$
$\quad 7p + 3q = -11$

8. $\quad 12x + 5y = 65$
$\quad 9x + 5y = 50$

9. $3x + 4y = 15$
$3x + 2y = 12$

11. $2x + 3y = -8$
$2x + y = -4$

10. $9c + 2d = 54$
$c + 2d = 6$

12. $9x + 5y = 45$
$4x + 5y = 45$

Not all pairs of simultaneous equations can be solved by subtracting one from the other.

Consider $4x + y = 6$ [1]
$2x - y = 0$ [2]

If we subtract we get $2x + 2y = 6$ which is no improvement.

On the other hand, if we add we get $6x = 6$ which eliminates y.

If the signs in front of the letter to be eliminated are the same we should *subtract*; if the signs are different we should *add*.

EXERCISE 7b

Solve the equations $x - 2y = 1$
$3x + 2y = 19$

$x - 2y = 1$ [1]

$3x + 2y = 19$ [2]

[1] + [2] gives $4x = 20$

$x = 5$

(It is easier to use the equation with the + sign to find y.)

Substitute 5 for x in [2] $15 + 2y = 19$

Take 15 from both sides $2y = 4$

$y = 2$

Check in [1]   Left-hand side $= 5 - 4$

$= 1 =$ right-hand side

Therefore the solution is $x = 5$, $y = 2$

Solve the following pairs of equations:

1. $x - y = 2$
$3x + y = 10$

2. $2x - y = 6$
$3x + y = 14$

3. $p + 2q = 11$
   $3p - 2q = 1$

7. $x + y = 2$
   $2x - y = 10$

4. $3a - b = 10$
   $a + b = 2$

8. $5p + 3q = 5$
   $4p - 3q = 4$

5. $6x + 2y = 19$
   $x - 2y = 2$

9. $3x - 4y = -24$
   $5x + 4y = 24$

6. $4x + y = 37$
   $2x - y = 17$

10. $5x - 2y = 4$
   $3x + 2y = 12$

To solve the following equations, first decide whether to add or subtract:

11. $3x + 2y = 12$
   $x + 2y = 8$

16. $2x + 3y = 13$
   $2x + 5y = 21$

12. $x - 2y = 6$
   $4x + 2y = 14$

17. $5x - 2y = 24$
   $x + 2y = 0$

13. $x + 3y = 12$
   $x + y = 8$

18. $x + 3y = 0$
   $x - y = -4$

14. $9x + 2y = 48$
   $x - 2y = 2$

19. $5p - 3q = 9$
   $4p + 3q = 9$

15. $4x + y = 19$
   $3x + y = 15$

20. $6a - b = 20$
   $6a + 5b = 8$

Solve the equations $4x - y = 10$
   $x - y = 1$

$$4x - y = 10 \qquad [1]$$
$$x - y = 1 \qquad [2]$$

(The signs in front of the y terms are the same so we subtract: $-y - (-y) = -y + y = 0$)

$[1] - [2]$ gives $3x = 9$
   $x = 3$

Substitute 3 for x in [2] $3 - y = 1$
Add y to both sides $3 = 1 + y$
Take 1 from both sides $2 = y$

Check in [1]   Left-hand side $= 12 - 2$
   $= 10 =$ right-hand side

Therefore the solution is $x = 3$, $y = 2$

Solve the following pairs of equations:

21. $2x - y = 4$
$\quad\quad x - y = 1$

24. $6x - y = 7$
$\quad\quad 2x - y = 1$

22. $2p - 3q = -7$
$\quad\quad 4p - 3q = 1$

25. $5x - 2y = -19$
$\quad\quad x - 2y = -7$

23. $x - y = 3$
$\quad\quad 3x - y = 9$

26. $2x - 3y = 14$
$\quad\quad 2x - y = 10$

27. $3x - 2y = 14$
$\quad\quad x + 2y = 10$

30. $3p - 5q = 7$
$\quad\quad 4p + 5q = -14$

28. $3p - 5q = -3$
$\quad\quad 4p - 5q = 1$

31. $3p + 5q = 35$
$\quad\quad 4p - 5q = 0$

29. $3p + 5q = 17$
$\quad\quad 4p + 5q = 16$

32. $3x - y = 10$
$\quad\quad x + y = -2$

HARDER ELIMINATION

Equations are not always as simple as the ones we have had so far.

Consider $\quad\quad\quad\quad\quad\quad 2x + 3y = 4 \quad\quad\quad\quad\quad$ [1]
$\quad\quad\quad\quad\quad\quad\quad\quad\quad 4x + y = -2 \quad\quad\quad\quad$ [2]

Whether we add or subtract neither letter will disappear, so it is necessary to do something else first. If we multiply the second equation by 3 to give $12x + 3y = -6$, we have the same number of ys in each equation. Then we can use the same method as before:

$[2] \times 3 \quad\quad\quad\quad\quad\quad 12x + 3y = -6 \quad\quad\quad\quad$ [3]

$\quad\quad\quad\quad\quad\quad\quad\quad\quad 2x + 3y = 4 \quad\quad\quad\quad\quad$ [1]

$[3] - [1]$ gives $\quad\quad\quad\quad\quad 10x = -10$

$\quad\quad\quad\quad\quad\quad\quad\quad\quad\quad x = -1$

Substitute -1 for x in [2] $\quad -4 + y = -2$

Add 4 to both sides $\quad\quad\quad\quad\quad y = 2$

Therefore the solution is $\quad x = -1, \ y = 2$

EXERCISE 7c

Solve the equations $\begin{aligned} 3x - 2y &= 1 \\ 4x + y &= 5 \end{aligned}$

$$\begin{aligned} 3x - 2y &= 1 \qquad [1] \\ 4x + y &= 5 \qquad [2] \end{aligned}$$

$[2] \times 2$ gives

$$\begin{aligned} 8x + 2y &= 10 \qquad [3] \\ 3x - 2y &= 1 \qquad [1] \end{aligned}$$

$[1] + [3]$ gives

$$\begin{aligned} 11x &= 11 \\ x &= 1 \end{aligned}$$

Substitute 1 for x in [2] $4 + y = 5$
Take 4 from both sides $y = 1$

Check in [1]   Left-hand side $= 3 - 2$
$= 1 =$ right-hand side

Therefore the solution is $x = 1$, $y = 1$

Solve the following pairs of equations:

1. $\begin{aligned} 2x + y &= 7 \\ 3x + 2y &= 11 \end{aligned}$

2. $\begin{aligned} 5x - 4y &= -3 \\ 3x + y &= 5 \end{aligned}$

3. $\begin{aligned} 9x + 7y &= 10 \\ 3x + y &= 2 \end{aligned}$

4. $\begin{aligned} 5x + 3y &= 21 \\ 2x + y &= 3 \end{aligned}$

5. $\begin{aligned} 6x - 4y &= -4 \\ 5x + 2y &= 2 \end{aligned}$

6. $\begin{aligned} 4x + 3y &= 25 \\ x + 5y &= 19 \end{aligned}$

Solve the equations $\begin{aligned} 5x + 3y &= 7 \\ 10x + 4y &= 16 \end{aligned}$

$$\begin{aligned} 5x + 3y &= 7 \qquad [1] \\ 10x + 4y &= 16 \qquad [2] \end{aligned}$$

$[1] \times 2$ gives

$$\begin{aligned} 10x + 6y &= 14 \qquad [3] \\ 10x + 4y &= 16 \qquad [2] \end{aligned}$$

$[3] - [2]$ gives

$$\begin{aligned} 2y &= -2 \\ y &= -1 \end{aligned}$$

Substitute -1 for y in [1] $5x - 3 = 7$
Add 3 to both sides $5x = 10$
$x = 2$

Check in [2]   Left-hand side $= 20 + (-4) = 16$

Therefore the solution is $x = 2$, $y = -1$

7. $5x + 3y = 11$
$4x + 6y = 16$

10. $9x + 5y = 15$
$3x - 2y = -6$

8. $2x - 3y = 1$
$5x + 9y = 19$

11. $4x + 3y = 1$
$16x - 5y = 21$

9. $2x + 5y = 1$
$4x + 3y = 9$

12. $7p + 2q = 22$
$3p + 4q = 11$

Sometimes we need to alter both equations before we add or subtract.

EXERCISE 7d

Solve the equations $3x + 5y = 6$
$2x + 3y = 5$

	$3x + 5y = 6$	[1]
	$2x + 3y = 5$	[2]
[1] × 3 gives	$9x + 15y = 18$	[3]
[2] × 5 gives	$10x + 15y = 25$	[4]
	$9x + 15y = 18$	[3]

[4] − [3] gives $x = 7$

Substitute 7 for x in [2] $14 + 3y = 5$
Take 14 from both sides $3y = -9$
Divide both sides by 3 $y = -3$

Check in [1]   Left-hand side $= 21 - 15$
$= 6 =$ right-hand side

Therefore the solution is $x = 7$, $y = -3$

Solve the following pairs of equations:

1. $2x + 3y = 12$
$5x + 4y = 23$

5. $14x - 3y = -18$
$6x + 2y = 12$

2. $3x - 2y = -7$
$4x + 3y = 19$

6. $6x - 7y = 25$
$7x + 6y = 15$

3. $2x - 5y = 1$
$5x + 3y = 18$

7. $5x + 4y = 21$
$3x + 6y = 27$

4. $6x + 5y = 9$
$4x + 3y = 6$

8. $9x + 8y = 17$
$2x - 6y = -4$

9. $9x - 2y = 14$
  $7x + 3y = 20$

14. $3p + 4q = 5$
  $2p + 10q = 18$

10. $5x + 4y = 11$
  $2x + 3y = 3$

15. $6x + 5y = 8$
  $3x + 4y = 1$

11. $4x + 5y = 26$
  $5x + 4y = 28$

16. $7x - 3y = 20$
  $2x + 4y = -4$

12. $2x - 6y = -6$
  $5x + 4y = -15$

17. $10x + 3y = 12$
  $3x + 5y = 20$

13. $5x - 6y = 6$
  $2x + 9y = 10$

18. $6x - 5y = 4$
  $4x + 2y = -8$

19. $5x + 3y = 8$
  $3x + 5y = 8$

24. $7x + 6y = 0$
  $5x - 8y = 43$

20. $7x + 2y = 23$
  $3x - 5y = 4$

25. $2x + 6y = 30$
  $3x + 10y = 49$

21. $6x - 5y = 17$
  $5x + 4y = 6$

26. $4x - 3y = -7$
  $3x + 2y = 16$

22. $3x + 8y = 56$
  $5x - 6y = 16$

27. $17x - 2y = 47$
  $5x - 3y = 9$

23. $7x + 3y = -9$
  $2x + 5y = 14$

28. $8x + 3y = -17$
  $7x - 4y = 5$

MIXED QUESTIONS

EXERCISE 7e Solve the following pairs of equations:

1. $x + 2y = 9$
  $2x - y = -2$

6. $3x + 2y = -5$
  $3x - 4y = 1$

2. $x + y = 4$
  $x + 2y = 9$

7. $x + y = 6$
  $x - y = 1$

3. $2x + 3y = 0$
  $3x + 2y = 5$

8. $3x - 5y = 13$
  $2x + 5y = -8$

4. $3x - y = -10$
  $4x - y = -4$

9. $7x + 3y = 35$
  $2x - 5y = 10$

5. $5x + 2y = 16$
  $2x - 3y = -5$

10. $9x + 2y = 8$
  $7x + 3y = 12$

11. $2x - 5y = 1$
     $3x + 4y = 13$

12. $3x - 2y = -2$
     $5x - y = -15$

Sometimes the equations are arranged in an awkward fashion and need to be rearranged before solving them.

EXERCISE 7f

Solve the equations $x = 4 - 3y$
             $2y - x = 1$

                $x = 4 - 3y$ [1]
             $2y - x = 1$ [2]

(We must first arrange the letters in the same order in both equations.
By adding $3y$ to both sides, equation [1] can be written $3y + x = 4$)

                $3y + x = 4$ [3]
             $2y - x = 1$ [2]

[3] + [2] gives $5y = 5$
             $y = 1$

Substitute 1 for y in [1] $x = 4 - 3$
                 $x = 1$

Check in [2]   Left-hand side $= 2 - 1$
                   $= 1 =$ right-hand side

Therefore the solution is $x = 1$, $y = 1$

Solve the following pairs of equations:

1. $y = 6 - x$
    $2x + y = 8$

4. $9 + x = y$
    $x + 2y = 12$

2. $x - y = 2$
    $2y = x + 1$

5. $2y = 16 - x$
    $x - 2y = -8$

3. $3 = 2x + y$
    $4x + 6 = 10y$

6. $3x + 4y = 7$
    $2x = 5 - 3y$

As long as the x and y and number terms are in corresponding positions in the two equations, they need not be in the order we have had so far.

Solve the equations $y = x + 5$
$y = 7 - x$

$$y = x + 5 \qquad [1]$$
$$y = 7 - x \qquad [2]$$

Rewrite [1] as $y = 5 + x \qquad [3]$

[2] + [3] gives $2y = 12$
$y = 6$

Substitute 6 for y in [1] $6 = x + 5$
$x = 1$

Check in [2]   Left-hand side $= 6$
Right-hand side $= 1 + 5 = 6$

Therefore the solution is $x = 1$, $y = 6$

---

7. $y = 9 + x$
$y = 11 - x$

8. $x = 3 + y$
$2x = 4 - y$

9. $y = 4 - x$
$y = x + 6$

10. $2y = 4 + x$
$y = x + 8$

11. $x + 4 = y$
$y = 10 - 2x$

12. $x + y = 12$
$y = 3 + x$

SPECIAL CASES

Some pairs of equations have no solution and some have an infinite number of solutions.

EXERCISE 7g Try solving the following pairs of equations. Comment on why the method breaks down:

1. $x + 2y = 6$
$x + 2y = 7$

2. $3x + 4y = 1$
$6x + 8y = 2$

3. $y = 4 + 2x$
$y - 2x = 6$

4. $9x = 3 - 6y$
$3x + 2y = 1$

5. Make up other pairs of equations which either have no solution or have an infinite set of solutions.

PROBLEMS ──

EXERCISE 7h

I think of two numbers. If I add three times the smaller number to the bigger number I get 14. If I subtract the bigger number from twice the smaller number I get 1. Find the two numbers

Let the smaller number be x and the bigger number be y.

$$3x + y = 14 \qquad [1]$$
$$2x - y = 1 \qquad [2]$$

[1] + [2] gives
$$5x = 15$$
$$x = 3$$

Substitute 3 for x in [1] $9 + y = 14$
$$y = 5$$

Therefore, the two numbers are 3 and 5

(Check by reading the original statements to see if the numbers fit.)

Solve the following problems by forming a pair of simultaneous equations:

1. The sum of two numbers is 20 and their difference is 4. Find the numbers.

2. The sum of two numbers is 16 and they differ by 6. What are the numbers?

3. I think of two numbers. If I double the first and add the second I get 18. If I double the first and subtract the second I get 14. What are the numbers?

4. Three times a number added to a second number is 33. The first number added to three times the second number is 19. Find the two numbers.

5. Find two numbers such that twice the first added to the second is 26 and the first added to three times the second is 28.

6. Find the two numbers such that twice the first added to the second gives 27 and twice the second added to the first gives 21.

A shop sells bread rolls. If five brown rolls and six white rolls cost 98 p while three brown rolls and four white rolls cost 62 p find the cost of each type of roll.

Let one brown roll cost x p and one white roll cost y p.

$$5x + 6y = 98 \qquad [1]$$
$$3x + 4y = 62 \qquad [2]$$

[1] × 2 gives $\qquad 10x + 12y = 196 \qquad$ [3]
[2] × 3 gives $\qquad 9x + 12y = 186 \qquad$ [4]
[3] − [4] gives $\qquad\qquad x = 10$
Substitute 10 for x in [1] $\quad 50 + 6y = 98$
Take 50 from both sides $\qquad 6y = 48$
$$y = 8$$

Therefore one brown roll costs 10 p and one white roll costs 8 p.

7. I buy x choc ices and y orange ices and spend 230 p. I buy ten ices altogether. The choc ices cost 30 p each and the orange ices cost 20 p each. How many of each do I buy?

8.

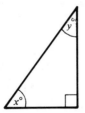

x is bigger than y. The difference between x and y is 18. Find x and y.

9. A cup and saucer cost £1.05 together. A cup and two saucers cost £1.50. Find the cost of a cup and of a saucer.

10. The cost of two Star Bars is the same as the cost of five Toffee Knobs. One Star Bar and one Toffee Knob together cost 35 p. What do they each cost?

11. In a test, the sum of Harry's marks and Adam's marks is 42. Sam has twice as many marks as Adam, and the sum of Harry's and Sam's marks is 52. What are the marks of each of the three boys?

12.

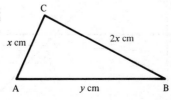

The perimeter of triangle ABC is 14 cm. AB is 2 cm longer than AC. Find x and y.

13.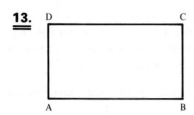

The perimeter of the rectangle is 31 cm. The difference between the lengths of AB and BC is $3\frac{1}{2}$ cm. Find the lengths of AB and BC.

14. The equation of a straight line is $y = mx + c$. When $x = 1$, $y = 6$ and when $x = 3$, $y = 10$. Form two equations for m and c and hence find the equation of the line.

GRAPHICAL SOLUTIONS OF SIMULTANEOUS EQUATIONS

We saw in a previous chapter that when we are given an equation we can draw a graph. Any of the equations which occur in this chapter gives us a straight line. Two equations give us two straight lines which usually cross one another.

Consider the two equations $x + y = 4$
$$y = 1 + x$$

Suppose we know that the x coordinate of the point of intersection is in the range $0 \leqslant x \leqslant 5$:

$x + y = 4$

x	0	4	5
y	4	0	-1

$y = 1 + x$

x	0	2	5
y	1	3	6

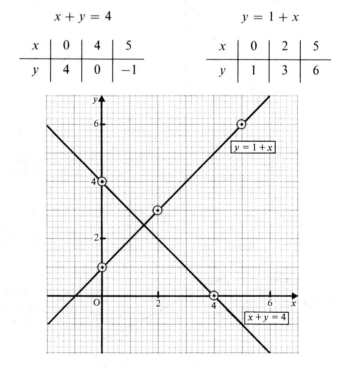

At the point where the two lines cross, the values of x and y are the same for both equations, so they are the solutions of the pair of equations.

From the graph we see that the solution is $x = 1\frac{1}{2}$, $y = 2\frac{1}{2}$.

EXERCISE 7i Solve the following equations graphically. In each case draw axes for x and y and use values in the ranges indicated, taking 2 cm to 1 unit:

1. $x + y = 6$ $0 \leqslant x \leqslant 6,\ 0 \leqslant y \leqslant 6$
    $y = 3 + x$

2. $x + y = 5$ $0 \leqslant x \leqslant 6,\ 0 \leqslant y \leqslant 6$
    $y = 2x + 1$

3. $y = 4 + x$ $0 \leqslant x \leqslant 6,\ 0 \leqslant y \leqslant 6$
    $y = 1 + 3x$

4. $x + y = 1$ $-3 \leqslant x \leqslant 2,\ -2 \leqslant y \leqslant 4$
    $y = x + 2$

5. $2x + y = 3$ $0 \leqslant x \leqslant 3,\ -3 \leqslant y \leqslant 3$
    $x + y = 2\frac{1}{2}$

6. $y = 5 - x$ $0 \leqslant x \leqslant 5,\ 0 \leqslant y \leqslant 7$
    $y = 2 + x$

7. $3x + 2y = 9$ $0 \leqslant x \leqslant 4,\ -2 \leqslant y \leqslant 5$
    $2x - 2y = 3$

8. $2x + 3y = 4$ $-2 \leqslant x \leqslant 2,\ 0 \leqslant y \leqslant 4$
    $y = x + 2$

9. $x + 3y = 6$ $0 \leqslant x \leqslant 5,\ 0 \leqslant y \leqslant 5$
    $3x - y = 6$

10. $x = 2y - 3$ $-2 \leqslant x \leqslant 3,\ 0 \leqslant y \leqslant 4$
    $y = 2x + 1$

SPECIAL CASES

EXERCISE 7j Try to solve the following equations graphically. Why do you think the method breaks down?

1. $x + y = 9$ $0 \leqslant x \leqslant 9$ 3. $2x + y = 3$ $0 \leqslant x \leqslant 3$
    $x + y = 4$ $0 \leqslant y \leqslant 9$ $4x + 2y = 7$ $-3 \leqslant y \leqslant 4$

2. $y = 2x + 3$ $0 \leqslant x \leqslant 4$ 4. $y = 2x - 4$ $0 \leqslant x \leqslant 4$
    $y = 2x - 1$ $-1 \leqslant y \leqslant 11$ $2x = y + 4$ $-4 \leqslant y \leqslant 4$

8 INVERSE AND SQUARE MATRICES

SQUARE MATRICES

A square matrix has the same number of rows and columns.

$$\begin{pmatrix} 1 & 2 \\ 3 & -1 \end{pmatrix} \text{ and } \begin{pmatrix} 1 & 6 & 3 \\ 1 & 0 & 1 \\ 4 & 4 & 3 \end{pmatrix} \text{ are square matrices.}$$

A square matrix has two diagonals:

$$\begin{pmatrix} 1 & 3 & 4 \\ 0 & 1 & 2 \\ 3 & 1 & 2 \end{pmatrix}$$

The diagonal that goes from top left to bottom right is the *leading* diagonal:

$$\begin{pmatrix} 1 & 2 \\ 3 & 4 \end{pmatrix}$$

EXERCISE 8a

State which of the two matrices $\begin{pmatrix} 1 \\ 2 \end{pmatrix}$ and $\begin{pmatrix} 4 & 1 \\ 3 & 2 \end{pmatrix}$ is a square matrix. Give its size and mark its leading diagonal

$\begin{pmatrix} 4 & 1 \\ 3 & 2 \end{pmatrix}$ is a 2 × 2 square matrix.

State whether or not each of the following matrices is a square matrix. If it is, give its size and mark its leading diagonal:

1. $\begin{pmatrix} 3 & 6 & 1 \\ 4 & 0 & 1 \\ 3 & 2 & 1 \end{pmatrix}$

2. $\begin{pmatrix} 4 & 3 & 2 \\ 1 & 3 & 1 \end{pmatrix}$

3. $\begin{pmatrix} 2 & 1 \\ 4 & 3 \end{pmatrix}$

5. $\begin{pmatrix} -1 & 4 \\ 4 & -1 \end{pmatrix}$

4. $\begin{pmatrix} 2 & 3 \\ 6 & 2 \\ 1 & 4 \end{pmatrix}$

6. $\begin{pmatrix} 6 & 1 & -3 \\ 4 & 1 & 2 \\ 6 & 8 & 1 \end{pmatrix}$

REMINDER OF MATRIX MULTIPLICATION

EXERCISE 8b

Find, where possible

a) $\begin{pmatrix} 2 & 3 \\ 1 & -3 \\ 4 & -2 \end{pmatrix} \begin{pmatrix} 2 & 3 \\ 1 & 4 \end{pmatrix}$

b) $\begin{pmatrix} 1 & 2 \end{pmatrix} \begin{pmatrix} 3 & 4 \end{pmatrix}$

a)
$$\begin{pmatrix} 2 & 3 \\ 1 & -3 \\ 4 & -2 \end{pmatrix} \begin{pmatrix} 2 & 3 \\ 1 & 4 \end{pmatrix} = \begin{pmatrix} 7 & 18 \\ -1 & -9 \\ 6 & 4 \end{pmatrix}$$

$3 \times \mathbf{2}$ $\quad$ $\mathbf{2} \times 2$ $\quad\quad$ 3×2

b) $\quad\quad\quad$ $\begin{pmatrix} 1 & 2 \end{pmatrix} \begin{pmatrix} 3 & 4 \end{pmatrix}$ $\quad\quad$ not possible

$1 \times 2 \quad\quad 1 \times 2$

Find where possible:

1. $\begin{pmatrix} 3 & 1 \\ 4 & 3 \end{pmatrix} \begin{pmatrix} 1 & 2 \\ 1 & 1 \end{pmatrix}$

4. $\begin{pmatrix} 1 & 6 & -1 \\ 3 & 2 & 1 \end{pmatrix} \begin{pmatrix} 1 & 2 \\ 3 & 4 \end{pmatrix}$

2. $\begin{pmatrix} 1 & 0 & 1 \end{pmatrix} \begin{pmatrix} 4 & -1 \\ 2 & 4 \\ 3 & -5 \end{pmatrix}$

5. $\begin{pmatrix} 4 & -1 \\ 2 & 4 \\ 3 & -5 \end{pmatrix} \begin{pmatrix} 1 & 0 & 1 \end{pmatrix}$

3. $\begin{pmatrix} 1 & 2 \\ 3 & 4 \end{pmatrix} \begin{pmatrix} 1 & 6 & -1 \\ 3 & 2 & 1 \end{pmatrix}$

6. $\begin{pmatrix} 3 & 5 \\ 4 & -3 \end{pmatrix} \begin{pmatrix} 2 & 1 \\ 4 & 2 \end{pmatrix}$

7. $\begin{pmatrix} 1 & 2 \\ 1 & 1 \end{pmatrix}\begin{pmatrix} 3 & 1 \\ 4 & 3 \end{pmatrix}$

10. $\begin{pmatrix} 1 & 6 & 1 \end{pmatrix}\begin{pmatrix} 4 \\ 3 \\ 2 \end{pmatrix}$

8. $\begin{pmatrix} 4 \\ 3 \\ 2 \end{pmatrix}\begin{pmatrix} 1 \\ 6 \\ 1 \end{pmatrix}$

11. $\begin{pmatrix} 1 & -1 \\ -1 & 1 \end{pmatrix}\begin{pmatrix} 4 \\ 3 \end{pmatrix}$

9. $\begin{pmatrix} 4 \\ 3 \\ 2 \end{pmatrix}\begin{pmatrix} 1 & 6 & 1 \end{pmatrix}$

12. $\begin{pmatrix} 2 & 6 \end{pmatrix}\begin{pmatrix} 5 & -3 \\ 4 & 2 \end{pmatrix}$

THE UNIT MATRIX AND THE ZERO MATRIX

EXERCISE 8c   Find:

1. $\begin{pmatrix} 4 & 2 \\ 3 & 4 \end{pmatrix}\begin{pmatrix} 1 & 0 \\ 0 & 1 \end{pmatrix}$

4. $\begin{pmatrix} 0 & 0 \\ 0 & 0 \end{pmatrix}\begin{pmatrix} 2 & 3 \\ 4 & 1 \end{pmatrix}$

2. $\begin{pmatrix} 1 & 0 \\ 0 & 1 \end{pmatrix}\begin{pmatrix} 4 \\ 5 \end{pmatrix}$

5. $\begin{pmatrix} 3 & 2 \end{pmatrix}\begin{pmatrix} 1 & 0 \\ 0 & 1 \end{pmatrix}$

3. $\begin{pmatrix} 6 & 2 \end{pmatrix}\begin{pmatrix} 0 & 0 \\ 0 & 0 \end{pmatrix}$

6. $\begin{pmatrix} 1 & 0 \\ 0 & 1 \end{pmatrix}\begin{pmatrix} 3 & 2 & -1 \\ 4 & 3 & 1 \end{pmatrix}$

The *unit matrix* $\begin{pmatrix} 1 & 0 \\ 0 & 1 \end{pmatrix}$ does not alter the matrix it multiplies or is multiplied by, e.g.:

$$\begin{pmatrix} 1 & 2 \\ 4 & 5 \end{pmatrix}\begin{pmatrix} 1 & 0 \\ 0 & 1 \end{pmatrix}=\begin{pmatrix} 1 & 2 \\ 4 & 5 \end{pmatrix} \text{ and } \begin{pmatrix} 1 & 0 \\ 0 & 1 \end{pmatrix}\begin{pmatrix} 1 & 2 \\ 4 & 5 \end{pmatrix}=\begin{pmatrix} 1 & 2 \\ 4 & 5 \end{pmatrix}$$

This is similar to multiplying by 1 in ordinary number multiplication. A unit matrix is usually represented by **I**.

The *zero matrix* $\begin{pmatrix} 0 & 0 \\ 0 & 0 \end{pmatrix}$, when multiplied by any compatible matrix, gives the zero matrix:

$$\begin{pmatrix} 1 & 2 \\ 4 & 5 \end{pmatrix}\begin{pmatrix} 0 & 0 \\ 0 & 0 \end{pmatrix} = \begin{pmatrix} 0 & 0 \\ 0 & 0 \end{pmatrix} \quad \text{and} \quad \begin{pmatrix} 0 & 0 \\ 0 & 0 \end{pmatrix}\begin{pmatrix} 1 & 2 \\ 4 & 5 \end{pmatrix} = \begin{pmatrix} 0 & 0 \\ 0 & 0 \end{pmatrix}$$

This is similar to multiplying by zero in ordinary number multiplication. A zero matrix is represented by **0**.

There are other unit and zero matrices, e.g.

$$\begin{pmatrix} 1 & 0 & 0 \\ 0 & 1 & 0 \\ 0 & 0 & 1 \end{pmatrix} \quad \text{and} \quad \begin{pmatrix} 0 & 0 & 0 \\ 0 & 0 & 0 \\ 0 & 0 & 0 \end{pmatrix}$$

but we are concerned mainly with 2×2 square matrices in this chapter.

EXERCISE 8d Find:

1. $\begin{pmatrix} 1 & 1 \\ 2 & 3 \end{pmatrix}\begin{pmatrix} 3 & -1 \\ -2 & 1 \end{pmatrix}$

5. $\begin{pmatrix} 6 & -3 \\ -3 & 2 \end{pmatrix}\begin{pmatrix} 2 & 3 \\ 3 & 6 \end{pmatrix}$

2. $\begin{pmatrix} 4 & 1 \\ 3 & 2 \end{pmatrix}\begin{pmatrix} 2 & -1 \\ -3 & 4 \end{pmatrix}$

6. $\begin{pmatrix} 2 & 3 \\ 3 & 6 \end{pmatrix}\begin{pmatrix} 6 & -3 \\ -3 & 2 \end{pmatrix}$

3. $\begin{pmatrix} 2 & -1 \\ -3 & 4 \end{pmatrix}\begin{pmatrix} 4 & 1 \\ 3 & 2 \end{pmatrix}$

7. $\begin{pmatrix} 9 & 7 \\ 4 & 3 \end{pmatrix}\begin{pmatrix} 3 & -7 \\ -4 & 9 \end{pmatrix}$

4. $\begin{pmatrix} 9 & 4 \\ 4 & 2 \end{pmatrix}\begin{pmatrix} 2 & -4 \\ -4 & 9 \end{pmatrix}$

8. $\begin{pmatrix} 3 & -2 \\ -4 & 3 \end{pmatrix}\begin{pmatrix} 3 & 2 \\ 4 & 3 \end{pmatrix}$

9. What do you notice about the arrangement of the numbers and signs in each pair of matrices? What do you notice about the answers?

In questions 10 to 18, find a second matrix **B** such that **B** × **A** is of the form $\begin{pmatrix} k & 0 \\ 0 & k \end{pmatrix}$, i.e. like the answers to questions 1 to 8.

10.
$$A = \begin{pmatrix} 3 & 2 \\ 8 & 6 \end{pmatrix}$$

15.
$$A = \begin{pmatrix} 5 & 3 \\ 9 & 6 \end{pmatrix}$$

11.
$$A = \begin{pmatrix} 2 & 1 \\ 2 & 3 \end{pmatrix}$$

16.
$$A = \begin{pmatrix} 5 & 4 \\ 9 & 8 \end{pmatrix}$$

12.
$$A = \begin{pmatrix} 6 & -1 \\ -20 & 3 \end{pmatrix}$$

17.
$$A = \begin{pmatrix} 6 & 2 \\ 14 & 4 \end{pmatrix}$$

13.
$$A = \begin{pmatrix} 1 & -3 \\ 1 & 5 \end{pmatrix}$$

18.
$$A = \begin{pmatrix} 1 & 2 \\ 2 & 1 \end{pmatrix}$$

14.
$$A = \begin{pmatrix} 5 & 7 \\ 3 & 4 \end{pmatrix}$$

19.
$$A = \begin{pmatrix} 4 & -3 \\ 3 & -2 \end{pmatrix}$$

If we are given a 2 × 2 matrix **A** and we want a matrix **B** such that **A** × **B** is of the form $\begin{pmatrix} k & 0 \\ 0 & k \end{pmatrix}$ we

a) interchange the numbers in the leading diagonal of **A**

b) leave the other two numbers of **A** where they are, but change their signs.

For example, if $A = \begin{pmatrix} 4 & 2 \\ 5 & 3 \end{pmatrix}$ to get **B**

we change $\begin{pmatrix} 4 & \\ & 3 \end{pmatrix}$ to $\begin{pmatrix} 3 & \\ & 4 \end{pmatrix}$ and $\begin{pmatrix} & 2 \\ 5 & \end{pmatrix}$ to $\begin{pmatrix} & -2 \\ -5 & \end{pmatrix}$

then $\begin{pmatrix} 4 & 2 \\ 5 & 3 \end{pmatrix} \begin{pmatrix} 3 & -2 \\ -5 & 4 \end{pmatrix} = \begin{pmatrix} 2 & 0 \\ 0 & 2 \end{pmatrix}$

THE INVERSE OF A MATRIX

In ordinary numbers we say $\frac{1}{7}$ is the inverse of 7 because $7 \times \frac{1}{7} = 1$
Similarly $\frac{2}{5}$ is the inverse of $\frac{5}{2}$ because $\frac{5}{2} \times \frac{2}{5} = 1$

In the same way $\begin{pmatrix} 3 & 5 \\ 1 & 2 \end{pmatrix} \times \begin{pmatrix} 2 & -5 \\ -1 & 3 \end{pmatrix} = \begin{pmatrix} 1 & 0 \\ 0 & 1 \end{pmatrix}$

so $\begin{pmatrix} 2 & -5 \\ -1 & 3 \end{pmatrix}$ is the inverse of $\begin{pmatrix} 3 & 5 \\ 1 & 2 \end{pmatrix}$

and $\begin{pmatrix} 3 & 5 \\ 1 & 2 \end{pmatrix}$ is the inverse of $\begin{pmatrix} 2 & -5 \\ -1 & 3 \end{pmatrix}$

EXERCISE 8e

Find the inverse of $\begin{pmatrix} 5 & 3 \\ 3 & 2 \end{pmatrix}$

Try $\begin{pmatrix} 2 & -3 \\ -3 & 5 \end{pmatrix}$

$\begin{pmatrix} 5 & 3 \\ 3 & 2 \end{pmatrix} \begin{pmatrix} 2 & -3 \\ -3 & 5 \end{pmatrix} = \begin{pmatrix} 1 & 0 \\ 0 & 1 \end{pmatrix}$

So $\begin{pmatrix} 2 & -3 \\ -3 & 5 \end{pmatrix}$ is the inverse of $\begin{pmatrix} 5 & 3 \\ 3 & 2 \end{pmatrix}$

Find the inverses of the following matrices:

1. $\begin{pmatrix} 4 & 1 \\ 7 & 2 \end{pmatrix}$ **3.** $\begin{pmatrix} 10 & -3 \\ 7 & -2 \end{pmatrix}$ **5.** $\begin{pmatrix} 7 & 4 \\ 12 & 7 \end{pmatrix}$

2. $\begin{pmatrix} 11 & 3 \\ 7 & 2 \end{pmatrix}$ **4.** $\begin{pmatrix} 3 & 5 \\ 4 & 7 \end{pmatrix}$ **6.** $\begin{pmatrix} 1 & 1 \\ 1 & 2 \end{pmatrix}$

Not all inverses can be produced in this way, e.g.

$\begin{pmatrix} 5 & 4 \\ 3 & 3 \end{pmatrix} \times \begin{pmatrix} 3 & -4 \\ -3 & 5 \end{pmatrix} = \begin{pmatrix} 3 & 0 \\ 0 & 3 \end{pmatrix}$ and not $\begin{pmatrix} 1 & 0 \\ 0 & 1 \end{pmatrix}$

But $\begin{pmatrix} 3 & 0 \\ 0 & 3 \end{pmatrix} = 3 \begin{pmatrix} 1 & 0 \\ 0 & 1 \end{pmatrix}$ so our attempt at the inverse is three times too big.

We therefore divide our attempt at the inverse by 3

The inverse of $\begin{pmatrix} 5 & 4 \\ 3 & 3 \end{pmatrix}$ is $\begin{pmatrix} 1 & -\frac{4}{3} \\ -1 & \frac{5}{3} \end{pmatrix}$

The number 3 is called the *determinant* of $\begin{pmatrix} 5 & 4 \\ 3 & 3 \end{pmatrix}$

In exercise 8e, questions 1 to 6, the determinant in each case was 1

EXERCISE 8f

Find the inverse of $\begin{pmatrix} 5 & 8 \\ 1 & 2 \end{pmatrix}$

Try $\begin{pmatrix} 2 & -8 \\ -1 & 5 \end{pmatrix}$

$$\begin{pmatrix} 5 & 8 \\ 1 & 2 \end{pmatrix}\begin{pmatrix} 2 & -8 \\ -1 & 5 \end{pmatrix} = \begin{pmatrix} 2 & 0 \\ 0 & 2 \end{pmatrix}$$

The determinant of $\begin{pmatrix} 5 & 8 \\ 1 & 2 \end{pmatrix}$ is 2,

so the inverse of $\begin{pmatrix} 5 & 8 \\ 1 & 2 \end{pmatrix}$ is $\begin{pmatrix} \frac{2}{2} & \frac{-8}{2} \\ \frac{-1}{2} & \frac{5}{2} \end{pmatrix}$

$$= \begin{pmatrix} 1 & -4 \\ -\frac{1}{2} & 2\frac{1}{2} \end{pmatrix}$$

Note that this inverse can be written $\frac{1}{2}\begin{pmatrix} 2 & -8 \\ -1 & 5 \end{pmatrix}$

Find the inverses of the following matrices:

1. $\begin{pmatrix} 6 & 2 \\ 8 & 3 \end{pmatrix}$ **4.** $\begin{pmatrix} 1 & -2 \\ 1 & 1 \end{pmatrix}$ **7.** $\begin{pmatrix} 4 & -3 \\ -5 & 4 \end{pmatrix}$

2. $\begin{pmatrix} 9 & 2 \\ 3 & 1 \end{pmatrix}$ **5.** $\begin{pmatrix} 2 & 0 \\ 0 & 3 \end{pmatrix}$ **8.** $\begin{pmatrix} 4 & 3 \\ 5 & 4 \end{pmatrix}$

3. $\begin{pmatrix} 4 & 2 \\ 10 & 6 \end{pmatrix}$ **6.** $\begin{pmatrix} 3 & 2 \\ 11 & 8 \end{pmatrix}$ **9.** $\begin{pmatrix} 3 & -4 \\ -2 & 4 \end{pmatrix}$

Sometimes the determinant is negative.

Find the inverse of $\begin{pmatrix} 1 & 5 \\ 1 & 2 \end{pmatrix}$

Try $\begin{pmatrix} 2 & -5 \\ -1 & 1 \end{pmatrix}$

$$\begin{pmatrix} 1 & 5 \\ 1 & 2 \end{pmatrix}\begin{pmatrix} 2 & -5 \\ -1 & 1 \end{pmatrix} = \begin{pmatrix} -3 & 0 \\ 0 & -3 \end{pmatrix}$$

The determinant of $\begin{pmatrix} 1 & 5 \\ 1 & 2 \end{pmatrix}$ is -3,

so the inverse of $\begin{pmatrix} 1 & 5 \\ 1 & 2 \end{pmatrix}$ is $\begin{pmatrix} \frac{2}{-3} & \frac{-5}{-3} \\ \frac{-1}{-3} & \frac{1}{-3} \end{pmatrix}$

$$= \begin{pmatrix} -\frac{2}{3} & \frac{5}{3} \\ \frac{1}{3} & -\frac{1}{3} \end{pmatrix}$$

i.e. $-\frac{1}{3}\begin{pmatrix} 2 & -5 \\ -1 & 1 \end{pmatrix}$

Find the inverses of the following matrices:

10. $\begin{pmatrix} 6 & 2 \\ 7 & 2 \end{pmatrix}$ **12.** $\begin{pmatrix} -4 & 1 \\ 1 & 1 \end{pmatrix}$ **14.** $\begin{pmatrix} 3 & -4 \\ -3 & 3 \end{pmatrix}$

11. $\begin{pmatrix} 9 & 4 \\ 5 & 2 \end{pmatrix}$ **13.** $\begin{pmatrix} 3 & 2 \\ 4 & 2 \end{pmatrix}$ **15.** $\begin{pmatrix} 5 & 4 \\ 8 & 6 \end{pmatrix}$

Sometimes the determinant is zero. In this case, as we cannot divide by zero, we cannot produce an inverse.

16. State whether the given matrix has a zero determinant and hence no inverse:

a) $\begin{pmatrix} 6 & 4 \\ 3 & 2 \end{pmatrix}$ b) $\begin{pmatrix} 11 & 3 \\ 4 & 1 \end{pmatrix}$ c) $\begin{pmatrix} 8 & 6 \\ 4 & 3 \end{pmatrix}$

17. Which of the following matrices are their own inverses?

a) $\begin{pmatrix} 1 & 0 \\ 0 & -1 \end{pmatrix}$.   b) $\begin{pmatrix} 0 & 1 \\ 1 & 0 \end{pmatrix}$ c) $\begin{pmatrix} -1 & 0 \\ 0 & -1 \end{pmatrix}$

Find, where possible, the inverses of the following matrices:

18. $\begin{pmatrix} 6 & 5 \\ 5 & 5 \end{pmatrix}$ **20.** $\begin{pmatrix} 5 & 0 \\ 0 & 5 \end{pmatrix}$ **22.** $\begin{pmatrix} 5 & 3 \\ 9 & 6 \end{pmatrix}$

19. $\begin{pmatrix} 6 & 4 \\ 6 & 4 \end{pmatrix}$ **21.** $\begin{pmatrix} 5 & 7 \\ 3 & 4 \end{pmatrix}$ **23.** $\begin{pmatrix} 8 & 3 \\ 16 & 6 \end{pmatrix}$

NOTATION

With ordinary numbers, $\frac{1}{3}$ is the inverse of 3 under multiplication. $\frac{1}{3}$ can be written as 3^{-1}, so 3^{-1} is the inverse of 3.
In the same way, the inverse of $\mathbf{A}$ is written $\mathbf{A}^{-1}$.

If $\mathbf{A} = \begin{pmatrix} 4 & 7 \\ 1 & 2 \end{pmatrix}$, then $\mathbf{A}^{-1} = \begin{pmatrix} 2 & -7 \\ -1 & 4 \end{pmatrix}$

$$\begin{pmatrix} 4 & 7 \\ 1 & 2 \end{pmatrix} \begin{pmatrix} 2 & -7 \\ -1 & 4 \end{pmatrix} = \begin{pmatrix} 1 & 0 \\ 0 & 1 \end{pmatrix}$$

and $\mathbf{A}\mathbf{A}^{-1} = \mathbf{I}$

The determinant of $\mathbf{A}$ is written $|\mathbf{A}|$. In this case $|\mathbf{A}| = 1$

EXERCISE 8g   The questions in this exercise refer to the following matrices:

$\mathbf{A} = \begin{pmatrix} 5 & 3 \\ 3 & 2 \end{pmatrix}$ $\mathbf{B} = \begin{pmatrix} 2 & 2 \\ 2 & 3 \end{pmatrix}$ $\mathbf{C} = \begin{pmatrix} 6 & 2 \\ 5 & 2 \end{pmatrix}$

$\mathbf{D} = \begin{pmatrix} 7 & 3 \\ 2 & 1 \end{pmatrix}$ $\mathbf{I} = \begin{pmatrix} 1 & 0 \\ 0 & 1 \end{pmatrix}$

Find   a) $|\mathbf{D}|$ b) $\mathbf{D}^{-1}$ c) $|\mathbf{C}|$ d) $\mathbf{C}^{-1}$ e) $\mathbf{D}^{-1}\mathbf{C}$

a) Try $\begin{pmatrix} 1 & -3 \\ -2 & 7 \end{pmatrix}$

$$\begin{pmatrix} 7 & 3 \\ 2 & 1 \end{pmatrix} \begin{pmatrix} 1 & -3 \\ -2 & 7 \end{pmatrix} = \begin{pmatrix} 1 & 0 \\ 0 & 1 \end{pmatrix}$$

$$|\mathbf{D}| = 1$$

b) $\mathbf{D}^{-1} = \begin{pmatrix} 1 & -3 \\ -2 & 7 \end{pmatrix}$

c) Try $\begin{pmatrix} 2 & -2 \\ -5 & 6 \end{pmatrix}$

$$\begin{pmatrix} 6 & 2 \\ 5 & 2 \end{pmatrix} \begin{pmatrix} 2 & -2 \\ -5 & 6 \end{pmatrix} = \begin{pmatrix} 2 & 0 \\ 0 & 2 \end{pmatrix}$$

$$|\mathbf{C}| = 2$$

d) $\mathbf{C}^{-1} = \frac{1}{2} \begin{pmatrix} 2 & -2 \\ -5 & 6 \end{pmatrix} = \begin{pmatrix} 1 & -1 \\ -2\frac{1}{2} & 3 \end{pmatrix}$

e) $\mathbf{D}^{-1}\mathbf{C} = \begin{pmatrix} 1 & -3 \\ -2 & 7 \end{pmatrix} \begin{pmatrix} 6 & 2 \\ 5 & 2 \end{pmatrix} = \begin{pmatrix} -9 & -4 \\ 23 & 10 \end{pmatrix}$

Write out the matrices $\mathbf{A}$, $\mathbf{B}$, $\mathbf{C}$, $\mathbf{D}$ and $\mathbf{I}$ and hence find:

1. $|\mathbf{A}|$ and $\mathbf{A}^{-1}$ **5.** $(\mathbf{AB})^{-1}$ **9.** $(\mathbf{A}^2)^{-1}$

2. $|\mathbf{B}|$ and $\mathbf{B}^{-1}$ **6.** $\mathbf{A}^{-1}\mathbf{B}^{-1}$ **10.** $(\mathbf{A}^{-1})^2$

3. $\mathbf{I}^{-1}$ **7.** $\mathbf{B}^{-1}\mathbf{A}^{-1}$ **11.** $\mathbf{C}^{-1}\mathbf{A}$

4. $\mathbf{AB}$ **8.** $\mathbf{A}^2$ **12.** $\mathbf{AC}^{-1}$

FINDING THE DETERMINANT

The determinant can be found without trying to find the inverse.

The determinant of the matrix $\begin{pmatrix} a & b \\ c & d \end{pmatrix}$ is $ad - bc$.

We find the product of the entries in the leading diagonal, ad, and then we subtract the product of the entries in the other diagonal, bc.

For example, the determinant of $\begin{pmatrix} 3 & 1 \\ 4 & 2 \end{pmatrix}$ is $3 \times 2 - 1 \times 4 = 2$

EXERCISE 8h

Find the determinant of $\begin{pmatrix} 3 & -1 \\ 5 & 2 \end{pmatrix}$

The determinant of $\begin{pmatrix} 3 & -1 \\ 5 & 2 \end{pmatrix} = 3 \times 2 - (-1) \times 5$

$$= 6 - (-5)$$

$$= 11$$

Find the determinants of the following matrices:

1. $\begin{pmatrix} 6 & 1 \\ 3 & 2 \end{pmatrix}$ **5.** $\begin{pmatrix} 5 & 1 \\ 4 & -2 \end{pmatrix}$ **9.** $\begin{pmatrix} 1 & 4 \\ -1 & 5 \end{pmatrix}$

2. $\begin{pmatrix} 4 & 1 \\ 3 & 5 \end{pmatrix}$ **6.** $\begin{pmatrix} 6 & -2 \\ 2 & 1 \end{pmatrix}$ **10.** $\begin{pmatrix} -2 & -3 \\ -1 & -4 \end{pmatrix}$

3. $\begin{pmatrix} 6 & 2 \\ 6 & 2 \end{pmatrix}$ **7.** $\begin{pmatrix} 3 & 4 \\ 1 & 1 \end{pmatrix}$ **11.** $\begin{pmatrix} 2 & 3 \\ 1 & 4 \end{pmatrix}$

4. $\begin{pmatrix} 4 & -3 \\ 1 & 4 \end{pmatrix}$ **8.** $\begin{pmatrix} 6 & -7 \\ -2 & 1 \end{pmatrix}$ **12.** $\begin{pmatrix} 6 & 1 \\ 3 & -1 \end{pmatrix}$

SOLUTION OF SIMULTANEOUS EQUATIONS USING MATRICES

A pair of simultaneous equations is sometimes given in matrix form.

Consider the matrix equation

$$\begin{pmatrix} 5 & 2 \\ 3 & 5 \end{pmatrix} \begin{pmatrix} x \\ y \end{pmatrix} = \begin{pmatrix} 1 \\ -7 \end{pmatrix}$$

Multiply the left-hand side

$$\begin{pmatrix} 5x + 2y \\ 3x + 5y \end{pmatrix} = \begin{pmatrix} 1 \\ -7 \end{pmatrix}$$

This represents the pair of equations

$$5x + 2y = 1$$
$$3x + 5y = -7$$

Similarly, a pair of given equations can be written in matrix form.

For example, $\begin{aligned} 4x + 2y &= 7 \\ x + 3y &= 8 \end{aligned}$ gives $\begin{pmatrix} 4x + 2y \\ x + 3y \end{pmatrix} = \begin{pmatrix} 7 \\ 8 \end{pmatrix}$

i.e. $$\begin{pmatrix} 4 & 2 \\ 1 & 3 \end{pmatrix} \begin{pmatrix} x \\ y \end{pmatrix} = \begin{pmatrix} 7 \\ 8 \end{pmatrix}$$

EXERCISE 8i

Find the pair of simultaneous equations which are equivalent to the matrix equation $\begin{pmatrix} 1 & 2 \\ 3 & 1 \end{pmatrix} \begin{pmatrix} x \\ y \end{pmatrix} = \begin{pmatrix} 7 \\ 5 \end{pmatrix}$

$$\begin{pmatrix} 1 & 2 \\ 3 & 1 \end{pmatrix} \begin{pmatrix} x \\ y \end{pmatrix} = \begin{pmatrix} 7 \\ 5 \end{pmatrix}$$

(Multiply the left-hand side)

$$\begin{pmatrix} 1x + 2y \\ 3x + 1y \end{pmatrix} = \begin{pmatrix} 7 \\ 5 \end{pmatrix}$$

$\therefore$ the pair of equations is $\begin{aligned} x + 2y &= 7 \\ 3x + y &= 5 \end{aligned}$

Find the pairs of simultaneous equations which are equivalent to the following matrix equations:

1. $\begin{pmatrix} 1 & 2 \\ 3 & 2 \end{pmatrix} \begin{pmatrix} x \\ y \end{pmatrix} = \begin{pmatrix} 3 \\ 5 \end{pmatrix}$

3. $\begin{pmatrix} 9 & 2 \\ 4 & 1 \end{pmatrix} \begin{pmatrix} x \\ y \end{pmatrix} = \begin{pmatrix} 24 \\ 11 \end{pmatrix}$

2. $\begin{pmatrix} 4 & 2 \\ 5 & 3 \end{pmatrix} \begin{pmatrix} x \\ y \end{pmatrix} = \begin{pmatrix} 12 \\ 15 \end{pmatrix}$

4. $\begin{pmatrix} 6 & -1 \\ 2 & 1 \end{pmatrix} \begin{pmatrix} p \\ q \end{pmatrix} = \begin{pmatrix} -8 \\ 0 \end{pmatrix}$

Find the matrix equation which is equivalent to the pair of equations

$$2x - 4y = 0$$
$$9x + y = 19$$

$$2x - 4y = 0$$
$$9x + y = 19$$

$$\begin{pmatrix} 2 & -4 \\ 9 & 1 \end{pmatrix} \begin{pmatrix} x \\ y \end{pmatrix} = \begin{pmatrix} 0 \\ 19 \end{pmatrix}$$

Find the matrix equations which are equivalent to the following pairs of equations:

5. $3x + 2y = 8$
$\quad x + y = 3$

8. $3x - 2y = 1$
$\quad x - y = 0$

6. $4x - 3y = 1$
$\quad 2x + y = 3$

9. $7x - 2y = 3$
$\quad 3x + 4y = 11$

7. $4x + 3y = 5$
$\quad 5x + 4y = 6$

10. $5x + y = -8$
$\quad 4x - 3y = -14$

Consider the equation $2x = 6$

If we multiply both sides by $\frac{1}{2}$, we get $\frac{1}{2} \times 2x = \frac{1}{2} \times 6$

i.e. $x = 3$

$\frac{1}{2}$ is the reciprocal of 2, i.e. $\frac{1}{2}$ is the inverse of 2 under multiplication.

In the same way, we can simplify a matrix equation if we multiply both sides by the inverse of the first matrix.

EXERCISE 8j

Solve the matrix equation $\begin{pmatrix} 3 & 1 \\ 1 & 2 \end{pmatrix}\begin{pmatrix} x \\ y \end{pmatrix} = \begin{pmatrix} 7 \\ 4 \end{pmatrix}$

(Find the inverse first.)

The determinant of $\begin{pmatrix} 3 & 1 \\ 1 & 2 \end{pmatrix} = 3 \times 2 - 1 \times 1 = 5$

∴ the inverse is $\frac{1}{5}\begin{pmatrix} 2 & -1 \\ -1 & 3 \end{pmatrix}$

Multiply both sides of the equation by $\frac{1}{5}\begin{pmatrix} 2 & -1 \\ -1 & 3 \end{pmatrix}$

(The order matters now. *The inverse must go in front on both sides.*)

$$\frac{1}{5}\begin{pmatrix} 2 & -1 \\ -1 & 3 \end{pmatrix}\begin{pmatrix} 3 & 1 \\ 1 & 2 \end{pmatrix}\begin{pmatrix} x \\ y \end{pmatrix} = \frac{1}{5}\begin{pmatrix} 2 & -1 \\ -1 & 3 \end{pmatrix}\begin{pmatrix} 7 \\ 4 \end{pmatrix}$$

$$\begin{pmatrix} 1 & 0 \\ 0 & 1 \end{pmatrix}\begin{pmatrix} x \\ y \end{pmatrix} = \frac{1}{5}\begin{pmatrix} 10 \\ 5 \end{pmatrix}$$

i.e. $\begin{pmatrix} x \\ y \end{pmatrix} = \begin{pmatrix} 2 \\ 1 \end{pmatrix}$

∴ $x = 2$ and $y = 1$

Solve the following matrix equations:

1. $\begin{pmatrix} 4 & 1 \\ 3 & 1 \end{pmatrix}\begin{pmatrix} x \\ y \end{pmatrix} = \begin{pmatrix} 6 \\ 5 \end{pmatrix}$ **4.** $\begin{pmatrix} 3 & -1 \\ 4 & 1 \end{pmatrix}\begin{pmatrix} x \\ y \end{pmatrix} = \begin{pmatrix} 7 \\ 7 \end{pmatrix}$

2. $\begin{pmatrix} 2 & 1 \\ 5 & 3 \end{pmatrix}\begin{pmatrix} x \\ y \end{pmatrix} = \begin{pmatrix} 7 \\ 19 \end{pmatrix}$ **5.** $\begin{pmatrix} 2 & 1 \\ 1 & 2 \end{pmatrix}\begin{pmatrix} x \\ y \end{pmatrix} = \begin{pmatrix} 6 \\ 3 \end{pmatrix}$

3. $\begin{pmatrix} 4 & 3 \\ 2 & 2 \end{pmatrix}\begin{pmatrix} x \\ y \end{pmatrix} = \begin{pmatrix} 1 \\ 0 \end{pmatrix}$ **6.** $\begin{pmatrix} 7 & -2 \\ 3 & 4 \end{pmatrix}\begin{pmatrix} x \\ y \end{pmatrix} = \begin{pmatrix} 3 \\ 11 \end{pmatrix}$

7. $\begin{pmatrix} 1 & -1 \\ 1 & 3 \end{pmatrix} \begin{pmatrix} x \\ y \end{pmatrix} = \begin{pmatrix} 2 \\ 10 \end{pmatrix}$

8. $\begin{pmatrix} 5 & 2 \\ 2 & 1 \end{pmatrix} \begin{pmatrix} x \\ y \end{pmatrix} = \begin{pmatrix} 1 \\ 0 \end{pmatrix}$

9. $\begin{pmatrix} 3 & 2 \\ 3 & 3 \end{pmatrix} \begin{pmatrix} x \\ y \end{pmatrix} = \begin{pmatrix} 16 \\ 18 \end{pmatrix}$

10. $\begin{pmatrix} 4 & -2 \\ -3 & 1 \end{pmatrix} \begin{pmatrix} p \\ q \end{pmatrix} = \begin{pmatrix} 2 \\ -2 \end{pmatrix}$

11. $\begin{pmatrix} -5 & 1 \\ 9 & -2 \end{pmatrix} \begin{pmatrix} s \\ t \end{pmatrix} = \begin{pmatrix} 13 \\ -24 \end{pmatrix}$

Give the following pairs of simultaneous equations in matrix form and hence solve them:

12. $x + y = 2$
$x + 2y = 3$

13. $4x - y = 5$
$x + y = 5$

14. $5x + 4y = 1$
$x + y = 0$

15. $2x + 3y = 15$
$3x + 5y = 23$

16. $9x + 2y = 11$
$3x + y = 5$

17. $2x + 3y = 7$
$3x + 2y = 8$

18. $5x + 2y = 16$
$3x - y = 3$

19. $x + 4y = 11$
$2x + 3y = 7$

20. Try using matrices to solve the equations $\begin{array}{l} x + 2y = 3 \\ x + 2y = 6 \end{array}$

Comment on why the method breaks down.

21. Repeat question 20 with $\begin{array}{l} 3x + 2y = 1 \\ 9x + 6y = 5 \end{array}$

MATRIX EQUATIONS

EXERCISE 8k The questions in this exercise refer to the following matrices:

$$A = \begin{pmatrix} 4 & 3 \\ 2 & 1 \end{pmatrix} \qquad B = \begin{pmatrix} 1 & -2 \\ 0 & 1 \end{pmatrix} \qquad C = \begin{pmatrix} 6 & 4 \\ -3 & 1 \end{pmatrix}$$

Find **X** if **A + X = B**

$$A + X = B$$
$$X = B - A \qquad \text{(Taking A from each side)}$$

$$= \begin{pmatrix} 1 & -2 \\ 0 & 1 \end{pmatrix} - \begin{pmatrix} 4 & 3 \\ 2 & 1 \end{pmatrix}$$

$$= \begin{pmatrix} -3 & -5 \\ -2 & 0 \end{pmatrix}$$

Find **X** in the following equations:

1. **B + X = C**

2. **X − C = B**

3. **X = AC**

4. **B − X = C**

5. **2X = A**

6. $\frac{1}{3}$**X = B**

7. **A + X = B + C**

8. **AX = I**

MIXED EXERCISES

EXERCISE 8l **1.** Find $\begin{pmatrix} 2 & 3 & -1 \\ 4 & 1 & 2 \end{pmatrix} + \begin{pmatrix} 3 & 1 & 4 \\ 6 & -9 & 2 \end{pmatrix}$

2. Find $\frac{1}{2}\begin{pmatrix} 2 & 7 \\ 3 & -1 \end{pmatrix}$

3. Find the determinant of $\begin{pmatrix} 16 & 3 \\ 8 & 3 \end{pmatrix}$

4. Find the inverse of $\begin{pmatrix} 4 & 3 \\ 3 & 4 \end{pmatrix}$

5. Find $\begin{pmatrix} 1 & 3 & -1 \end{pmatrix} \begin{pmatrix} 1 \\ -3 \\ 1 \end{pmatrix}$

6. Find $\begin{pmatrix} 1 & 4 \\ 3 & 1 \end{pmatrix} \begin{pmatrix} 2 & -1 \\ 2 & 1 \end{pmatrix} \begin{pmatrix} 1 & 3 \\ 1 & 1 \end{pmatrix}$

EXERCISE 8m The questions in this exercise refer to the following matrices:

$$A = \begin{pmatrix} 2 & 4 \\ 1 & 3 \end{pmatrix} \qquad B = \begin{pmatrix} 3 & -1 \\ -2 & 1 \end{pmatrix} \qquad C = \begin{pmatrix} 2 \\ 3 \end{pmatrix} \qquad D = \begin{pmatrix} 4 & -2 \end{pmatrix}$$

1. Find $A + B$.

2. Find the determinant of A.

3. Find the inverse of B.

4. Find BC.

5. Find DA.

6. Find $\frac{3}{4}D$.

9 AREAS

AREAS OF FAMILIAR SHAPES

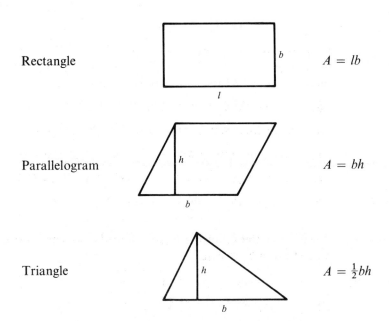

Rectangle $\qquad\qquad A = lb$

Parallelogram $\qquad\qquad A = bh$

Triangle $\qquad\qquad A = \frac{1}{2}bh$

Remember that when we talk about the height of a figure we mean the *perpendicular height*, not the slant height.

Remember also that both the lengths we use must be measured in the *same* unit.

EXERCISE 9a   Find the areas of the following figures:

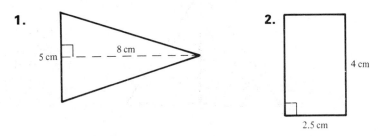

1.   5 cm   8 cm

2.   4 cm   2.5 cm

3.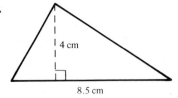

4.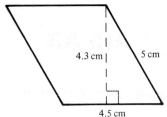

In questions 5 to 9 use squared paper and draw axes for x and y in the ranges $-6 \leqslant x \leqslant 6$, $-6 \leqslant y \leqslant 6$ using 1 square to 1 unit. Draw the figure and find its area in square units:

5. Triangle ABC with A $(0, 6)$, B $(6, 6)$ and C $(5, 2)$.

6. Parallelogram ABCD with A $(0, 1)$, B $(0, 6)$, C $(6, 4)$ and D $(6, -1)$.

7. Rectangle ABCD with A $(-4, 2)$, B $(0, 2)$ and C $(0, -1)$.

8. Square ABCD with A $(0, 0)$, B $(0, 4)$ and C $(4, 4)$.

9. Triangle ABC with A $(-5, -4)$, B $(2, -4)$, C $(-2, 3)$.

For each of the following figures, find the missing measurement. Draw a diagram in each case:

	Figure	*Base*	*Height*	*Area*
10.	Triangle	8 cm		16 cm^2
11.	Rectangle	3 cm	15 mm	
12.	Parallelogram	4 cm		20 cm^2
13.	Square	5 m		
14.	Triangle	70 mm		14 cm^2

In questions 15 to 18 give answers correct to three significant figures where necessary.

15.

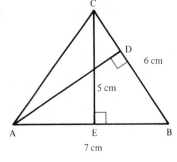

In $\triangle$ABC, AB $= 7$ cm, CB $= 6$ cm and CE $= 5$ cm. Find a) the area of $\triangle$ABC b) the length of AD.

16.

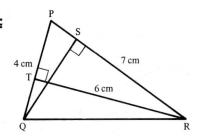

In △PQR, PQ = 4cm,
PR = 7cm and RT = 6cm.
Find a) the area of △PQR
b) the length of QS.

17.

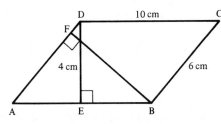

In parallelogram ABCD,
DC = 10cm, BC = 6cm
and DE = 4cm.
Find a) the area of ABCD
b) the length of BF.

18.

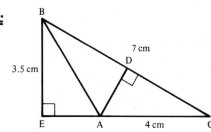

In △ABC, AC = 4cm,
BC = 7cm and BE = 3.5cm.
Find a) the area of △ABC
b) the length of AD.

AREAS OF COMPOUND SHAPES

EXERCISE 9b

ABCD is a rhombus. AC = 8cm and BD = 12cm.
Find the area of ABCD

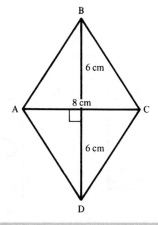

(The diagonals of a rhombus
bisect each other at right angles.)

Area △ABC = $\frac{1}{2}$ base × height

$= \frac{1}{2} \times 8 \times 6 \, cm^2$

$= 24 \, cm^2$

Area △ACD = area △ABC

(AC is an axis of symmetry)

∴ total area = 48 cm²

Find the area of each of the following shapes. Draw a diagram for each question and mark in all the measurements:

1.

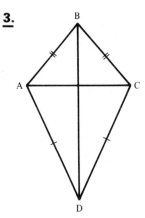

ABCD is a rhombus.
AC = 15 cm and BD = 8 cm

3.

ABCD is a kite.
AC = 6 cm and BD = 10 cm

2.

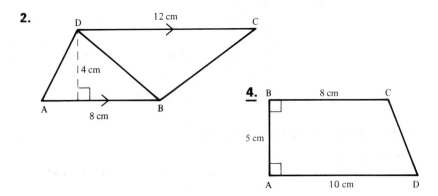

4. B   8 cm   C

5 cm

A   10 cm   D

In questions 5 and 6, find the area of the shaded figure (find the area of the complete figure, then subtract the areas of the unshaded parts).

5.

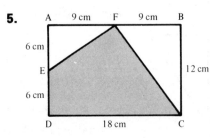

ABCD is a rectangle.

6.

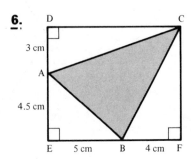

AREA OF A TRAPEZIUM

In the last exercise we found the areas of several trapeziums. A trapezium is a shape that occurs often enough to justify finding a formula for its area.

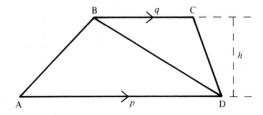

$$\text{Area of } \triangle ABD = \tfrac{1}{2} \text{ base} \times \text{height} = \tfrac{1}{2}p \times h$$

$$\text{Area of } \triangle BCD = \tfrac{1}{2} \text{ base} \times \text{height} = \tfrac{1}{2}q \times h$$

The heights of both triangles are the same, as each is the distance between the parallel sides of the trapezium.

$\therefore$ total area of ABCD $= \tfrac{1}{2}ph + \tfrac{1}{2}qh = \tfrac{1}{2}(p + q) \times h$

i.e.   **the area of a trapezium is equal to**

$\tfrac{1}{2}$(sum of parallel sides) × (distance between them)

EXERCISE 9c

Find the area of the trapezium in the diagram

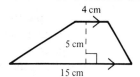

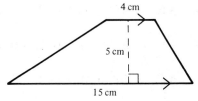

Area $= \tfrac{1}{2}$(sum of parallel sides) × (distance between them)
$= \tfrac{1}{2}(4 + 15) \times 5\,\text{cm}^2$
$= \tfrac{1}{2} \times 19 \times 5\,\text{cm}^2$
$= 47.5\,\text{cm}^2$

Find the area of each of the following trapeziums:

1.
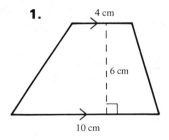

3.
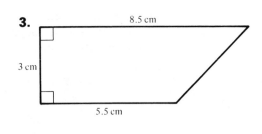

2.
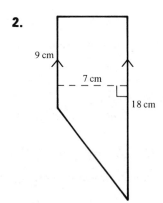

4.
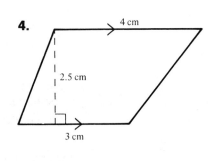

For questions 5 to 10 use squared paper and draw axes for x and y using ranges $-6 \leqslant x \leqslant 6$ and $-6 \leqslant y \leqslant 6$ and a scale of one square to 1 unit. Plot the points and join them up in alphabetical order. Find, in square units, the area of the resulting shape:

5. A $(6, 1)$, B $(4, -3)$, C $(-2, -3)$, D $(-3, 1)$

6. A $(4, 4)$, B $(-2, 2)$, C $(-2, -2)$, D $(4, -3)$

7. A $(3, 5)$, B $(-4, 4)$, C $(-4, -2)$, D $(3, -5)$

8. A $(1, 0)$, B $(5, 0)$, C $(5, 3)$, D $(3, 5)$, E $(1, 3)$

9. A $(6, -4)$, B $(6, 1)$, C $(2, 5)$, D $(-5, 3)$, E $(-5, -4)$

10. A $(2, 0)$, B $(6, 4)$, C $(-4, 4)$, D $(-4, -2)$, E $(5, -2)$

SHAPES THAT HAVE EQUAL AREAS

EXERCISE 9d   1.

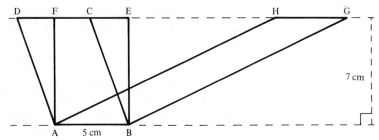

There are three parallelograms in the diagram. Write down the area of each one.

2.

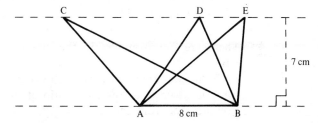

Write down the area of each of the three triangles ABC, ABD and ABE.

In questions 3 and 4, take the side of a square as the unit of length.

3.

What is the length of the base and the height of each of these parallelograms? What can you conclude about the areas of the three parallelograms?

4.

What is the length of the base and the height of each of these triangles? What can you conclude about the areas of the three triangles?

5. Using squared paper draw x and y axes for $-8 \leqslant x \leqslant 8$ and $-4 \leqslant y \leqslant 12$. Draw parallelogram ABCD where A (2, 2), B (2, 7), C (7, 4), D (7, −1). Using AB as the base in each case, draw three other parallelograms whose areas are equal to the area of ABCD.

6. Use the same set of axes as you used for question 5. Draw the triangle LMN where L(−6, 1), M(−1, 1), N(−5, 5). Using LM as the base draw two other triangles whose areas are equal to the area of triangle LMN.

7. Draw again the diagram that you used for question 6. Draw a triangle which is equal in area to triangle LMN and which has LN for one of its sides.

8. There are four triangles in the diagram all with the same base AB.
Find, in ascending order, the ratio of their heights.
Find, in ascending order, the ratio of their areas.
Comment on your results.

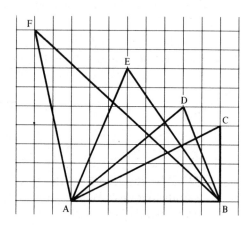

9. Using squared paper draw x and y axes for $0 \leqslant x \leqslant 10$ and $0 \leqslant y \leqslant 12$. Plot the points A (2, 1), B (10, 1) and C (8, 5). Draw a triangle ABD whose area is twice that of triangle ABC.
Give the y coordinate of D.

10. Using the same set of axes as in question 9, draw a triangle ABE whose area is half the area of $\triangle$ABC.
Give the y coordinate of E.

From the last exercise we see that:

parallelograms with equal bases and equal heights have the same area.

triangles with equal bases and equal heights have the same area.

triangles on equal bases but with different heights have areas proportional to their heights.

EXERCISE 9e

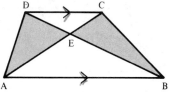

ABCD is a trapezium. Show that the shaded triangles have the same area.

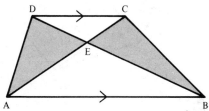

Area △ADB = area △ACB

    (Same base AB and same height, as DC∥AB)

Area △AEB is common to both △ADB and △ACB. Removing it from each triangle in turn leaves the shaded areas.

∴   area △AED = area △BEC

1.

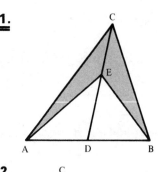

D is the midpoint of AB. Show that the shaded triangles have the same area.

2.

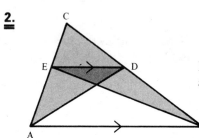

ED is parallel to AB. Show that area △ACD = area △BCE.

3.

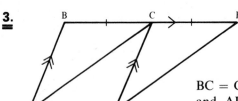

BC = CE, AB is parallel to DC and AD is parallel to BE. Show that area △ADC = area △DCE.

4.

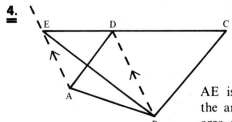

AE is parallel to BD. Show that the area of △BCE is equal to the area of the quadrilateral ABCD.

5.

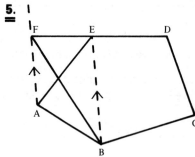

AF is parallel to BE. Show that the area of the pentagon ABCDE is equal to the area of the quadrilateral BCDF.

6.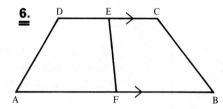

ABCD is a trapezium. E is the midpoint of DC and F is the midpoint of AB. Show that EF divides the area of ABCD into two equal parts.

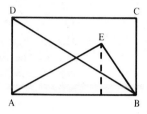

ABCD is a rectangle.
The height of $\triangle$ABE (i.e. the distance of E above AB) is two thirds of the height, AD, of the rectangle.
The area of ABCD is $12\,\text{cm}^2$.
Find the area of $\triangle$AEB.

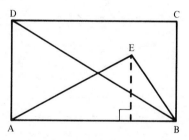

Area of $\triangle$ADB $= 6\,\text{cm}^2$ (half area of ABCD)

$\triangle$s ABD, ABE are on the same base AB, so their areas are in the same ratio as their heights.

$$\therefore \quad \frac{\text{area } \triangle ABE}{\text{area } \triangle ADB} = \frac{2}{3}$$

$$\frac{\text{area } \triangle ABE}{6} = \frac{2}{3}$$

$$6 \times \frac{\text{area } \triangle ABE}{6} = 6 \times \frac{2}{3}$$

$$\therefore \quad \text{area } \triangle ABE = 4\,\text{cm}^2$$

7.

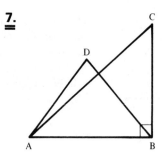

The area of △ABD is $\frac{3}{5}$ of the area of △ABC. BC = 20 cm. Find the height of D above AB.

8.

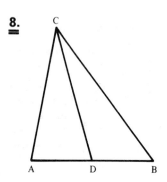

D is the midpoint of AB. Find the ratio of the area of △ABC to the area of △ADC.

9.

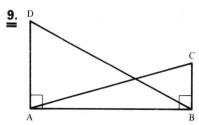

The area of △ABC = $\frac{7}{12}$ of the area of △ADB. AD = 24 cm. Find the length of BC.

10.

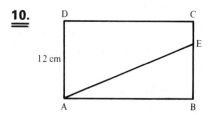

In the rectangle ABCD, E is a point on BC such that area of △ABE is $\frac{1}{3}$ of the area of the rectangle ABCD.
Find the length of BE.

11.

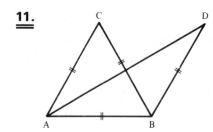

AB = AC = CB = BD.
Area △ABC = area △ABD.
Find the size of $A\widehat{D}B$.

12. 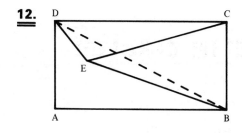 E is a point inside the rectangle ABCD such that the distance of E from BC is $\frac{3}{4}$ of the distance of A from BC and the distance of E from DC is $\frac{1}{3}$ of the distance of A from DC. The area of the rectangle ABCD is $72\,\text{cm}^2$.

Find the areas of $\triangle$BEC and $\triangle$DEC.

In questions 13 and 14, before beginning the construction draw a rough sketch and on it mark any extra lines that you will need:

13. Construct a parallelogram ABCD with $AB = 4\,\text{cm}$, $BC = 6\,\text{cm}$ and $A\widehat{B}C = 60°$. Construct a parallelogram ABEF that is equal in area to ABCD such that $BE = 7\,\text{cm}$, and such that E and D are on opposite sides of BC. Measure $A\widehat{B}E$.

14. Construct $\triangle$ABC with $AB = 12\,\text{cm}$, $\widehat{A} = 30°$ and $\widehat{B} = 30°$. On AC as base, construct a triangle ADC that is equal in area to $\triangle$ABC, such that $C\widehat{A}D = 90°$. Measure AD.

10 ANGLES IN CIRCLES

These facts were introduced in Book 1A:

angles on a straight line add up to 180°;

angles at a point add up to 360°;

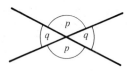

vertically opposite angles are equal.

When a transversal cuts a pair of parallel lines, various angles are formed and:

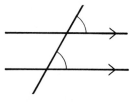

corresponding angles are equal;

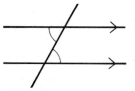

alternate angles are equal;

interior angles add up to 180°.

In *any* triangle, whatever its shape or size, the sum of the three angles is 180°,

e.g.

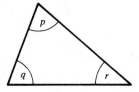

$$p + q + r = 180°$$

The exercise that follows will help you revise these facts. However, you must state which fact you have used as a reason for statements that you make. The facts do not have to be quoted in full but can be shortened using a number of standard abbreviations.

For example, if you state that $x = 60°$ and the reason is that x and 60° are vertically opposite angles then you could write

$$x = 60° \qquad \text{(vert. opp. } \angle\text{s)}$$

EXERCISE 10a

Find the size of the angle marked x, giving brief reasons to justify your statements.

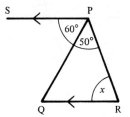

(Fill in the size of any angles that you find.)

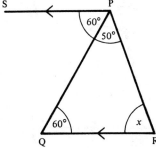

From the diagram

$$P\widehat{Q}R = 60° \qquad \text{(alt. } \angle\text{s)}$$
$$50° + 60° + x = 180° \qquad (\angle \text{ sum of } \triangle PQR)$$
$$\therefore \qquad\qquad x = 70°$$

In each of the following diagrams find the size of the angle marked x, giving brief reasons for your answer:

1.

5.

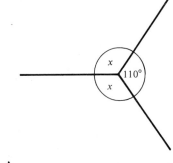

2.

6.

3.

7.

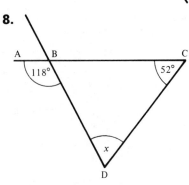

4.

8.

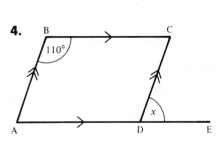

Find the angles denoted by the letters:

9.

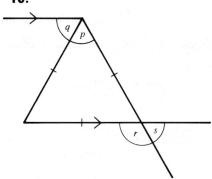

12.

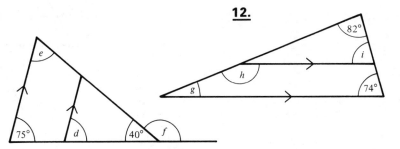

10.

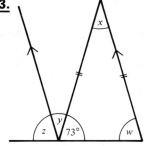

13.

11.

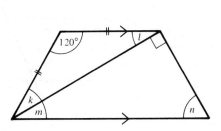

14.

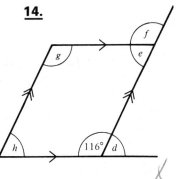

Prove that $\widehat{ABC} = \widehat{CDE}$

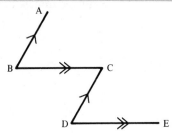

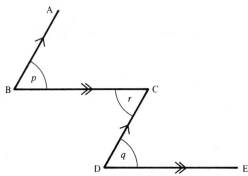

(Mark with a letter the angles you need to refer to.)

From the diagram

$$p = r \quad \text{(alt. } \angle\text{s)}$$
$$r = q \quad \text{(alt. } \angle\text{s)}$$
$$\therefore \qquad p = q$$

i.e. $\qquad \widehat{ABC} = \widehat{CDE}$

15. Prove that

$$\widehat{ABF} = \widehat{CDE}$$

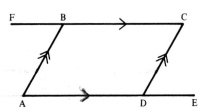

16. Prove that

$$\widehat{TVQ} + \widehat{RWV} = 180°$$

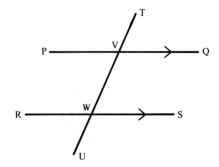

17. Prove that

$$A\widehat{P}E = B\widehat{A}P + F\widehat{E}P$$

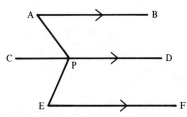

18. Prove that

$$D\widehat{C}E + C\widehat{A}B + C\widehat{B}A = 180°$$

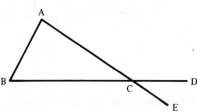

19. Prove that

$$x + y + z = 180°$$

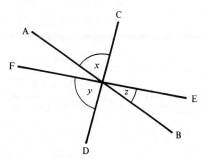

20. Prove that

$$x = y + z$$

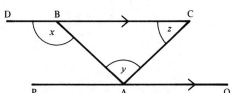

CONGRUENT SHAPES

Two shapes are *congruent* if they are exactly the same shape and size, i.e. one shape is an exact copy of the other shape, although not necessarily drawn in the same position.

In each of these diagrams the two figures are congruent:

In each case the second shape is an exact copy of the first shape although it may be turned round or over.

In each of these diagrams the two figures are not congruent (they may be similar; i.e. have the same shape but different sizes):

EXERCISE 10b In the following questions state whether or not the two shapes are congruent. If you are not sure, trace one shape and see if it fits exactly over the other shape:

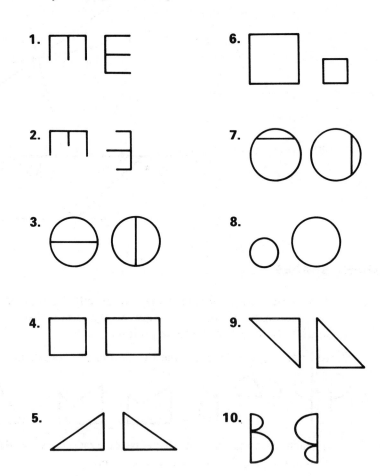

TRANSFORMATIONS AND CONGRUENT FIGURES

The shape and size of a figure are not altered by certain transformations. Reflection produces congruent shapes:

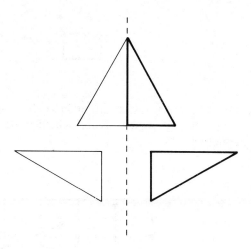

Rotation produces congruent shapes:

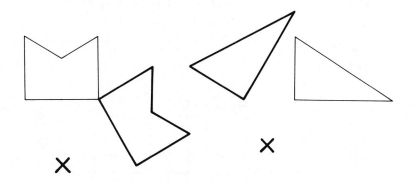

Translation produces congruent shapes:

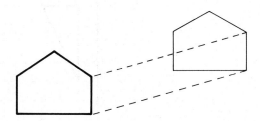

But enlargement does *not* produce congruent shapes, it produces similar shapes:

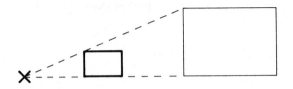

EXERCISE 10c Describe the transformation in each of the following cases. The grey shape is the image. State whether the object and the image are congruent:

1.

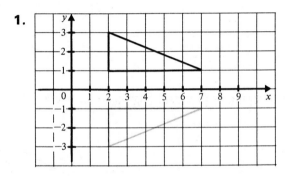

2.

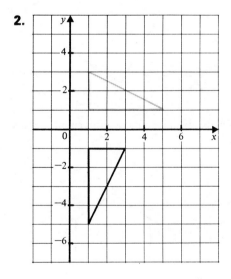

3.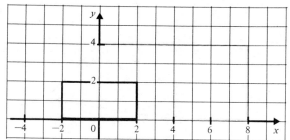

4. 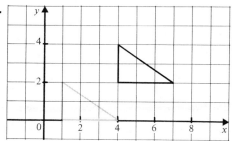   **5.**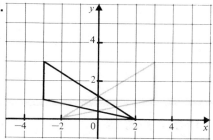

6.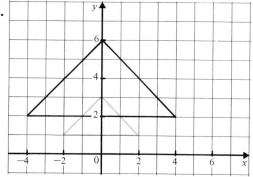

7.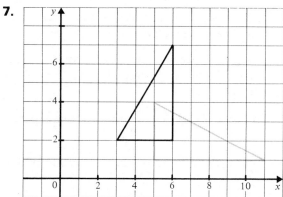

8.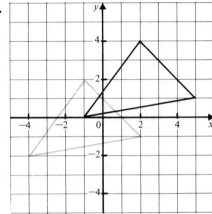

CIRCLE FACTS

First we will revise some of the facts we already know about the circle.

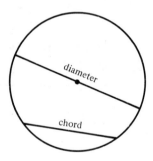

Every point on a circle is the same distance from its centre. This distance is called the *radius* of the circle.

Sometimes we use the word "circle" to include the space inside the curve. When we do this we call the curve itself the *circumference* of the circle.

A straight line joining any two points on the circumference is called a *chord*.

Any chord passing through the centre of a circle is called a *diameter*.

We will now learn some new facts and definitions.

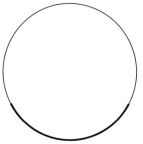

Any part of the circumference is called an arc. If the arc is less than half the circumference it is called a *minor arc*; if it is greater than half the circumference it is called a *major arc*.

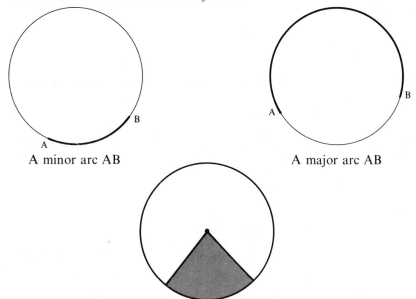

A minor arc AB   A major arc AB

The shaded area is enclosed by two radii and an arc. It looks like a slice of cake and is called a *sector*.

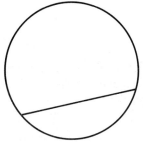

A chord divides a circle into two regions called segments. The larger region is called a *major segment* and the smaller region is called a *minor segment*.

EXERCISE 10d **1.** Name six chords in this diagram. Is any one of these chords a diameter? If so, name it.

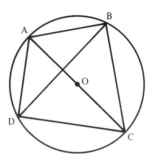

Copy the following diagrams. They do not have to be identical:

2. DC divides the circle into two segments. Shade the minor segment.

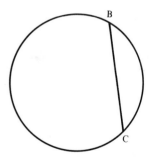

3. BC divides the circle into two segments. Shade the major segment.

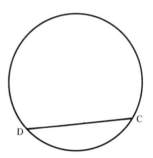

4. AB divides the circle into two segments. Shade the major segment.

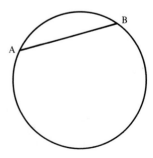

5. AD divides the circle into two segments. Shade the minor segment.

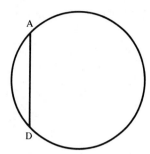

THE ANGLE SUBTENDED BY AN ARC

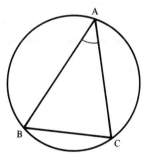

Consider a triangle ABC whose vertices A, B and C lie on a circle. The angle BAC is said to stand on the minor arc BC. We say that BC *subtends* an angle BAC at A which is on the circumference.

Similarly the angle ABC stands on the arc AC, or AC subtends an angle ABC at B, and the angle ACB stands on the arc AB, or AB subtends an angle ACB at C.

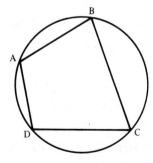

If the four vertices of a quadrilateral ABCD all lie on a circle, we say that the quadrilateral is *cyclic*, i.e. ABCD is a *cyclic quadrilateral*.

EXERCISE 10e Questions 1 to 6 refer to the following diagram:

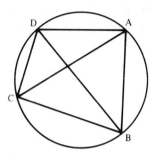

1. What arc does $D\hat{B}C$ stand on?

2. What arc does $B\hat{D}C$ stand on?

3. Name the two angles at the circumference standing on arc AB.

4. Name the two angles at the circumference standing on arc BC.

5. What arc subtends $D\hat{B}A$?

6. What arc subtends $A\hat{C}B$?

Questions 7 to 12 refer to the following diagram:

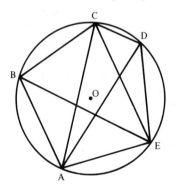

7. What arc does $B\hat{C}E$ stand on?   **9.** What arc subtends $C\hat{A}E$?

8. What arc does $C\hat{A}D$ stand on?   **10.** What arc subtends $D\hat{A}B$?

11. What angles stand on   a) the arc AB?   b) the arc BC?

12. What angles stand on   a) the arc AE?   b) the arc CE?

DISCOVERING RELATIONSHIPS BETWEEN ANGLES

EXERCISE 10f Copy the following diagrams making them at least twice as large. For each diagram measure the angles denoted by the letters:

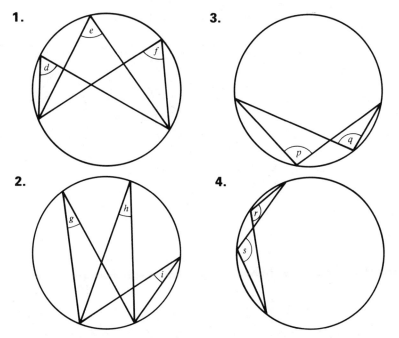

1.

2.

3.

4.

First Fact
The results for questions 1 to 4 show that

angles standing on the same arc of a circle and in the same segment are equal.

EXERCISE 10g

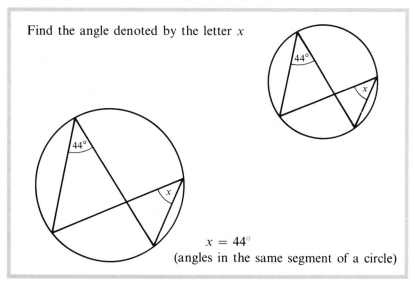

Find the angle denoted by the letter x

$x = 44°$
(angles in the same segment of a circle)

Find the angles denoted by the letters:

1.

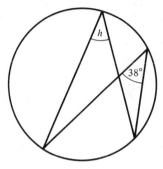

3.

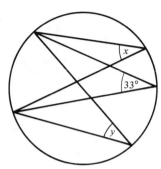

2.

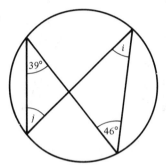

4.

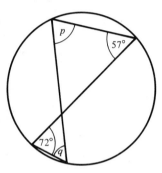

Find the angle denoted by the letter *k*

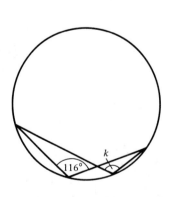

$k = 116°$

(angles in the same segment)

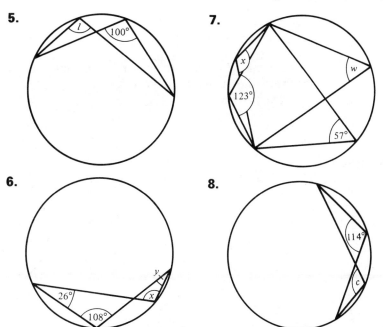

5.

7.

6.

8.

EXERCISE 10h Copy the following diagrams making them at least twice as large. They need not be identical. For each diagram measure the angles denoted by the letters. Hence find a relationship between x and y:

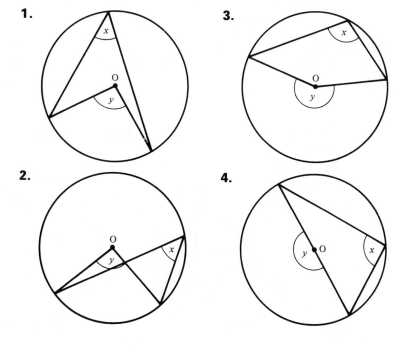

1.

3.

2.

4.

Second Fact
These results show that

> the angle which the arc of a circle subtends at the centre is equal to twice the angle it subtends at any point on the remaining circumference.

EXERCISE 10i

Find the angle denoted by the letter d

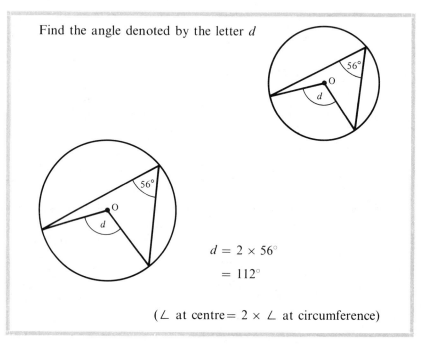

$$d = 2 \times 56°$$
$$= 112°$$

($\angle$ at centre $= 2 \times \angle$ at circumference)

Find the angles denoted by the letters. O denotes the centre of the circle:

1.

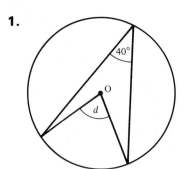

2.

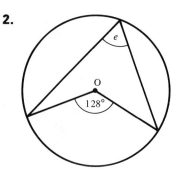

3.

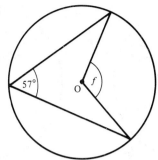

4.

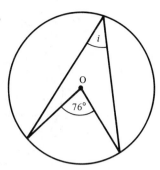

Find the angle denoted by the letter *f*

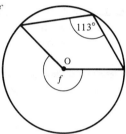

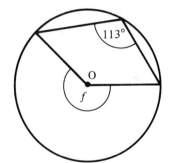

$$f = 2 \times 113°$$
$$= 226°$$

(∠ at centre = 2 × ∠ at circumference)

5.

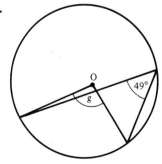

6.

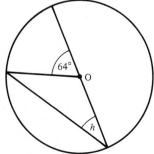

7.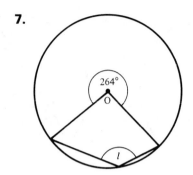

8.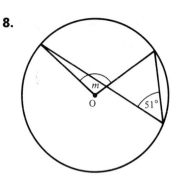

EXERCISE 10j Copy the following diagrams making them at least twice as large. For each diagram measure the angles denoted by p and q. What do you notice about their sum?

1.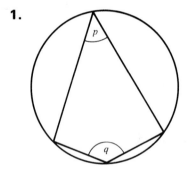

2.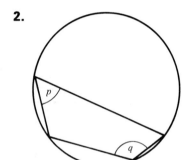

3.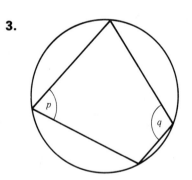

Third Fact
These results show that

the opposite angles of a cyclic quadrilateral are supplementary.

EXERCISE 10k Find the angles denoted by the letters.

Find the angle denoted by the letter *p*

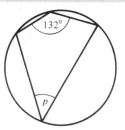

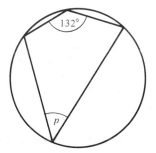

$p + 132° = 180°$ (opp. ∠s cyclic quad. supplementary)

∴ $p = 180° - 132°$

            $= 48°$

1.

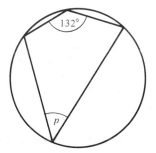

3.

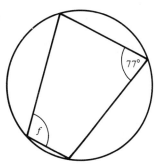

2.

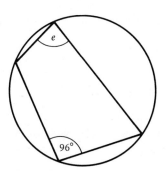

4.

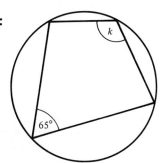

5.

7.

6.

8.

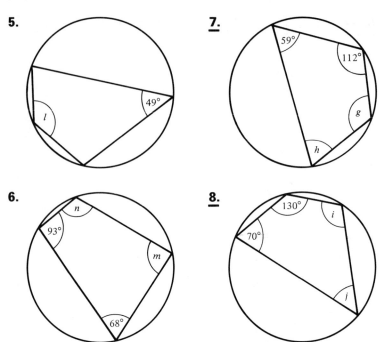

Fourth Fact

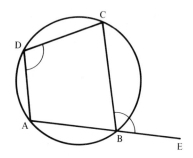

If the side AB of the cyclic quadrilateral ABCD is produced to E, the angle $C\hat{B}E$ is called an exterior angle of quadrilateral ABCD.

Then $A\hat{B}C + C\hat{B}E = 180°$ ($\angle$s on a str. line)

and $A\hat{B}C + A\hat{D}C = 180°$ (opp. $\angle$s cyclic quadrilateral)

Hence $C\hat{B}E = A\hat{D}C$

i.e.   **any exterior angle of a cyclic quadrilateral is equal to the opposite interior angle.**

EXERCISE 10I Copy the following diagrams making them at least twice as large. For each diagram measure the angles denoted by the letters. What result do they confirm?

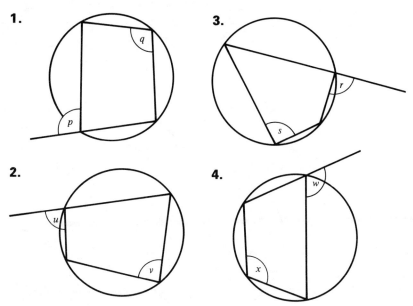

1.

q

p

3.

r

s

2.

u

v

4.

w

x

In questions 5 to 8 find the angles denoted by the letters:

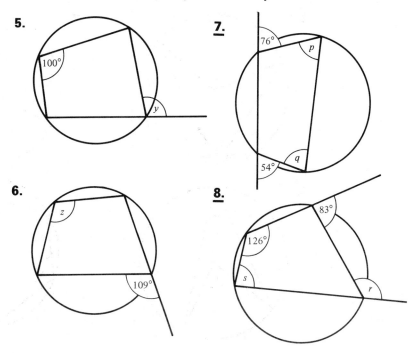

5.

100°

y

7.

76°

p

54°

q

6.

z

109°

8.

83°

126°

s

r

EXERCISE 10m This exercise brings together the results used in the last three exercises. In some questions more than one of those results is required.

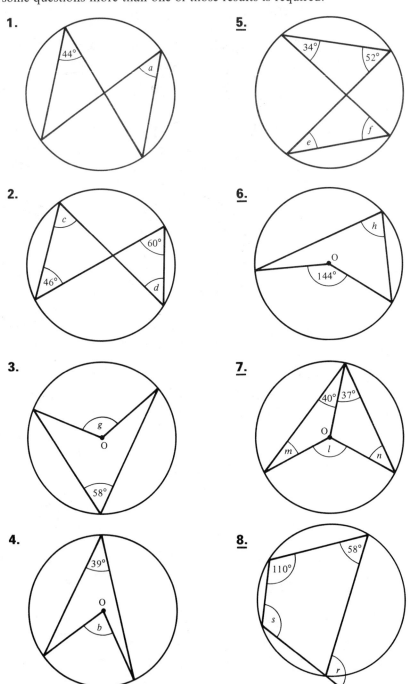

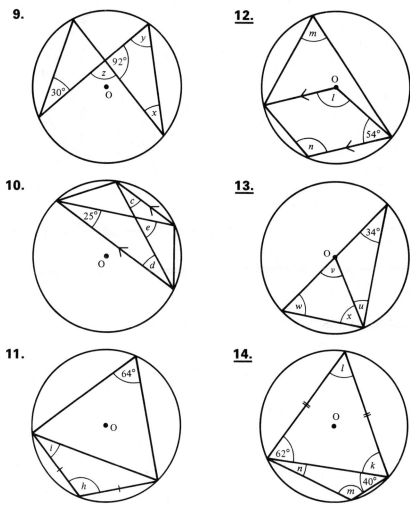

9.

10.

11.

12.

13.

14.

EXERCISE 10n Copy the following diagrams making yours at least twice as big. Measure the angles denoted by the letters. What result do these values show?

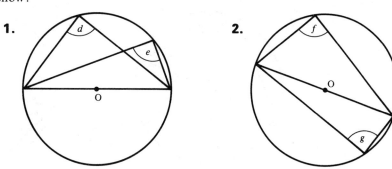

1.

2.

3.

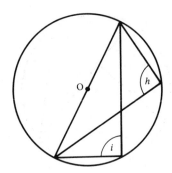

Fifth Fact

The angle in a semicircle is a right angle.

EXERCISE 10p Find the angles denoted by the letters. The centre of the circle is marked O.

1.

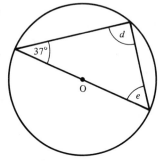

3.

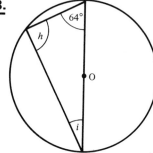

2.

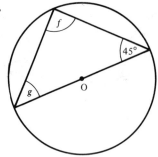

4.

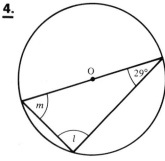

5.

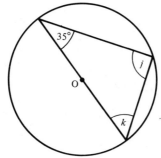

6.

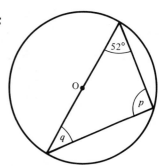

Find the angles denoted by the letters

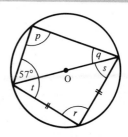

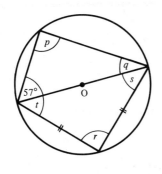

$$p = 90° \quad (\angle \text{ in a semicircle})$$
$$q = 33° \quad (\angle\text{s of a triangle})$$
$$r = 90° \quad (\angle\text{s in a semicircle})$$
$$s + t = 90° \quad (\angle\text{s of a triangle})$$

But $\quad s = t \quad$ (isosceles triangle)

$\therefore \quad s = t = 45°$

7.

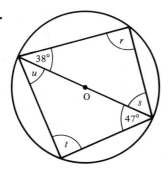

8.

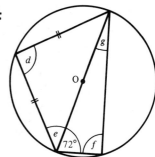

190   *ST(P) Mathematics 3A*

9.

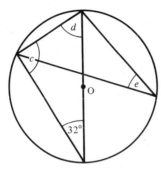

11.

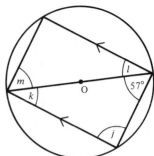

10.

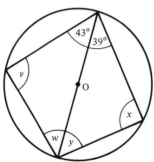

12.

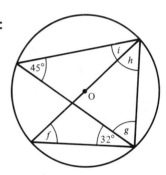

MIXED EXERCISE

EXERCISE 10q   Find the angles denoted by the letters:

1.

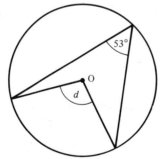

3.

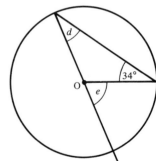

2.

4.

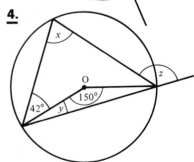

5.

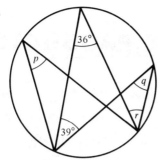

7.

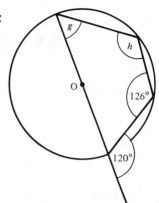

6.

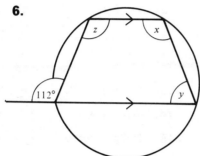

8.

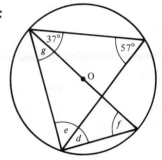

11 ALGEBRAIC PRODUCTS

BRACKETS

Remember that $5(x + 1) = 5x + 5$

and that $4x(y + z) = 4xy + 4xz$

EXERCISE 11a Expand:

1. $2(x + 1)$	**4.** $5(a + 4)$	**7.** $5(1 - b)$
2. $3(x - 1)$	**5.** $3(b + 7)$	**8.** $2(3a - 1)$
3. $4(x + 3)$	**6.** $3(1 - a)$	**9.** $4(2 + 3b)$
10. $5a(b - c)$	**13.** $5x(3y + z)$	**16.** $8r(2t - s)$
11. $4a(b - 2c)$	**14.** $4y(4x + 3z)$	**17.** $3a(b - 5c)$
12. $3a(2a + b)$	**15.** $2n(3p - 5q)$	**18.** $4x(3y + 2z)$

THE PRODUCT OF TWO BRACKETS

Frequently, we wish to find the product of two brackets, each of which contains two terms, e.g. $(a + b)(c + d)$. The meaning of this product is that each term in the first bracket has to be multiplied by each term in the second bracket.

Always multiply the brackets together in the following order:

1. the first terms in the brackets
2. the outside terms
3. the inside terms
4. the second terms in the brackets.

Thus
$$(a + b)(c + d) = ac + ad + bc + bd$$

192

EXERCISE 11b

Expand $(x + 2y)(2y - z)$

$(x + 2y)(2y - z) = 2xy - xz + 4y^2 - 2yz$

Expand:

1. $(a + b)(c + d)$	**6.** $(a - b)(c + d)$
2. $(p + q)(s + t)$	**7.** $(x + y)(y + z)$
3. $(2a + b)(c + 2d)$	**8.** $(2a + b)(3c + d)$
4. $(5x + 2y)(z + 3)$	**9.** $(5x + 4y)(z + 2)$
5. $(x + y)(z - 4)$	**10.** $(3x - 2y)(5 - z)$
11. $(p + q)(2s - 3t)$	**16.** $(3p - q)(4r - 3s)$
12. $(a - 2b)(c - d)$	**17.** $(3a - 4b)(3c + 4d)$
13. $(6u - 5v)(w - 5r)$	**18.** $(7x - 2y)(3 - 2z)$
14. $(3a + 4b)(2c - 3d)$	**19.** $(2a + b)(5c - 2)$
15. $(3x + 2y)(3z + 2)$	**20.** $(5a - 4b)(3 - 2d)$

We get a slightly simpler form when we find the product of two brackets such as $(x + 2)$ and $(x + 3)$,
i.e. using the order we chose earlier

$$(x + 2)(x + 3) = x^2 + 3x + 2x + 6$$

$$= x^2 + 5x + 6 \quad \text{(since ② and ③ are like terms)}$$

i.e. $(x + 2)(x + 3) = x^2 + 5x + 6$

EXERCISE 11c Expand:

1. $(x + 3)(x + 4)$	**6.** $(a + 4)(a + 5)$
2. $(x + 2)(x + 4)$	**7.** $(b + 2)(b + 7)$
3. $(x + 1)(x + 6)$	**8.** $(c + 4)(c + 6)$
4. $(x + 5)(x + 2)$	**9.** $(p + 3)(p + 12)$
5. $(x + 8)(x + 3)$	**10.** $(q + 7)(q + 10)$

Expand $(x - 4)(x - 6)$

$$(x - 4)(x - 6) = x^2 - 6x - 4x + 24$$

$$= x^2 - 10x + 24$$

Expand:

11. $(x - 2)(x - 3)$ **16.** $(x - 3)(x - 4)$

12. $(x - 5)(x - 7)$ **17.** $(x - 4)(x - 8)$

13. $(a - 2)(a - 8)$ **18.** $(b - 4)(b - 2)$

14. $(x - 10)(x - 3)$ **19.** $(a - 4)(a - 4)$

15. $(b - 5)(b - 5)$ **20.** $(p - 7)(p - 8)$

Expand $(x + 3)(x - 6)$

$$(x + 3)(x - 6) = x^2 - 6x + 3x - 18$$

$$= x^2 - 3x - 18$$

Expand:

21. $(x + 3)(x - 2)$ **26.** $(x + 7)(x - 2)$

22. $(x - 4)(x + 5)$ **27.** $(x - 5)(x + 6)$

23. $(x - 7)(x + 4)$ **28.** $(x + 10)(x - 1)$

24. $(a + 3)(a - 10)$ **29.** $(b - 8)(b - 7)$

25. $(p + 5)(p - 5)$ **30.** $(z - 1)(z - 12)$

FINDING THE PATTERN

You may have noticed in the previous exercise, that when you expanded the brackets and simplified the answers, there was a definite pattern,

e.g. $$(x + 5)(x + 9) = x^2 + 9x + 5x + 45$$
$$= x^2 + 14x + 45$$

We could have written it

$$(x + 5)(x + 9) = x^2 + (9 + 5)x + (5) \times (9)$$
$$= x^2 + 14x + 45$$

Similarly $$(x + 4)(x - 7) = x^2 + (-7 + 4)x + (4) \times (-7)$$
$$= x^2 - 3x - 28$$

and $$(x - 3)(x - 8) = x^2 + (-8 - 3)x + (-3) \times (-8)$$
$$= x^2 - 11x + 24$$

In each case there is a pattern:
the *product* of the two numbers in the brackets gives the number term in the expansion, while *collecting* them gives the number of xs.

EXERCISE 11d Use the pattern given above to expand the following products:

1.	$(x + 4)(x + 5)$	**5.**	$(x + 8)(x + 6)$
2.	$(a + 2)(a + 5)$	**6.**	$(a + 10)(a + 7)$
3.	$(x - 4)(x - 5)$	**7.**	$(x - 8)(x - 6)$
4.	$(a - 2)(a - 5)$	**8.**	$(a - 10)(a - 7)$
9.	$(a + 2)(a - 5)$	**13.**	$(a - 10)(a + 7)$
10.	$(y - 6)(y + 3)$	**14.**	$(y + 10)(y - 2)$
11.	$(z + 4)(z - 10)$	**15.**	$(z - 12)(z + 1)$
12.	$(p + 5)(p - 8)$	**16.**	$(p + 2)(p - 13)$
17.	$(x - 5)(x - 1)$	**21.**	$(p + 12)(p + 2)$
18.	$(b + 9)(b + 7)$	**22.**	$(t + 5)(t - 12)$
19.	$(a + 4)(a - 4)$	**23.**	$(c - 5)(c + 8)$
20.	$(r - 14)(r + 2)$	**24.**	$(x + 5)(x - 5)$

The pattern is similar when the brackets are slightly more complicated.

EXERCISE 11e

Expand the product $(2x + 3)(x + 2)$

$$\overset{①}{(2x + 3)}\overset{④}{(x + 2)} = 2x^2 + 4x + 3x + 6$$

$$= 2x^2 + 7x + 6$$

Expand the following products:

1. $(2x + 1)(x + 1)$ **5.** $(3x + 2)(x + 1)$

2. $(x + 2)(5x + 2)$ **6.** $(x + 3)(3x + 2)$

3. $(5x + 2)(x + 3)$ **7.** $(4x + 3)(x + 1)$

4. $(3x + 4)(x + 5)$ **8.** $(7x + 2)(x + 3)$

Expand the product $(3x - 2)(2x + 5)$

$$\overset{①}{(3x - 2)}\overset{④}{(2x + 5)} = 6x^2 + 15x - 4x - 10$$

$$= 6x^2 + 11x - 10$$

Expand:

9. $(3x + 2)(2x + 3)$ **13.** $(5x + 3)(2x + 5)$

10. $(4x - 3)(3x - 4)$ **14.** $(7x - 2)(3x - 2)$

11. $(5x + 6)(2x - 3)$ **15.** $(3x - 2)(4x + 1)$

12. $(7a - 3)(3a - 7)$ **16.** $(3b + 5)(2b - 5)$

17. $(2a + 3)(2a - 3)$ **21.** $(4x + 3)(4x - 3)$

18. $(3b - 7)(3b + 7)$ **22.** $(5y - 2)(5y + 2)$

19. $(7y - 5)(7y + 5)$ **23.** $(3x - 1)(3x + 1)$

20. $(5a + 4)(4a - 3)$ **24.** $(4x - 7)(4x + 5)$

Expand $(3x - 2)(5 - 2x)$

$$
\begin{aligned}
(3x - 2)(5 - 2x) &= 15x - 6x^2 - 10 + 4x \\
&= 19x - 6x^2 - 10 \\
&= -6x^2 + 19x - 10
\end{aligned}
$$

Expand:

25. $(2x + 1)(1 + 3x)$ **31.** $(5x + 2)(4 + 3x)$

26. $(5x + 2)(2 - x)$ **32.** $(7x + 4)(3 - 2x)$

27. $(6x - 1)(3 - x)$ **33.** $(4x - 3)(3 - 5x)$

28. $(5a - 2)(3 - 7a)$ **34.** $(3 - p)(4 + p)$

29. $(3x + 2)(4 - x)$ **35.** $(x - 5)(2 + x)$

30. $(4x - 5)(3 + x)$ **36.** $(4x - 3)(3 + x)$

IMPORTANT PRODUCTS

Three very important products are:

$$
(x + a)^2 = (x + a)(x + a)
$$

$$
= x^2 + xa + ax + a^2
$$

$$
= x^2 + 2ax + a^2 \quad \text{(since } xa \text{ is the same as } ax\text{)}
$$

i.e.
$$
\boxed{(x + a)^2 = x^2 + 2ax + a^2}
$$

so
$$
(x + 3)^2 = x^2 + 6x + 9
$$

$$
(x - a)^2 = (x - a)(x - a)
$$

$$
= x^2 - xa - ax + a^2
$$

i.e.
$$
\boxed{(x - a)^2 = x^2 - 2ax + a^2}
$$

so
$$
(x - 4)^2 = x^2 - 8x + 16
$$

$$(x + a)(x - a) = x^2 - xa + ax - a^2$$
$$= x^2 - a^2$$

i.e.

$$(x+a)(x-a) = x^2 - a^2$$

and

$$(x-a)(x+a) = x^2 - a^2$$

so

$$(x + 5)(x - 5) = x^2 - 25$$

and

$$(x - 3)(x + 3) = x^2 - 9$$

You should learn these three results thoroughly, for they will appear time and time again. Given the left-hand side you should know the right-hand side and vice versa.

EXERCISE 11f

Expand $(x + 5)^2$

$$(x + 5)^2 = x^2 + 10x + 25$$

Expand:

1. $(x + 1)^2$

2. $(x + 2)^2$

3. $(a + 3)^2$

4. $(b + 4)^2$

5. $(t + 10)^2$

6. $(x + 12)^2$

7. $(x + 8)^2$

8. $(p + 7)^2$

9. $(x + y)^2$

10. $(y + z)^2$

11. $(c + d)^2$

12. $(m + n)^2$

13. $(p + q)^2$

14. $(a + b)^2$

15. $(e + f)^2$

16. $(u + v)^2$

Expand $(2x + 3)^2$

$$(2x + 3)^2 = (2x)^2 + 2(2x)(3) + (3)^2$$

i.e.

$$(2x + 3)^2 = 4x^2 + 12x + 9$$

Expand:

17. $(2x + 1)^2$ **21.** $(3a + 1)^2$

18. $(4b + 1)^2$ **22.** $(2x + 5)^2$

19. $(5x + 2)^2$ **23.** $(3a + 4)^2$

20. $(6c + 1)^2$ **24.** $(4y + 3)^2$

Expand $(2x + 3y)^2$

$$(2x + 3y)^2 = (2x)^2 + 2(2x)(3y) + (3y)^2$$

i.e. $(2x + 3y)^2 = 4x^2 + 12xy + 9y^2$

Expand:

25. $(x + 2y)^2$ **29.** $(3a + b)^2$

26. $(3x + y)^2$ **30.** $(p + 4q)^2$

27. $(2x + 5y)^2$ **31.** $(7x + 2y)^2$

28. $(3a + 2b)^2$ **32.** $(3s + 4t)^2$

Expand $(x - 5)^2$

$$(x - 5)^2 = x^2 - 10x + 25$$

Expand:

33. $(x - 2)^2$ **37.** $(x - 3)^2$

34. $(x - 6)^2$ **38.** $(x - 7)^2$

35. $(a - 10)^2$ **39.** $(a - b)^2$

36. $(x - y)^2$ **40.** $(u - v)^2$

Expand $(2x - 7)^2$

$$(2x - 7)^2 = (2x)^2 + 2(2x)(-7) + (-7)^2$$

i.e. $(2x - 7)^2 = 4x^2 - 28x + 49$

Expand:

41. $(3x - 1)^2$ **45.** $(2a - 1)^2$

42. $(5z - 1)^2$ **46.** $(4y - 1)^2$

43. $(10a - 9)^2$ **47.** $(7b - 2)^2$

44. $(4x - 3)^2$ **48.** $(5x - 3)^2$

Expand $(7a - 4b)^2$

$$(7a - 4b)^2 = (7a)^2 + 2(7a)(-4b) + (-4b)^2$$

i.e. $$(7a - 4b)^2 = 49a^2 - 56ab + 16b^2$$

Expand:

49. $(2y - x)^2$ **53.** $(a - 3b)^2$

50. $(5x - y)^2$ **54.** $(m - 8n)^2$

51. $(3m - 2n)^2$ **55.** $(5a - 2b)^2$

52. $(7x - 3y)^2$ **56.** $(3p - 5q)^2$

THE DIFFERENCE BETWEEN TWO SQUARES

EXERCISE 11g

Expand   a) $(a + 2)(a - 2)$ b) $(2x + 3)(2x - 3)$

a) $(a + 2)(a - 2) = a^2 - 4$

b) $(2x + 3)(2x - 3) = 4x^2 - 9$

Expand:

1. $(x + 4)(x - 4)$ **5.** $(x + 5)(x - 5)$

2. $(b + 6)(b - 6)$ **6.** $(a - 7)(a + 7)$

3. $(c - 3)(c + 3)$ **7.** $(q + 10)(q - 10)$

4. $(x + 12)(x - 12)$ **8.** $(x - 8)(x + 8)$

9. $(2x - 1)(2x + 1)$ **13.** $(5x + 1)(5x - 1)$

10. $(3x + 1)(3x - 1)$ **14.** $(2a - 3)(2a + 3)$

11. $(7a + 2)(7a - 2)$ **15.** $(10m - 1)(10m + 1)$

12. $(5a - 4)(5a + 4)$ **16.** $(6a + 5)(6a - 5)$

Expand $(3x + 2y)(3x - 2y)$

$$(3x + 2y)(3x - 2y) = (3x)^2 - (2y)^2$$
$$= 9x^2 - 4y^2$$

Expand:

17. $(3x + 4y)(3x - 4y)$ **22.** $(5a - 4b)(5a + 4b)$

18. $(2a - 5b)(2a + 5b)$ **23.** $(1 + 3x)(1 - 3x)$

19. $(1 - 2a)(1 + 2a)$ **24.** $(3 - 5x)(3 + 5x)$

20. $(7y + 3z)(7y - 3z)$ **25.** $(5m + 8n)(5m - 8n)$

21. $(10a - 9b)(10a + 9b)$ **26.** $(6p - 7q)(6p + 7q)$

The results from this exercise are very important when written the other way around,

i.e. $$a^2 - b^2 = (a + b)(a - b)$$

We refer to this as "factorising the difference between two squares" and we will deal with it in detail in the next chapter.

HARDER EXPANSIONS

EXERCISE 11h

Simplify $(x + 2)(x + 5) + 2x(x + 7)$

$$(x + 2)(x + 5) + 2x(x + 7) = x^2 + 5x + 2x + 10 + 2x^2 + 14x$$
$$= 3x^2 + 21x + 10$$

Simplify:

1. $(x + 3)(x + 4) + x(x + 2)$

2. $x(x + 6) + (x + 1)(x + 2)$

3. $(x + 4)(x + 5) + 6(x + 2)$

4. $(a - 6)(a - 5) + 2(a + 3)$

5. $(a - 5)(2a + 3) - 3(a - 4)$

6. $(x + 3)(x + 5) + 5(x + 2)$

7. $(x - 3)(x + 4) - 3(x + 3)$

8. $(x + 7)(x - 5) - 4(x - 3)$

9. $(2x + 1)(3x - 4) + (2x + 3)(5x - 2)$

10. $(5x - 2)(3x + 5) - (3x + 5)(x + 2)$

Expand $(xy - z)^2$

$$(xy - z)^2 = (xy)^2 - 2(xy)(z) + z^2$$

i.e. $$(xy - z)^2 = x^2y^2 - 2xyz + z^2$$

Expand:

11. $(xy - 3)^2$ **14.** $(3pq + 8)^2$ **17.** $(6 - pq)^2$

12. $(5 - yz)^2$ **15.** $(a - bc)^2$ **18.** $(mn + 3)^2$

13. $(xy + 4)^2$ **16.** $(ab - 2)^2$ **19.** $(uv - 2w)^2$

SUMMARY

The following is a summary of the most important types of examples considered in this chapter that will be required in future work.

1. $2(3x + 4) = 6x + 8$

2. $(x + 2)(x + 3) = x^2 + 5x + 6$

3. $(x - 2)(x - 3) = x^2 - 5x + 6$

4. $(x - 2)(x + 3) = x^2 + x - 6$

5. $(2x + 1)(3x + 2) = 6x^2 + 7x + 2$

6. $(2x - 1)(3x - 2) = 6x^2 - 7x + 2$

7. $(2x + 1)(3x - 2) = 6x^2 - x - 2$

Note that a) if the signs in the brackets are the same, i.e. both + or both −, then the number term is + (numbers **2**, **3**, **5** and **6**)

whereas b) if the signs in the brackets are different, i.e. one + and one −, then the number term is − (numbers **4** and **7**)

c) the middle term is given by collecting the product of the outside terms in the brackets and the product of the inside terms in the brackets,

i.e. in **2** the middle term is $3x + 2x$ or $5x$

in **3** the middle term is $-3x - 2x$ or $-5x$

in **4** the middle term is $3x - 2x$ or x

in **5** the middle term is $4x + 3x$ or $7x$

in **6** the middle term is $-4x - 3x$ or $-7x$

in **7** the middle term is $-4x + 3x$ or $-x$.

Most important of all we must remember the general expansions:

$$(x+a)^2 = x^2 + 2ax + a^2$$
$$(x-a)^2 = x^2 - 2ax + a^2$$
$$(x+a)(x-a) = x^2 - a^2$$

MIXED EXERCISES

EXERCISE 11i Expand:

1. $5(x + 2)$

2. $8p(3q - 2r)$

3. $(3a + b)(2a - 5b)$

4. $(4x + 1)(3x - 5)$

5. $(x + 6)(x + 10)$

6. $(x - 8)(x - 12)$

7. $(4y + 3)(4y - 7)$

8. $(4y - 9)(4y + 9)$

9. $(5x + 2)^2$

10. $(2a - 7b)^2$

EXERCISE 11j Expand:

1. $4(2 - 5x)$

2. $8a(2 - 3a)$

3. $(4a + 3)(3a - 11)$

4. $(x + 11)(x - 9)$

5. $(2x + 5)(1 - 10x)$

6. $(y + 2z)^2$

7. $(6y - z)(6y + 5z)$

8. $(4a + 1)^2$

9. $(5a - 7)^2$

10. $(6z - 13y)^2$

EXERCISE 11k Expand:

1. $3(2 - a)$

2. $4a(2b + c)$

3. $(5a + 2b)(2c + 5d)$

4. $(x - 7)(x - 12)$

5. $(a + 7)(a + 9)$

6. $(a + 4)(a - 5)$

7. $(3x + 1)(2x + 3)$

8. $(5x - 2)(5x + 2)$

9. $(3x - 7)^2$

10. $(5x + 2y)(5x - 2y)$

EXERCISE 11l Expand:

1. $5(3 - x)$

2. $12x(3x - 2)$

3. $3x(2y - 5z)$

4. $(a - b)(c + d)$

5. $(x + 7)(x - 4)$

6. $(x - 9)(x - 2)$

7. $(4x + 1)(3x + 2)$

8. $(x - 4y)^2$

9. $(2x + 7z)^2$

10. $(1 - 5a)(1 + 5a)$

12 ALGEBRAIC FACTORS

FINDING FACTORS

In a previous chapter we removed brackets and expanded expressions. Frequently we need to be able to do the reverse, i.e. to find the factors of an expression. This is called *factorising*.

COMMON FACTORS

In the expression $7a + 14b$ we could write the first term as $7 \times a$ and the second term as $7 \times 2b$,

i.e. $$7a + 14b = 7 \times a + 7 \times 2b$$

The 7 is a common factor.
However we already know that $7(a + 2b) = 7 \times a + 7 \times 2b$

∴ $$7a + 14b = 7 \times a + 7 \times 2b = 7(a + 2b)$$

EXERCISE 12a

> Factorise $3x - 12$
>
> $$3x - 12 = 3(x - 4)$$

Factorise:

1.	$4x + 4$	**4.**	$5a - 10b$	**7.**	$12a + 4$
2.	$12x - 3$	**5.**	$3t - 9$	**8.**	$2a + 4b$
3.	$6a + 2$	**6.**	$10a - 5$	**9.**	$14x - 7$

> Factorise $x^2 - 7x$
>
> $$x^2 - 7x = x \times x - 7 \times x$$
> $$= x(x - 7)$$

Factorise:

10. $x^2 + 2x$ **13.** $2x^2 + x$ **16.** $x^2 - 4x$

11. $x^2 - 7x$ **14.** $4t - 2t^2$ **17.** $b^2 + 4b$

12. $a^2 + 6a$ **15.** $x^2 + 5x$ **18.** $4a^2 - a$

Factorise $9ab + 12bc$

$$9ab + 12bc = 3b \times 3a + 3b \times 4c$$
$$= 3b(3a + 4c)$$

Factorise:

19. $2x^2 - 6x$ **22.** $12x^2 + 16x$ **25.** $2a^2 - 12a$

20. $2z^3 + 4z$ **23.** $5ab - 10bc$ **26.** $6p^2 + 2p$

21. $25a^2 - 5a$ **24.** $3y^2 + 27y$ **27.** $9y^2 - 6y$

Factorise $ab + 2bc + bd$

$$ab + 2bc + bd = b(a + 2c + d)$$

Factorise:

28. $2x^2 + 4x + 6$ **33.** $3x^2 - 6x + 9$

29. $10a^2 - 5a + 20$ **34.** $4a^2 + 8a - 4$

30. $ab + 4bc - 3bd$ **35.** $5xy + 4xz + 3x$

31. $8x - 4y + 12z$ **36.** $5ab + 10bc + 5bd$

32. $9ab - 6ac - 3ad$ **37.** $2xy - 4yz + 8yw$

Factorise $8x^3 - 4x^2$

$$8x^3 - 4x^2 = 4x^2(2x - 1)$$

Factorise:

38. $x^3 + x^2$

39. $x^2 - x^3$

40. $20a^2 - 5a^3$

41. $12x^3 - 16x^2$

42. $4x^4 + 12x^2$

43. $a^2 + a^3$

44. $b^3 - b^2$

45. $4x^3 - 2x^2$

46. $27a^2 - 18a^3$

47. $10x^2 - 15x^4$

Factorise:

48. $12x + 8$

49. $8x^2 + 12x$

50. $9x^2 - 6x + 12$

51. $5x^3 - 10x$

52. $8pq + 4qr$

53. $x^2 - 8x$

54. $12 + 9y^2$

55. $12xy + 16xz + 8x$

56. $4x^3 + 6x$

57. $12abc - 8bcd$

Factorise   a) $2\pi r^2 + 2\pi rh$ b) $\frac{1}{2}Mu^2 - \frac{1}{2}mu^2$

a) $$2\pi r^2 + 2\pi rh = 2\pi r(r + h)$$

b) $$\frac{1}{2}Mu^2 - \frac{1}{2}mu^2 = \frac{1}{2}u^2(M - m)$$

Factorise:

58. $\frac{1}{2}ah + \frac{1}{2}bh$

59. $mg - ma$

60. $\frac{1}{2}mv^2 + \frac{1}{2}mu^2$

61. $P + \dfrac{PRT}{100}$

62. $2\pi r^2 + \pi rh$

63. $\pi R^2 + \pi r^2$

64. $2gh_1 - 2gh_2$

65. $\frac{1}{2}mv^2 - mgh$

66. $\frac{4}{3}\pi r^3 - \frac{1}{3}\pi r^2 h$

67. $3\pi r^2 + 2\pi rh$

68. $\frac{1}{2}mu^2 + \frac{1}{2}mu$

69. $\frac{1}{2}bc - \frac{1}{4}ca$

FACTORISING QUADRATIC EXPRESSIONS

The type of expression we are most likely to want to factorise is one of the form $ax^2 + bx + c$ where a, b and c are numbers.

To factorise such an expression we look for two brackets whose product is the original expression.

When we expanded $(x + 2)(x + 4)$ we had

$$(x + 2)(x + 4) = x^2 + 6x + 8$$

If we write $x^2 + 6x + 8 = (x + 2)(x + 4)$ we say we have factorised $x^2 + 6x + 8$,
i.e. just as 10 is 2×5 so $x^2 + 6x + 8$ is $(x + 2) \times (x + 4)$.

To factorise an expression of the form $x^2 + 7x + 10$, i.e. where all the terms are positive, we remind ourselves of the patterns we observed in the previous chapter and summarised on pages 202–3.

We found when expanding brackets that:
a) if the sign in each bracket is $+$ then the number term in the expansion is $+$
b) the x^2 term comes from $x \times x$
c) the number term in the expansion comes from multiplying the numbers in the brackets together
d) the middle term, or x term in the expansion, comes from collecting the product of the outside terms in the brackets and the product of the inside terms in the brackets.

Using these ideas in reverse order

$$x^2 + 7x + 10 = (x + \quad)(x + \quad)$$
$$= (x + 2)(x + 5)$$

(choosing two numbers whose product is 10 and whose sum is 7).

EXERCISE 12b

Factorise $x^2 + 8x + 15$

$$x^2 + 8x + 15 = (x + 3)(x + 5) \text{ or } (x + 5)(x + 3)$$

(The product of 3 and 5 is 15, and their sum is 8.)

Factorise:

1. $x^2 + 3x + 2$ **3.** $x^2 + 7x + 12$

2. $x^2 + 6x + 5$ **4.** $x^2 + 8x + 15$

5. $x^2 + 21x + 20$

6. $x^2 + 8x + 7$

7. $x^2 + 8x + 12$

8. $x^2 + 13x + 12$

9. $x^2 + 16x + 15$

10. $x^2 + 12x + 20$

11. $x^2 + 8x + 16$

12. $x^2 + 15x + 36$

13. $x^2 + 19x + 18$

14. $x^2 + 22x + 40$

15. $x^2 + 9x + 8$

16. $x^2 + 6x + 9$

17. $x^2 + 20x + 36$

18. $x^2 + 9x + 18$

19. $x^2 + 11x + 30$

20. $x^2 + 14x + 40$

To factorise an expression of the form $x^2 - 6x + 8$ remember the pattern:

a) the numbers in the brackets must multiply to give $+8$, i.e. they must have the same sign. Since the middle term in the expression is $-$ they must both be $-$

b) the x^2 term comes from $x \times x$

c) the middle term, or x term, comes from collecting the product of the outside terms and the product of the inside terms.

Thus $\qquad x^2 - 6x + 8 = (x - 2)(x - 4)$

Since $\qquad (-2) \times (-4) = +8$

and $\qquad x \times (-4) + (-2) \times x = -4x - 2x = -6x$

EXERCISE 12c

Factorise $x^2 - 7x + 12$

$$x^2 - 7x + 12 = (x - 3)(x - 4)$$

(The product of -3 and -4 is $+12$.
The outside product is $-4x$ and the inside product is $-3x$.
Collecting these gives $-7x$.)

Factorise:

1. $x^2 - 9x + 8$

2. $x^2 - 7x + 12$

3. $x^2 - 17x + 30$

4. $x^2 - 11x + 28$

5. $x^2 - 13x + 42$

6. $x^2 - 5x + 6$

7. $x^2 - 16x + 15$

8. $x^2 - 6x + 9$

9. $x^2 - 18x + 32$

10. $x^2 - 16x + 63$

Similarly $\qquad\qquad x^2 + x - 12 = (x + 4)(x - 3)$

If the number term in the expansion is negative the signs in the brackets are different.

Thus $\qquad\qquad\qquad (+4) \times (-3) = -12$

Working as before, the product of the outside terms is $-3x$
$\qquad\qquad\qquad$ and the product of the inside terms is $+4x$

Therefore the total is $+x$.

Similarly $\qquad\qquad x^2 + 2x - 15 = (x + 5)(x - 3)$

or $\qquad\qquad\qquad x^2 + 2x - 15 = (x - 3)(x + 5)$

EXERCISE 12d Factorise:

1. $x^2 - x - 6$ $\qquad\qquad$ **6.** $x^2 - 2x - 24$

2. $x^2 + x - 20$ $\qquad\qquad$ **7.** $x^2 + 6x - 27$

3. $x^2 - x - 12$ $\qquad\qquad$ **8.** $x^2 - 9x - 22$

4. $x^2 + 3x - 28$ $\qquad\qquad$ **9.** $x^2 - 2x - 35$

5. $x^2 + 2x - 15$ $\qquad\qquad$ **10.** $x^2 - 8x - 20$

Most of the values in the previous three exercises have been easy to spot. Should you have difficulty, set out all possible pairs of numbers, as shown below, until you find the pair that gives the original expression when you multiply back.

Factorise

a) $x^2 - 11x + 24$

(Because the number term is $+$ the two numbers in the brackets must have the same sign.)

Possible numbers		Sum
-1	-24	-25
-2	-12	-14
-3	-8	-11

$\therefore \qquad x^2 - 11x + 24 = (x - 3)(x - 8)$

b) $x^2 + 5x - 24$

(Because the number term is $-$ the two numbers in the brackets have different signs)

Possible numbers		Sum
-1	$+24$	$+23$
-2	$+12$	$+10$
-3	$+8$	$+5$

$\therefore \qquad x^2 + 5x - 24 = (x - 3)(x + 8)$

Remember that a $+$ before the number term means that the signs in the brackets are the same, whereas a $-$ before the number term means that they are different.

EXERCISE 12e Factorise:

1. $x^2 + 9x + 14$

5. $x^2 + 9x + 8$

2. $x^2 - 10x + 21$

6. $x^2 - 10x + 25$

3. $x^2 + 5x - 14$

7. $x^2 + 8x - 9$

4. $x^2 + x - 30$

8. $x^2 - 15x + 26$

9. $x^2 + x - 56$

13. $x^2 + 14x + 13$

10. $x^2 + 32x + 60$

14. $x^2 + 12x - 28$

11. $x^2 - 6x - 27$

15. $x^2 + 2x - 80$

12. $x^2 + 16x - 80$

16. $x^2 - 11x + 30$

17. $x^2 + 8x - 48$

21. $x^2 + 11x + 24$

18. $x^2 + 18x + 72$

22. $x^2 - 11x - 42$

19. $x^2 + 17x + 52$

23. $x^2 - 18x + 32$

20. $x^2 - 12x - 28$

24. $x^2 + 7x - 60$

Sometimes the terms need rearranging before we try to factorise.

EXERCISE 12f

Factorise $6 + x^2 - 5x$

$$6 + x^2 - 5x = x^2 - 5x + 6$$
$$= (x - 2)(x - 3)$$

Factorise:

1. $8 + x^2 + 9x$

5. $9 + x^2 + 6x$

2. $9 + x^2 - 6x$

6. $8 + x^2 - 9x$

3. $11x + 28 + x^2$

7. $17x + 30 + x^2$

4. $20 - x^2 - x$

8. $27 - 6x - x^2$

9. $x^2 + 22 + 13x$ **13.** $x^2 - 5x - 24$

10. $x^2 - 11x - 26$ **14.** $14 + x^2 - 9x$

11. $7 + x^2 - 8x$ **15.** $28x + 27 + x^2$

12. $x + x^2 - 42$ **16.** $2x - 63 + x^2$

Factorise $x^2 + 6x + 9$

$$x^2 + 6x + 9 = (x + 3)(x + 3)$$
$$= (x + 3)^2$$

Factorise:

17. $x^2 + 10x + 25$ **21.** $x^2 + 12x + 36$

18. $x^2 - 10x + 25$ **22.** $x^2 - 12x + 36$

19. $x^2 + 4x + 4$ **23.** $x^2 - 4x + 4$

20. $x^2 - 14x + 49$ **24.** $x^2 + 16x + 64$

EXERCISE 12g

Factorise $6 - 5x - x^2$

$$6 - 5x - x^2 = (6 + x)(1 - x)$$

Factorise:

1. $2 - x - x^2$ **5.** $6 - x - x^2$

2. $6 + x - x^2$ **6.** $2 + x - x^2$

3. $4 - 3x - x^2$ **7.** $8 - 2x - x^2$

4. $8 + 2x - x^2$ **8.** $5 - 4x - x^2$

9. $10 - 3x - x^2$ **13.** $6 + 5x - x^2$

10. $12 + 4x - x^2$ **14.** $20 - x - x^2$

11. $5 + 4x - x^2$ **15.** $15 - 2x - x^2$

12. $14 - 5x - x^2$ **16.** $12 + x - x^2$

THE DIFFERENCE BETWEEN TWO SQUARES

In the last chapter, one of the expansions we listed was

$$(x + a)(x - a) = x^2 - a^2$$

If we reverse this we have

$$x^2 - a^2 = (x + a)(x - a)$$
$$\text{or } x^2 - a^2 = (x - a)(x + a)$$

(the order of multiplication of two brackets makes no difference to the result).

This result is known as *factorising the difference between two squares* and is *very important*.

When factorising do not confuse $x^2 - 4$ with $x^2 - 4x$.

$$x^2 - 4 = (x + 2)(x - 2)$$

whereas $x^2 - 4x = x(x - 4)$ (4x is *not* a square)

EXERCISE 12h

Factorise $x^2 - 9$

$$x^2 - 9 = x^2 - 3^2$$
$$= (x + 3)(x - 3) \text{ or } (x - 3)(x + 3)$$

Factorise:

1. $x^2 - 25$ **4.** $x^2 - 1$ **7.** $x^2 - 36$

2. $x^2 - 4$ **5.** $x^2 - 64$ **8.** $x^2 - 81$

3. $x^2 - 100$ **6.** $x^2 - 16$ **9.** $x^2 - 49$

Factorise $4 - x^2$

$$4 - x^2 = 2^2 - x^2$$
$$= (2 + x)(2 - x) \text{ or } (2 - x)(2 + x)$$

Factorise:

10. $9 - x^2$ **11.** $36 - x^2$ **12.** $100 - x^2$

13. $a^2 - b^2$ **15.** $16 - x^2$ **17.** $81 - x^2$

14. $9y^2 - z^2$ **16.** $25 - x^2$ **18.** $x^2 - y^2$

We began this chapter by considering common factors. A little revision is now necessary followed by expressions of the form $ax^2 + bx + c$ where a is a common factor.

EXERCISE 12i

> Factorise $12x - 6$
>
> $$12x - 6 = 6(2x - 1)$$

Factorise:

1. $3x + 12$ **6.** $21x - 7$

2. $25x^2 + 10x$ **7.** $9x^2 - 18x$

3. $12x^2 - 8$ **8.** $20x + 12$

4. $14x + 21$ **9.** $4x - 14$

5. $4x^2 + 2$ **10.** $8x^2 - 4x$

> Factorise $3x^2 + 9x + 6$
>
> $$3x^2 + 9x + 6 = 3(x^2 + 3x + 2)$$
> $$= 3(x + 1)(x + 2)$$

Factorise:

11. $2x^2 + 14x + 24$ **16.** $3x^2 + 24x + 36$

12. $3x^2 - 27x + 24$ **17.** $4x^2 - 24x + 36$

13. $7x^2 + 14x + 7$ **18.** $5x^2 - 5x - 30$

14. $4x^2 - 4x - 48$ **19.** $2x^2 - 18x - 44$

15. $5x^2 + 40x + 35$ **20.** $3x^2 + 9x - 120$

In the previous exercises we considered the factors of the expression $ax^2 + bx + c$ when either a was 1 or a was a common factor.

We must now consider opther values of a which mean that the two brackets do not both start with x.

For example, to factorise $3x^2 + 7x + 2$ the only sensible first terms for the brackets are $3x$ and x.

Since all the signs in the given expression are $+$, all the signs in the brackets will be $+$.

The possible number values will be $+2$ and $+1$, or $+1$ and $+2$

Trying $+2$ and $+1$,

$(3x + 2)(x + 1)$ gives a middle term of $5x$ which is incorrect.

Trying $+1$ and $+2$,

$(3x + 1)(x + 2)$ gives a middle term of $7x$ which is correct.

$\therefore$ $$3x^2 + 7x + 2 = (3x + 1)(x + 2)$$

Similarly to factorise $5x^2 - 7x + 2$ the brackets start $(5x \quad)(x \quad)$. The number term is $+$, therefore the brackets have the same sign and since the middle term is $-$ the signs must both be $-$,

i.e. we have $(5x - \quad)(x - \quad)$

The last term is $+2$, therefore the numbers must be either -1 and -2, or -2 and -1.

Trying -1 and -2,

$(5x - 1)(x - 2)$ gives a middle term of $-11x$ which is incorrect.

Trying -2 and -1,

$(5x - 2)(x - 1)$ gives a middle term of $-7x$ which is correct.

$\therefore$ $$5x^2 - 7x + 2 = (5x - 2)(x - 1)$$

EXERCISE 12j Factorise:

1. $2x^2 + 3x + 1$ **6.** $3x^2 - 8x + 4$

2. $3x^2 - 5x + 2$ **7.** $2x^2 + 9x + 4$

3. $4x^2 + 7x + 3$ **8.** $5x^2 - 17x + 6$

4. $2x^2 - 7x + 3$ **9.** $2x^2 + 11x + 12$

5. $3x^2 + 13x + 4$ **10.** $7x^2 - 29x + 4$

11. $2x^2 - 3x - 2$ **16.** $7x^2 - 19x - 6$

12. $3x^2 + x - 4$ **17.** $6x^2 - 7x - 10$

13. $5x^2 - 13x - 6$ **18.** $5x^2 - 19x + 12$

14. $4x^2 + 5x - 6$ **19.** $3x^2 - 11x - 20$

15. $3x^2 + 10x - 8$ **20.** $4x^2 + 17x - 15$

THE GENERAL CASE

We investigate next the most general case of all; that is when neither bracket begins with x.

Suppose that we wish to factorise $15x^2 + 26x + 8$.

The first terms in the brackets may be

$$(15x \quad)(x \quad) \quad \text{or} \quad (5x \quad)(3x \quad)$$

All the signs must be $+$.

The brackets may end with 8 and 1, 4 and 2, 2 and 4, or 1 and 8.

Try each in turn, until the correct middle term is found.

$(15x + 8)(x + 1)$	middle term $23x$, incorrect
$(15x + 4)(x + 2)$	middle term $34x$, incorrect
$(15x + 2)(x + 4)$	middle term $62x$, incorrect
$(15x + 1)(x + 8)$	middle term $121x$, incorrect
$(5x + 8)(3x + 1)$	middle term $29x$, incorrect
$(5x + 4)(3x + 2)$	middle term $22x$, incorrect
$(5x + 2)(3x + 4)$	middle term $26x$, *correct*

$$\therefore \qquad 15x^2 + 26x + 8 = (5x + 2)(3x + 4)$$

After a little practice, you should be able to find the factors without going into so much detail.

EXERCISE 12k Factorise:

1. $6x^2 + 7x + 2$

2. $6x^2 + 19x + 15$

3. $15x^2 + 11x + 2$

4. $12x^2 + 28x + 15$

5. $35x^2 + 24x + 4$

6. $6x^2 - 11x + 3$

7. $9x^2 - 18x + 8$

8. $16x^2 - 10x + 1$

9. $15x^2 - 44x + 21$

10. $20x^2 - 23x + 6$

11. $8x^2 - 10x - 3$

12. $15x^2 - x - 2$

13. $21x^2 + 2x - 8$

14. $80x^2 - 6x - 9$

15. $24x^2 + 17x - 20$

16. $6a^2 - a - 15$

17. $6t^2 - t - 2$

18. $9b^2 - 12b + 4$

19. $5x^2 - 7xy - 6y^2$

20. $4x^2 - 11x + 6$

EXERCISE 12I

> Factorise $4x^2 - 9$
>
> $$4x^2 - 9 = (2x)^2 - 3^2$$
> $$= (2x + 3)(2x - 3)$$

Factorise:

1. $4x^2 - 25$

2. $9x^2 - 4$

3. $36a^2 - 1$

<u>4.</u> $16a^2 - b^2$

<u>5.</u> $9x^2 - 25$

<u>6.</u> $4a^2 - 1$

> Factorise $4x^2 - 9y^2$
>
> $$4x^2 - 9y^2 = (2x)^2 - (3y)^2$$
> $$= (2x + 3y)(2x - 3y)$$

Factorise:

7. $16a^2 - 9b^2$

8. $25s^2 - 9t^2$

9. $100x^2 - 49y^2$

10. $9y^2 - 16z^2$

<u>11.</u> $4x^2 - 49y^2$

<u>12.</u> $81x^2 - 100y^2$

<u>13.</u> $9a^2 - 4b^2$

<u>14.</u> $64p^2 - 81q^2$

> Factorise $2 - 18a^2$
>
> $$2 - 18a^2 = 2(1 - 9a^2)$$
> $$= 2(1^2 - (3a)^2)$$
> $$= 2(1 + 3a)(1 - 3a)$$

Factorise:

15. $3a^2 - 27b^2$

16. $18t^2 - 50s^2$

17. $27x^2 - 3y^2$

<u>18.</u> $45x^2 - 20$

<u>19.</u> $5a^2 - 20$

<u>20.</u> $45 - 5b^2$

<u>21.</u> $\frac{1}{2}a^2 - 2b^2$

<u>22.</u> $\frac{a^2}{4} - \frac{b^2}{9}$

<u>23.</u> $27x^2 - \frac{1}{3}y^2$

<u>24.</u> $\frac{x^2}{16} - \frac{y^2}{25}$

CALCULATIONS USING FACTORISING

Find $1.7^2 + 0.3 \times 1.7$

$$1.7^2 + 0.3 \times 1.7 = 1.7(1.7 + 0.3)$$
$$= 1.7 \times 2$$
$$= 3.4$$

Find, without using a calculator:

1. $2.5^2 + 0.5 \times 2.5$

2. $1.3 \times 3.7 + 3.7^2$

3. $5.9^2 - 2.9 \times 5.9$

4. $8.76^2 - 4.76 \times 8.76$

5. $5.2^2 + 0.8 \times 5.2$

6. $2.6 \times 3.4 + 3.4^2$

7. $4.3^2 - 1.3 \times 4.3$

8. $16.27^2 - 5.27 \times 16.27$

Find $100^2 - 98^2$

$$100^2 - 98^2 = (100 + 98)(100 - 98)$$
$$= 198 \times 2$$
$$= 396$$

9. $55^2 - 45^2$

10. $20.6^2 - 9.4^2$

11. $7.82^2 - 2.82^2$

12. $2.667^2 - 1.333^2$

13. $10.2^2 - 9.8^2$

14. $13.5^2 - 6.5^2$

15. $8.79^2 - 1.21^2$

16. $0.763^2 - 0.237^2$

Finally in this chapter we take out a common factor that does not leave the first term as x^2; and factorise expressions that do not begin with the x^2 term.

Factorise $8x^2 + 28x + 12$

$$8x^2 + 28x + 12 = 4(2x^2 + 7x + 3)$$
$$= 4(2x + 1)(x + 3)$$

Factorise:

1. $15x^2 + 25x + 10$
2. $4x^2 - 6x - 4$
3. $6x^2 + 9x + 3$
4. $18x^2 - 21x - 30$
5. $8x^2 + 34x - 30$

6. $8x^2 + 14x + 6$
7. $25x^2 - 65x - 30$
8. $9x^2 + 3x - 12$
9. $6x^2 + 26x + 8$
10. $15x^2 + 50x - 40$

11. $18x^2 - 36x + 16$
12. $48x^2 - 30x + 3$
13. $12x^2 + 14x + 4$

14. $100x^2 - 115x + 30$
15. $24x^2 - 4x - 8$
16. $21x^2 + 70x - 56$

Factorise $12 + 7x - 10x^2$

$$12 + 7x - 10x^2 = (3 - 2x)(4 + 5x)$$

Factorise:

17. $4 - 5x - 6x^2$
18. $12 + 7x - 12x^2$
19. $21 + 25x - 4x^2$
20. $24 - 16x + 2x^2$
21. $16 - 20x - 6x^2$

22. $9 + 8x - x^2$
23. $12 - 11x - x^2$
24. $8 + 24x + 18x^2$
25. $45 - 30x + 5x^2$
26. $20 + 40x + 15x^2$

MIXED QUESTIONS

Some quadratic expressions such as $x^2 + 9$ and $x^2 + 3x + 1$ will not factorise. The next exercise in this chapter includes some expressions that will not factorise.

EXERCISE 12p Factorise where possible:

1. $x^2 + 13x + 40$
2. $6x^2 + 5x + 1$
3. $x^2 - 36$
4. $x^2 + 4$

5. $x^2 - 8x + 12$
6. $2x^2 + 7x - 15$
7. $x^2 + 6x - 7$
8. $5x^2 + 3x - 2$

9. $x^2 - 11x + 24$

10. $3x^2 + 11x + 6$

11. $x^2 + 14x - 15$

12. $12x^2 - 7x + 1$

13. $x^2 + 8x + 12$

14. $8x^2 - 2x - 1$

15. $x^2 - 49$

16. $x^2 - 7x + 2$

17. $6x^2 - 11x - 10$

18. $x^2 + 13x + 42$

19. $4x^2 - 9y^2$

20. $15x^2 - 22x + 8$

21. $6x^2 - 5x - 6$

22. $x^2 + 11x - 26$

23. $30x^2 - 2x - 4$

24. $28 + 3x - x^2$

25. $6x^2 + 5x - 4$

26. $30x^2 + 35x + 10$

27. $x^2 + 11x + 18$

28. $x^2 - 10x + 24$

29. $4x^2 - 16y^2$

30. $x^2 - 11x - 10$

31. $12x^2 - 22x - 20$

32. $x^2 + 13x - 30$

33. $28 - 12x - x^2$

34. $a^2 - 16a + 63$

35. $6 - 16x + 8x^2$

36. $1 + 2x + 4x^2 + 8x^3$

37. $x^2 + 13x - 68$

38. $2x^4 - x^3 + 4x - 2$

39. $6x^2 - 9x - 6$

40. $p^3 + p^2 + p + 1$

41. $(a + b)^2 - c^2$

42. $116x^2 - 25x - 1$

43. $a^2 + 23a + 112$

44. $x^4 - y^2 - 2y - 1$

45. $3a^2 + 56 - 31a$

46. $2x^2 - 8x - 154$

47. $4x^2 - (y - z)^2$

48. $a^2b^2 - ab - 342$

MIXED EXERCISES

EXERCISE 12q

1. Expand a) $7(a + 3)$ b) $3(x - 2y)$

2. Expand a) $(x + 4)(x + 10)$ b) $(2x - 3)(3x - 5)$

3. Expand a) $(5 + x)^2$ b) $(5 - x)^2$ c) $(5 + x)(5 - x)$

4. Factorise a) $10a + 20$ b) $15p^2 - 10p$

5. Factorise a) $a^3 + a^2 + a + 1$ b) $2km - kn + 2lm - ln$

6. Factorise a) $x^2 + 6x - 27$ b) $5x^2 - 42x + 49$ c) $a^2 - \dfrac{b^2}{4}$

7. Factorise a) $10x^2 - 11x - 6$ b) $100a^2 - 81b^2$

EXERCISE 12r **1.** Expand a) $5a(a + 3)$ b) $4x(3x - 2y)$

2. Expand a) $(y - 4)(y - 5)$ b) $(3x - 4y)(5x + 2y)$

3. Expand a) $(2p + 3q)^2$ b) $(2p - 3q)^2$ c) $(2p + 3q)(2p - 3q)$

4. Factorise a) $8z^3 - 4z^2$ b) $5xy - 20yz$

5. Factorise a) $2m + 3mn + 3n + 2$ b) $ac - 2ad + 2bc - 4bd$

6. Factorise a) $x^2 - 6x - 27$ b) $4x^2 + 27x - 7$ c) $4m^2 - 81n^2$

7. Factorise a) $15x^2 - 54x + 27$ b) $15 + 25x - 20x^2$

EXERCISE 12s **1.** Expand a) $4(a + 7)$ b) $3x(2x - 3y)$

2. Expand a) $(x + 3)(x + 9)$ b) $(5x - 2)(3x + 1)$

3. Expand a) $(5x + 2)^2$ b) $(5x - 2)^2$ c) $(5x + 2)(5x - 2)$

4. Factorise a) $12z^2 - 6z$ b) $8xy - 12yz$

5. Factorise a) $z^3 + 2z^2 + z + 2$ b) $3ac + bc + 6a + 2b$

6. Factorise a) $x^2 - 2x - 24$ b) $4a^2 + 4a - 15$ c) $9m^2 - \dfrac{n^2}{9}$

7. Factorise a) $15x^2 + x - 6$ b) $6 + x - 15x^2$

13 QUADRATIC EQUATIONS

MULTIPLICATION BY ZERO

EXERCISE 13a

Find the value of $(x - 3)(x - 7)$ if
a) $x = 8$ b) $x = 7$ c) $x = 3$

a) If $x = 8$ $(x - 3)(x - 7) = (8 - 3)(8 - 7)$
$$= (5)(1)$$
$$= 5$$

b) If $x = 7$ $(x - 3)(x - 7) = (4)(0)$
$$= 0$$

c) If $x = 3$ $(x - 3)(x - 7) = (0)(-4)$
$$= 0$$

1. Find the value of $(x - 4)(x - 2)$ if
a) $x = 6$ b) $x = 4$ c) $x = 2$

2. Find the value of $(x - 5)(x - 9)$ if
a) $x = 5$ b) $x = 10$ c) $x = 9$

3. Find the value of $(x - 7)(x - 1)$ if
a) $x = 1$ b) $x = 8$ c) $x = 7$

4. Find the value of $(x - 4)(x - 6)$ if
a) $x = 4$ b) $x = 6$ c) $x = 3$

5. Find the value of $(x - 6)(x - 7)$ if
a) $x = 2$ b) $x = 6$ c) $x = 9$

Find the value of $(x - 2)(x + 4)$ if
a) $x = 2$ b) $x = 4$ c) $x = -4$

a) If $x = 2$ $(x - 2)(x + 4) = (0)(6)$
$$= 0$$

b) If $x = 4$ $(x - 2)(x + 4) = (2)(8)$
$$= 16$$

c) If $x = -4$ $(x - 2)(x + 4) = (-6)(0)$
$$= 0$$

6. Find the value of $(x - 3)(x + 5)$ if
  a) $x = 6$ b) $x = 3$ c) $x = -5$

7. Find the value of $(x - 4)(x + 6)$ if
  a) $x = 0$ b) $x = -6$ c) $x = 4$

8. Find the value of $(x - 7)(x + 2)$ if
  a) $x = -7$ b) $x = -2$ c) $x = 7$

9. Find the value of $(x + 4)(x + 5)$ if
  a) $x = -4$ b) $x = -5$ c) $x = 0$

10. Find the value of $(x + 7)(x + 1)$ if
  a) $x = -4$ b) $x = -1$ c) $x = -7$

The results of this exercise show that if the product of two factors is 0, then either one or both of these factors must be 0

In general we can say

$$\text{if} \qquad A \times B = 0$$
$$\text{then either} \quad A = 0 \quad \text{or/and} \quad B = 0$$

EXERCISE 13b In questions 1 to 12 find, if possible, the value or values of A. Note that if $A \times 0 = 0$ then A can have any value.

1. $A \times 6 = 0$ **7.** $A \times 10 = 0$

2. $A \times 7 = 0$ **8.** $A \times 9 = 18$

3. $A \times 4 = 0$ **9.** $A \times 20 = 0$

4. $A \times 0 = 0$ **10.** $A \times 3 = 21$

5. $3 \times A = 12$ **11.** $0 \times A = 0$

6. $8 \times A = 8$ **12.** $4 \times A = 0$

13. If $AB = 0$ find   a) A if $B = 2$ b) B if $A = 10$

14. If $AB = 0$ find   a) A if $B = 5$ b) B if $A = 5$

15. If $AB = 0$ find   a) A if $B = 10$ b) B if $A = 3$

16. If $AB = 0$ find   a) B if $A = 6$ b) A if $B = 0$

Find a and b if $a(b-3) = 0$

Either $a = 0$ or/and $b - 3 = 0$

i.e., either $a = 0$ or/and $b = 3$

Find a and b if:

17. $a(b-1) = 0$ **22.** $a(b-4) = 0$

18. $a(b-5) = 0$ **23.** $a(b-10) = 0$

19. $a(b-2) = 0$ **24.** $(a-1)b = 0$

20. $(a-3)b = 0$ **25.** $(a-7)b = 0$

21. $(a-9)b = 0$ **26.** $(a-12)b = 0$

QUADRATIC EQUATIONS

Previously we have considered equations such as $x - 1 = 0$ and $3x + 2 = 0$. These are examples of *linear equations*. The first equation is true only for $x = 1$ and the second only for $x = -\frac{2}{3}$.

If, however, we consider the equation

$$(x-1)(x-2) = 0$$

we find that it is true either when $x - 1 = 0$ or when $x - 2 = 0$, i.e. either when $x = 1$ or when $x = 2$

There are, therefore, two values of x that satisfy the equation $(x-1)(x-2) = 0$

Expanding the left-hand side gives

$$x^2 - 3x + 2 = 0$$

Equations like this, which contain an x^2 term, are called *quadratic equations*.

When we are given a quadratic equation we can often factorise the left-hand side into two linear factors,

e.g. $x^2 - 5x + 4 = 0$

gives $(x-4)(x-1) = 0$

It is this technique that concerns us in the present chapter.

EXERCISE 13c

> What values of x satisfy the equation $x(x - 9) = 0$?
>
> $$x(x - 9) = 0$$
>
> Either $\quad\quad x = 0 \quad$ or $\quad x - 9 = 0$
>
> i.e., either $\quad\quad\quad x = 0$ or $x = 9$

What values of x satisfy the following equations?

1. $x(x - 3) = 0$ **6.** $x(x - 6) = 0$

2. $x(x - 5) = 0$ **7.** $x(x - 10) = 0$

3. $(x - 3)x = 0$ **8.** $(x - 7)x = 0$

4. $x(x + 4) = 0$ **9.** $x(x + 7) = 0$

5. $(x + 5)x = 0$ **10.** $(x + 9)x = 0$

> What values of x satisfy the equation $(x - 3)(x + 5) = 0$?
>
> $$(x - 3)(x + 5) = 0$$
>
> Either $\quad\quad x - 3 = 0 \quad$ or $\quad x + 5 = 0$
>
> i.e., either $\quad\quad\quad x = 3$ or $x = -5$

What values of x satisfy the following equations?

11. $(x - 1)(x - 2) = 0$ **16.** $(x - 8)(x + 11) = 0$

12. $(x - 5)(x - 9) = 0$ **17.** $(x - 3)(x + 5) = 0$

13. $(x - 10)(x - 7) = 0$ **18.** $(x + 7)(x - 2) = 0$

14. $(x - 4)(x - 7) = 0$ **19.** $(x + 2)(x + 3) = 0$

15. $(x - 6)(x - 1) = 0$ **20.** $(x + 4)(x + 9) = 0$

21. $(x + 1)(x + 8) = 0$ **26.** $(x + 6)(x + 7) = 0$

22. $(x - p)(x - q) = 0$ **27.** $(x + 10)(x + 11) = 0$

23. $(x + a)(x + b) = 0$ **28.** $(x - a)(x - b) = 0$

24. $(x - 4)(x + 1) = 0$ **29.** $(x + a)(x - b) = 0$

25. $(x + 9)(x - 8) = 0$ **30.** $(x - c)(x + d) = 0$

EXERCISE 13d

> Solve the equation $(2x - 3)(3x + 1) = 0$
>
> $$(2x - 3)(3x + 1) = 0$$
> Either $2x - 3 = 0$ or $3x + 1 = 0$
>
> i.e., either $2x = 3$ or $3x = -1$
>
> i.e., either $x = \frac{3}{2} = 1\frac{1}{2}$ or $x = -\frac{1}{3}$

Solve the following equations:

1. $(2x - 5)(x - 1) = 0$ **11.** $(3x - 7)(x - 2) = 0$

2. $(x - 4)(3x - 2) = 0$ **12.** $(3x - 5)(2x - 1) = 0$

3. $(5x - 4)(4x - 3) = 0$ **13.** $x(3x - 1) = 0$

4. $x(4x - 5) = 0$ **14.** $x(7x - 3) = 0$

5. $x(10x - 3) = 0$ **15.** $(2x + 3)(x - 3) = 0$

6. $(5x + 2)(x - 7) = 0$ **16.** $(4x + 3)(2x - 5) = 0$

7. $(6x + 5)(3x - 2) = 0$ **17.** $(10x + 9)(5x - 4) = 0$

8. $(8x - 3)(2x + 5) = 0$ **18.** $(3x - 2)(4x + 9) = 0$

9. $(7x - 8)(4x + 15) = 0$ **19.** $(5x - 12)(2x + 7) = 0$

10. $(4x + 3)(2x + 3) = 0$ **20.** $(5x + 8)(4x + 3) = 0$

SOLUTION BY FACTORISATION

The previous two exercises suggest that if the left-hand side of a quadratic equation can be expressed as two linear factors, we can use these factors to solve the equation.

EXERCISE 13e

> Solve the equation $x^2 - 10x + 9 = 0$
>
> If $x^2 - 10x + 9 = 0$
>
> then $(x - 1)(x - 9) = 0$
>
> $\therefore$ either $x - 1 = 0$ or $x - 9 = 0$
>
> i.e. $x = 1$ or 9

Solve the equations:

1. $x^2 - 3x + 2 = 0$ **6.** $x^2 - 6x + 5 = 0$

2. $x^2 - 8x + 7 = 0$ **7.** $x^2 - 12x + 11 = 0$

3. $x^2 - 5x + 6 = 0$ **8.** $x^2 - 6x + 8 = 0$

4. $x^2 - 7x + 10 = 0$ **9.** $x^2 - 8x + 12 = 0$

5. $x^2 - 7x + 12 = 0$ **10.** $x^2 - 13x + 12 = 0$

Solve the equation $x^2 + 2x - 8 = 0$

$$x^2 + 2x - 8 = 0$$
$$(x + 4)(x - 2) = 0$$

$\therefore$ either $x + 4 = 0$ or $x - 2 = 0$

i.e. $x = -4$ or 2

Solve the equations:

11. $x^2 + 6x - 7 = 0$ **16.** $x^2 - 12x - 13 = 0$

12. $x^2 - 2x - 8 = 0$ **17.** $x^2 + x - 6 = 0$

13. $x^2 + x - 12 = 0$ **18.** $x^2 - 4x - 12 = 0$

14. $x^2 - 2x - 15 = 0$ **19.** $x^2 + x - 20 = 0$

15. $x^2 + 7x - 18 = 0$ **20.** $x^2 - 5x - 24 = 0$

Solve the equation $x^2 + 9x + 8 = 0$

$$x^2 + 9x + 8 = 0$$
$$(x + 1)(x + 8) = 0$$

Either $x + 1 = 0$ or $x + 8 = 0$

i.e. $x = -1$ or -8

Solve the equations:

21. $x^2 + 3x + 2 = 0$ **26.** $x^2 + 7x + 6 = 0$

22. $x^2 + 8x + 7 = 0$ **27.** $x^2 + 7x + 10 = 0$

23. $x^2 + 8x + 15 = 0$ **28.** $x^2 + 14x + 13 = 0$

24. $x^2 + 8x + 12 = 0$ **29.** $x^2 + 16x + 15 = 0$

25. $x^2 + 11x + 18 = 0$ **30.** $x^2 + 9x + 18 = 0$

Solve the equation $x^2 - 49 = 0$

$$x^2 - 49 = 0$$
$$(x + 7)(x - 7) = 0$$

Either $x + 7 = 0$ or $x - 7 = 0$

i.e. $x = -7$ or 7

Solve the equations:

31. $x^2 - 1 = 0$ **36.** $x^2 - 4 = 0$

32. $x^2 - 9 = 0$ **37.** $x^2 - 25 = 0$

33. $x^2 - 16 = 0$ **38.** $x^2 - 100 = 0$

34. $x^2 - 81 = 0$ **39.** $x^2 - 144 = 0$

35. $x^2 - 169 = 0$ **40.** $x^2 - 36 = 0$

The equations we have solved by factorising have all been examples of the equation $ax^2 + bx + c = 0$ when $a = 1$. We consider next the case when $c = 0$,

e.g. solve the equation $3x^2 + 2x = 0$

Since x is common to both terms on the left-hand side we can rewrite this equation

$$x(3x + 2) = 0$$

Then, either $x = 0$ or $3x + 2 = 0$

i.e. $x = 0$ or $3x = -2$

i.e. $x = 0$ or $-\frac{2}{3}$

EXERCISE 13f Solve the equations:

1. $x^2 - 2x = 0$ **6.** $x^2 - 5x = 0$

2. $x^2 - 10x = 0$ **7.** $x^2 + 3x = 0$

3. $x^2 + 8x = 0$ **8.** $x^2 + x = 0$

4. $2x^2 - x = 0$ **9.** $3x^2 - 5x = 0$

5. $4x^2 - 5x = 0$ **10.** $5x^2 - 7x = 0$

11. $2x^2 + 3x = 0$

12. $8x^2 + 5x = 0$

13. $x^2 - 7x = 0$

14. $3x^2 + 5x = 0$

15. $7x^2 - 12x = 0$

16. $6x^2 + 7x = 0$

17. $12x^2 + 7x = 0$

18. $x^2 + 4x = 0$

19. $7x^2 - 2x = 0$

20. $14x^2 + 3x = 0$

Sometimes a quadratic equation has two answers, or *roots*, that are exactly the same.

Consider $\qquad\qquad x^2 - 4x + 4 = 0$

then $\qquad\qquad\qquad (x - 2)(x - 2) = 0$

i.e., either $\qquad\quad x - 2 = 0 \qquad$ or $\qquad x - 2 = 0$

i.e. $\qquad\qquad\qquad x = 2 \qquad$ or $\qquad x = 2$

i.e. $\qquad\qquad\qquad x = 2 \qquad$ (twice)

Such an equation involves a *perfect square*. As with any quadratic equation, it has two answers, or roots, but they are equal. We say that such an equation has a repeated root.

EXERCISE 13g

Solve the equation $x^2 + 14x + 49 = 0$

$$x^2 + 14x + 49 = 0$$
$$(x + 7)(x + 7) = 0$$

Either $\qquad x + 7 = 0 \qquad$ or $\qquad x + 7 = 0$

i.e. $\qquad\qquad\qquad x = -7 \quad$ (twice)

Solve the equations:

1. $x^2 - 2x + 1 = 0$

2. $x^2 - 10x + 25 = 0$

3. $x^2 - 20x + 100 = 0$

4. $x^2 + 8x + 16 = 0$

5. $x^2 + 6x + 9 = 0$

6. $x^2 - 6x + 9 = 0$

7. $x^2 - 8x + 16 = 0$

8. $x^2 - 18x + 81 = 0$

9. $x^2 + 2x + 1 = 0$

10. $x^2 + 20x + 100 = 0$

11. $x^2 + 18x + 81 = 0$

12. $x^2 - 14x + 49 = 0$

13. $x^2 - 22x + 121 = 0$

14. $x^2 + 12x + 36 = 0$

15. $x^2 - x + \frac{1}{4} = 0$

16. $x^2 + 10x + 25 = 0$

17. $x^2 - 12x + 36 = 0$

18. $x^2 - 40x + 400 = 0$

19. $x^2 - 16x + 64 = 0$

20. $x^2 + \frac{4}{3}x + \frac{4}{9} = 0$

Frequently $a \neq 1$ in the equation $ax^2 + bx + c = 0$. The next exercise considers quadratic equations for values of a other than 1.

EXERCISE 13h

Solve the equation $5x^2 + 13x - 6 = 0$

$$5x^2 + 13x - 6 = 0$$
$$(5x - 2)(x + 3) = 0$$

Either $5x - 2 = 0$ or $x + 3 = 0$

i.e. $5x = 2$ or $x = -3$

i.e. $x = \frac{2}{5}$ or -3

Solve the equations:

1. $2x^2 - 5x + 2 = 0$

2. $2x^2 - 11x + 12 = 0$

3. $2x^2 - 13x + 20 = 0$

4. $3x^2 + 5x + 2 = 0$

5. $2x^2 + 9x - 35 = 0$

6. $3x^2 - 11x + 6 = 0$

7. $3x^2 - 7x + 2 = 0$

8. $2x^2 + 5x - 12 = 0$

9. $3x^2 + 11x + 6 = 0$

10. $5x^2 + 27x + 10 = 0$

11. $6x^2 - x - 2 = 0$

12. $15x^2 + 14x - 8 = 0$

13. $12x^2 - 7x + 1 = 0$

14. $6x^2 - 13x - 5 = 0$

15. $20x^2 + 19x + 3 = 0$

16. $8x^2 - 18x + 9 = 0$

17. $12x^2 - 20x - 25 = 0$

18. $4x^2 + 8x + 3 = 0$

19. $12x^2 + 17x + 6 = 0$

20. $10x^2 - 29x - 21 = 0$

Solve the equation $4x^2 - 9 = 0$

$$4x^2 - 9 = 0$$
$$(2x - 3)(2x + 3) = 0$$

Either $2x - 3 = 0$ or $2x + 3 = 0$

i.e. $2x = 3$ or $2x = -3$

i.e. $x = \frac{3}{2}$ or $-\frac{3}{2}$

Solve the equations:

21.	$16x^2 - 25 = 0$	**26.**	$9x^2 - 4 = 0$
22.	$100x^2 - 81 = 0$	**27.**	$81x^2 - 25 = 0$
23.	$4x^2 - 25 = 0$	**28.**	$25x^2 - 4 = 0$
24.	$9x^2 - 16 = 0$	**29.**	$36x^2 - 25 = 0$
25.	$25x^2 - 144 = 0$	**30.**	$4x^2 - 81 = 0$

EXERCISE 13i

> Solve the equation $x^2 - x = 12$
>
> $$x^2 - x = 12$$
>
> Subtracting 12 from each side gives
>
> $$x^2 - x - 12 = 0$$
> $$(x - 4)(x + 3) = 0$$
>
> Either $x - 4 = 0$ or $x + 3 = 0$
>
> i.e. $x = 4$ or -3

Solve the equations:

1.	$x^2 - x = 30$	**5.**	$x^2 - x = 6$
2.	$x^2 - 6x = 16$	**6.**	$x^2 + 6x = 7$
3.	$x^2 + 9x = 36$	**7.**	$2x^2 + 5x = 3$
4.	$3x^2 + 4x = 4$	**8.**	$5x^2 - 12x = 9$

> Solve the equations $x^2 = 2x + 3$
>
> $$x^2 = 2x + 3$$
>
> Subtract $2x$ and 3 from each side:
>
> $$x^2 - 2x - 3 = 0$$
> $$(x - 3)(x + 1) = 0$$
>
> Either $x - 3 = 0$ or $x + 1 = 0$
>
> i.e. $x = 3$ or -1

Solve the equations:

9. $x^2 = 2x + 8$

10. $x^2 = 2x + 24$

11. $x^2 = 12x - 35$

12. $10x^2 = 13x + 3$

13. $x^2 = 3x + 10$

14. $x^2 = 6x - 8$

15. $6x^2 = x + 1$

16. $3x^2 = 13x - 4$

17. $10 = 7x - x^2$

18. $7 = 8x - x^2$

19. $8 = 6x - x^2$

20. $21 = 10x - x^2$

21. $12 = 8x - x^2$

22. $20 = 9x - x^2$

23. $35 = 12x - x^2$

24. $15 = 8x - x^2$

Solve the equation $12x^2 + 10x - 12 = 0$

$$12x^2 + 10x - 12 = 0$$

Taking out the common factor 2 gives

$$(2)(6x^2 + 5x - 6) = 0$$

Since 2 is not zero, $6x^2 + 5x - 6$ must be zero

∴ $6x^2 + 5x - 6 = 0$

i.e. $(3x - 2)(2x + 3) = 0$

Either $3x - 2 = 0$ or $2x + 3 = 0$

i.e. $3x = 2$ or $2x = -3$

i.e. $x = \frac{2}{3}$ or $-\frac{3}{2}$

Solve the equations:

25. $8x^2 - 4x = 0$

26. $2x^2 - 10x + 12 = 0$

27. $3x^2 - 24x + 36 = 0$

28. $12x^2 + 20x + 8 = 0$

29. $8x^2 + 20x = 12$

30. $3x^2 - 9x = 0$

31. $5x^2 - 15x + 10 = 0$

32. $6x^2 + 18x + 12 = 0$

33. $15x^2 - 35x + 10 = 0$

34. $30x^2 = 39x + 9$

SUMMARY

To solve a quadratic equation by factorising:
a) collect all terms on one side of the equation, i.e. arrange it in the form
$$ax^2 + bx + c = 0$$
b) take out any common factors (these may or may not include x)
c) factorise.

EXERCISE 13j Solve the equations:

1. $x^2 - x - 20 = 0$

2. $x^2 = 4x - 4$

3. $9x^2 - 1 = 0$

4. $2x^2 + 7x = 0$

5. $x^2 + 13x + 12 = 0$

6. $1 - 16x^2 = 0$

7. $x^2 - 6x = 0$

8. $x^2 = 2x + 35$

9. $2x^2 + 3x - 14 = 0$

10. $2x^2 + 12x + 18 = 0$

11. $x^2 = 7 - 6x$

12. $4 = 25x^2$

13. $4x^2 = 25$

14. $x^2 + 11x + 18 = 0$

15. $2 - x = 6x^2$

16. $5x - 2x^2 = 0$

17. $5x = 3x^2 - 2$

18. $4 + 11x + 6x^2 = 0$

19. $7x = 4x^2$

20. $14x - 2 = 24x^2$

21. $6x^2 + 13x - 5 = 0$

22. $5x + 2 = 3x^2$

23. $3 + 8x + 4x^2 = 0$

24. $3 - 12x^2 = 0$

Solve the equation $x(x - 2) = 15$

$$x(x - 2) = 15$$
$$x^2 - 2x = 15$$
$$x^2 - 2x - 15 = 0$$
$$(x - 5)(x + 3) = 0$$

Either $\qquad x - 5 = 0 \qquad$ or $\qquad x + 3 = 0$

i.e. $\qquad\qquad x = 5$ or -3

Solve the equations:

25. $x(x + 1) = 12$

26. $x(x - 1) = x + 3$

27. $3x(2x + 1) = 4x + 1$

28. $5x(x - 1) = 4x^2 - 4$

29. $x(x - 5) = 24$

30. $x(x + 3) = 5(3x - 7)$

31. $3x(x + 3) = 5x + 4$

32. $2x(2x - 1) = x^2 + 3x + 2$

Solve the equation $(x - 3)(x + 2) = 6$

$$(x - 3)(x + 2) = 6$$
$$x^2 - x - 6 = 6$$
$$x^2 - x - 12 = 0$$
$$(x - 4)(x + 3) = 0$$

Either $x - 4 = 0$ or $x + 3 = 0$

i.e. $x = 4$ or -3

Solve the equations:

33. $(x + 2)(x + 3) = 56$

34. $(x + 9)(x - 6) = 34$

35. $(x - 2)(x + 6) = 33$

36. $(x + 3)(x - 8) + 10 = 0$

37. $(x - 5)(x + 2) = 18$

38. $(x + 8)(x - 2) = 39$

39. $(x + 1)(x + 8) + 12 = 0$

40. $(x - 1)(x + 10) + 30 = 0$

Solve the equation $x(x - 3)(x - 7) = 0$

$$x(x - 3)(x - 7) = 0$$

Either $x = 0$ or $x - 3 = 0$ or $x - 7 = 0$

i.e. $x = 0$ or 3 or 7

Solve the equations:

41. $x(x - 1)(x - 2) = 0$

42. $x(x - 3)(x + 4) = 0$

43. $x(2x - 5)(x - 2) = 0$

44. $x^3 - 2x^2 + x = 0$

45. $2x^3 + 9x^2 + 4x = 0$

46. $x(x - 6)(x - 7) = 0$

47. $x(x + 2)(x - 5) = 0$

48. $x(3x + 7)(x - 5) = 0$

49. $4x^3 - 9x = 0$

50. $8x = 6x^2 - x^3$

PROBLEMS

> I think of a positive number x, square it and then add three times the number I first thought of. If the answer is 54, form an equation in x and solve it to find the number I first thought of.
>
> If the given number is x, its square is x^2 and three times the number is $3x$.
>
> i.e.
> $$x^2 + 3x = 54$$
> $$x^2 + 3x - 54 = 0$$
> $$(x - 6)(x + 9) = 0$$
>
> Either $\qquad x - 6 = 0 \qquad$ or $\qquad x + 9 = 0$
>
> i.e. $\qquad\qquad\qquad x = 6$ or -9
>
> The number I first thought of was therefore 6

1. The square of a number x is 16 more than six times the number. Form an equation in x and solve it.

2. When five times a number x is subtracted from the square of the same number, the answer is 14. Form an equation in x and solve it.

3. I think of a number x. If I square it and add it to the number I first thought of the total is 42. Find the number I first thought of.

4. Peter had x marbles. The number of marbles Fred had was six fewer than the square of the number Peter had. Together they had 66 marbles. Form an equation in x and solve it. How many marbles did Fred have?

5. Ahmed is x years old and his father is x^2 years old. If the sum of their ages is 56 years, form an equation in x and solve it to find the age of each.

6. Kathryn is x years old. If her mother's age is two years more than the square of Kathryn's age, and the sum of their ages is 44 years, form an equation in x and solve it to find the ages of Kathryn and her mother.

7. Peter is x years old and his sister is 5 years older. If the product of their ages is 84, form an equation in x and solve it to find Peter's age.

8. Sally is x years old and her sister Ann is 4 years younger. If the product of their ages is 140, form an equation in x and solve it to find Ann's age.

A rectangle is 4 cm longer than it is wide. If it is x cm wide and has an area of 77 cm², form an equation in x and solve it to find the dimensions of the rectangle.

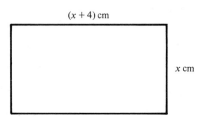

$(x + 4)$ cm

x cm

The area of a rectangle is length × breadth.

$$\text{Area} = (x + 4) \times x \text{ cm}^2$$

But the area is 77 cm²,

$\therefore$ $\quad\quad\quad\quad\quad\quad (x + 4)x = 77$

i.e. $\quad\quad\quad\quad\quad\quad x^2 + 4x = 77$

$$x^2 + 4x - 77 = 0$$

$$(x - 7)(x + 11) = 0$$

Either $\quad\quad x - 7 = 0 \quad\quad$ or $\quad\quad x + 11 = 0$

i.e. $\quad\quad\quad\quad\quad\quad x = 7 \text{ or } -11$

The breadth of a rectangle cannot be negative, so we use $x = 7$.

Therefore, the rectangle measures $(7 + 4)$ cm by 7 cm, i.e. 11 cm by 7 cm.

---

9. A rectangle is x cm wide and is 3 cm longer than it is wide. If its area is 28 cm², form an equation in x and solve it to find the dimensions of the rectangle.

10. A rectangle is 5 cm longer than it is wide. If its width is x cm and its area is 66 cm² form an equation in x and solve it. Hence find the dimensions of the rectangle.

11. The base of a triangle is x cm long and its perpendicular height is half the length of its base. If the triangle has an area of 25 cm², form an equation in x and solve it. What is the height of the triangle?

12. A rectangular lawn measuring 30 m by 20 m is bordered on two adjacent sides by a uniform path of width x m as shown in the diagram.

    a) Express in terms of x each of the areas denoted by the letters A, B and C.

    b) If the area of the path is 104 m² form an equation in x and solve it to find the width of the path.

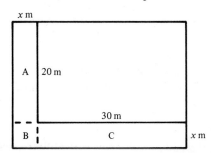

SOLVING EQUATIONS BY TRIAL AND IMPROVEMENT

The methods examined so far in this chapter apply to solving quadratic equations. There are many other types of equation for which there is no method of finding an exact solution.

In such cases we start by giving x a value that looks reasonable and *trying* it in the equation. Then we adjust the value of x to give an *improvement*.

Consider $x^3 + x = 9$. (This is a *cubic* equation because the highest power of x it contains is 3).

Whole number values of x can be tested easily.

We can see that if $x = 1$, then the value of $x^3 + x$ is clearly too small $(1^3 + 1 = 2)$.
When we try 2 for x, we find that $2^3 + 2 = 10$, so 2 is just too big.
As $x^3 + x = 9$, we can expect that the true value of x is nearer to 2 than to 1, so we will try 1.7 as a possible value of x.
$1.7^3 + 1.7 = 4.913 + 1.7 = 6.613$ which is too small.

Therefore the value of x lies between 1.7 and 2.

We can illustrate this on a number line.

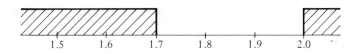

Again we expect the value of x to be nearer to 2 than to 1.7 so we try 1.9.

If $x = 1.9$, then $x^3 + x = 6.859 + 1.9 = 8.759$.

This is nearer but still too small so we will try 1.95.

If $x = 1.95$, $x^3 + x = 7.415 + 1.95 = 9.365$.

However, this is too big.

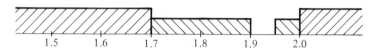

We can see that the next number to try is between 1.9 and 1.95, perhaps 1.92 or 1.93, so we can go on improving.

The solution to this equation cannot be found exactly so we must decide when to stop our search. In this case we might decide that we will be satisfied if we can find two numbers differing by 0.1, each to 1 decimal place, between which the solution lies and we already know that it is between 1.9 and 2.0. Alternatively, we might decide that the two numbers should differ by only 0.01; although we already know that the solution lies between 1.90 and 1.95 we need to investigate further.

EXERCISE 13I

Given the equation $x^2 - \dfrac{2}{x} = 6$, find two numbers differing by 0.1, each correct to 1 decimal place, between which a solution lies.

(1 is clearly too small, so try 2)

Try $x = 2$ $x^2 - \dfrac{2}{x} = 4 - \dfrac{2}{2} = 3$ (Too small)

Try $x = 3$ $x^2 - \dfrac{2}{x} = 9 - \dfrac{2}{3} = 8.333$ (Too big)

Try $x = 2.5$ $x^2 - \dfrac{2}{x} = 6.25 - \dfrac{2}{2.5} = 5.45$

(Nearer but too small)

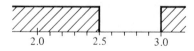

Try $x = 2.7$ $x^2 - \dfrac{2}{x} = 7.29 - \dfrac{2}{2.7}$

$= 6.549$ (Too big)

If $x = 2.6$ $x^2 - \dfrac{2}{x} = 6.76 - \dfrac{2}{2.6}$

$= 5.99$ (Very close!)

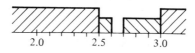

A solution lies very close to 2.6. It lies between 2.6 and 2.7.

(We could if necessary go on and find two numbers differing by 0.01, each to 2 decimal places, between which the solution lies, probably trying 2.61 first.)

For each equation, find two numbers differing by 0.1, each to 1 decimal place, between which a solution to the equation lies.

1. $x^3 + 7x = 12$ (Start by trying $x = 1$)

2. $\dfrac{10}{x} - x^2 = 6$ (Start by trying $x = 2$)

3. $x^3 - 2x - 5 = 0$ (Start by trying $x = 2$)

4. $x^2 + x + \dfrac{1}{x} = 11$ (Start by trying $x = 2$)

5. $x^3 + 3x = 7$ **7.** $24 - x^3 + 2x = 0$

6. $x + \dfrac{8}{x} = 12$ **8.** $x^2 - \dfrac{2}{x} = 6$

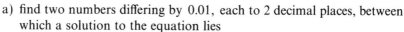

For each question from 9 to 12,

a) find two numbers differing by 0.01, each to 2 decimal places, between which a solution to the equation lies

b) give this solution correct to 1 decimal place.

9. $x^3 + x^2 + x = 45$ **11.** $8x^2 - x^3 = 10$

10. $\dfrac{12}{x} + x = 9$ **12.** $x^2 - \dfrac{6}{x} = 16$

MIXED EXERCISES

EXERCISE 13m **1.** Find the value of $(x + 4)(x - 3)$ if

  a) $x = 1$ b) $x = 3$ c) $x = 4$

2. Solve   a) $x(x + 7) = 0$ b) $4x(2x - 1) = 0$

3. Solve   a) $(x - 3)(x - 8) = 0$ b) $(x - 2)(5x + 3) = 0$

4. Solve   a) $x^2 - 2x - 35 = 0$ b) $x^2 - 13x + 40 = 0$

5. Solve   a) $10x^2 - 13x + 4 = 0$ b) $15x^2 - x - 2 = 0$
  c) $9x^2 - 4 = 0$

6. Solve   a) $7x^2 - 14x = 0$ b) $4x^2 - 3x = 0$

7. Solve   a) $x(x + 4) = 45$ b) $x(x + 2) = x + 30$

8. For the equation $x^3 + x = 20$, find two numbers differing by 0.1, each to 1 decimal place, between which a solution lies.

EXERCISE 13n **1.** Find the value of $(2x - 1)(x + 2)$ if

  a) $x = 0$ b) $x = \frac{1}{2}$ c) $x = 2$ d) $x = -2$

2. Solve   a) $3x(x - 2) = 0$ b) $x(7x + 3) = 0$

3. Solve   a) $(x - 2)(x + 5) = 0$ b) $(3x - 4)(x + 2) = 0$
  c) $(2x + 3)(2x - 3) = 0$

4. Solve   a) $x^2 + x - 6 = 0$ b) $x^2 + 11x + 30 = 0$

5. Solve   a) $20x^2 + 11x - 3 = 0$ b) $15x^2 + 41x + 14 = 0$

6. Solve   a) $15x^2 + 10x = 0$ b) $7x^2 + 3x = 0$

7. Solve   a) $x(x - 4) = 32$ b) $x(2x - 1) = 3x + 16$

8. For the equation $x + \frac{6}{x} = 20$, find two numbers differing by 0.1, each to 1 decimal place, between which a solution lies.

EXERCISE 13p **1.** Find the value of $(5x + 1)(x - 3)$ if

  a) $x = 2$ b) $x = 3$ c) $x = -\frac{1}{5}$

2. Solve   a) $5x(x + 7) = 0$ b) $3x(4x - 3) = 0$

3. Solve   a) $(x + 4)(x - 5) = 0$ b) $(4x - 7)(x + 3) = 0$

c) $(5x - 3)(5x + 3) = 0$

4. Solve   a) $x^2 - 2x - 15 = 0$ b) $x^2 + 12x + 32 = 0$

5. Solve   a) $20x^2 + 19x + 3 = 0$ b) $28x^2 + x - 2 = 0$

6. Solve   a) $12x^2 + 16x = 0$ b) $3x^2 + 5x = 0$

7. Solve   a) $x(x + 6) = 3x + 10$ b) $x(x + 8) = x + 30$

8. For the equation $x^3 + x - 4 = 0$, find two numbers differing by 0.1, each to 1 decimal place, between which a solution lies.

14 GRAPHS

GRAPHS FROM TABLES

Graphs are used to give a visual representation of information about two, related, varying quantities.

You can see examples of graphs in many newspapers and periodicals. The best of them give a clear visual impression of the way in which the related quantities vary and allow us to get more information from the graph. The worst graphs are misleading, often because the axes are incorrectly labelled or because the scales are distorted.

Before we can draw a graph we have to know how the quantities vary. Sometimes we are given this information in a table, while at other times we are told that the two quantities are connected by a formula. The following example shows how to draw a clear, informative graph.

The Forestry Commission collected data for a particular type of tree. This data is given in the following table:

Age of tree in years (A)	0.8	1.4	2.2	3.5	5.4	6.6	7.8
Girth of tree in cm (G)	15.8	20.9	26.5	33.2	41.2	45.9	49.5

To draw a graph to represent this data we must:

a) draw axes that intersect at right angles in the bottom left-hand corner of the graph paper

b) state the scale used on each axis

c) name the axes
   (We put time along the horizontal axis and the distance around the tree, or *girth*, along the vertical axis.)

d) plot carefully the points representing the data given in the table
   (We shall use a dot for each point.)

e) draw a smooth curve to pass through the points

f) give the graph a title.

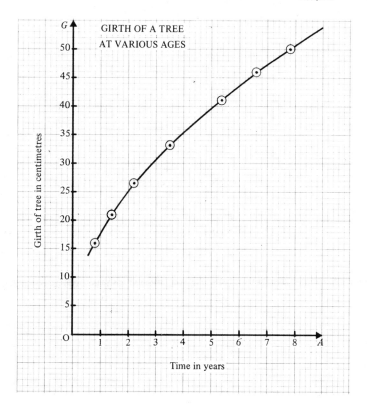

We can use this graph to find the girth of the tree at any given age, or the age of the tree when it has a particular girth, provided that we keep within the range of values on our axes.

From the graph:
a) when the tree is four years old its girth is 35.5 cm
b) when the girth is 40 cm the tree is five years old.

EXERCISE 14a   **1.** The table shows the weight (*W*) in tonnes which a chain of diameter (*D*) centimetres will support before breaking.

D	1.2	1.8	2.2	2.6	2.8	3	3.4	3.8
W	72	162	242	338	392	450	578	722

Draw a graph for values of *D* from 0 to 4. Take 4 cm as 1 unit for *D* and 1 cm as 50 units for *W*. Use your graph to find
a) the greatest weight that a chain of diameter 2 cm will support
b) the smallest diameter of chain that will support a load of 500 t.

2. When a sum of money is invested, the interest added to the original value gives an increased value called the amount. The table shows the amount (V) when £100 has been invested for T years at 10% compound interest.

T	0	1	2	3	4	5	6	7	8	9
V	100	110	121	133	146	161	177	195	214	236

Draw the graph connecting T and V. Take 2 cm as 1 unit for T and 4 cm as 50 units for V. From your graph find
a) the amount after $5\frac{1}{2}$ years
b) the time, in years, in which £100 will double in value.

3. The table shows the connection between two quantities X and Y as X takes different values from 1 to 8

X	1	1.5	2	3	4	5	6	8
Y	14	10	8	6	5	4.4	4	3.5

Draw a graph connecting X and Y taking 2 cm as 1 unit for X and 1 cm as 1 unit for Y. Start at 0 for each scale. From your graph find
a) the value of Y when X is 7.2
b) the value of X when Y is 7.2

4. The table gives values for the surface area, A cm^2, of a closed rectangular box of fixed volume, which has square ends of side x cm.

x (cm)	1	1.5	2	3	4	5	6
A (cm^2)	110	76.5	62	54	59	71.6	90

Draw a graph connecting A and x taking 4 cm to represent 1 unit on the x-axis and 1 cm to represent 5 units on the A-axis. Let 40 be the lowest value on the A-axis. Use your graph to find
a) the value of x that gives the lowest value of A
b) the value of A when x is 1.8

5. The table shows the capacity (*C*), in litres, for jugs that are mathematically similar, but of different heights (*H* cm).

Height of jug in cm (*H*)	4	5.25	7.5	10.25	12.75	15.5	18.25
Capacity of jug in litres (*C*)	0.05	0.11	0.33	0.84	1.67	2.90	4.74

Draw a graph to represent this data using 4 cm to represent 5 units on the *H*-axis and 1 unit on the *C*-axis. use your graph to find

a) the height of a similar jug with a capacity of 3.5 litres

b) the capacity of a similar jug that is 14 cm high.

CONSTRUCTING A TABLE FROM A FORMULA

In the exercise we have just completed, all the data was given in tables. However, in many questions we are given a formula and have to construct our own table for given values of the variables.

Consider a car that starts from rest and travels a distance *D* metres in *T* seconds. Suppose that the formula connecting *D* and *T* for the first 10 seconds of the journey is

$$D = 50\sqrt{T}$$

and that we wish to draw a graph to show this relationship. Before we can draw the graph, we must work out some corresponding values of *D* and *T*.

If we take the whole number values of *T* from 0 to 10 we get:

when

$T = 1 \quad D = 50\sqrt{1} \ = 50 \times 1 = 50$
$T = 2 \quad D = 50\sqrt{2} \ = 50 \times 1.414 = 71$
$T = 3 \quad D = 50\sqrt{3} \ = 50 \times 1.732 = 87$
$T = 4 \quad D = 50\sqrt{4} \ = 50 \times 2 = 100$
$T = 5 \quad D = 50\sqrt{5} \ = 50 \times 2.236 = 112$
$T = 6 \quad D = 50\sqrt{6} \ = 50 \times 2.449 = 122$
$T = 7 \quad D = 50\sqrt{7} \ = 50 \times 2.646 = 132$
$T = 8 \quad D = 50\sqrt{8} \ = 50 \times 2.828 = 141$
$T = 9 \quad D = 50\sqrt{9} \ = 50 \times 3 = 150$
$T = 10 \quad D = 50\sqrt{10} = 50 \times 3.162 = 158$

Each value of *D* is calculated correct to the nearest whole number. These values of *T* in seconds and *D* in metres can be set out in table form:

T	0	1	2	3	4	5	6	7	8	9	10
D	0	50	71	87	100	112	122	132	141	150	158

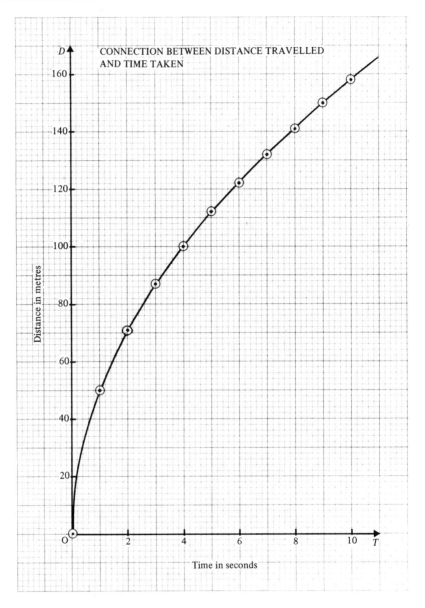

We plot time (*T*) along the horizontal axis and distance (*D*) along the vertical axis.

When all the points have been plotted, draw a smooth curve to pass through them. The resulting graph is shown in the diagram. From this graph we can find *D* for any given value of *T* from 0 to 10, e.g. the distance travelled in 4.4 s is 105 m. Similarly, the time to travel 146 m is 8.5 s.

EXERCISE 14b

1. A stone is dropped down a vertical mine shaft. After t seconds it has fallen s metres, where $s = 5t^2$. Construct a table to show the relation between s and t, for values of t from 0 to 6, at half-unit intervals.

Use these values to draw a graph, and use your graph to find

a) how far the stone has fallen in 3.4 s

b) how long the stone takes to fall 100 m.

2. Two quantities P and Q are connected by the formula
$$P = \frac{36}{Q}$$
Construct a table to show the relation between P and Q for values of Q from 1 to 8

Plot these points on a graph and from this graph find

a) the value of P when Q is 7.5

b) the value of Q when P is 4.8

3. Draw the graph of $y = \dfrac{6}{x}$ for values of x from 1 to 6 inclusive, plotting points at whole number values of x and at $x = 1\frac{1}{2}$. Take 2 cm as 1 unit on each axis. Use your graph to find

a) the value of x when $y = 3.6$

b) the value of y when $x = 5.5$

4. Taking 2 cm as the unit on both axes, draw the graph of $y = x + \dfrac{7}{x}$ for values of x between 1 and 7 at unit intervals. Use your graph to find the lowest value of y and the corresponding value of x.

5. Draw the graph of $xy = 8$ for values of x from -8 to -1 and from 1 to 8. (We cannot take $x = 0$ since there is no value of y which enables the product xy to be 8.) Let each axis range from -8 to 8, and take 1 cm as 1 unit on both axes. Use your graph to find

a) the value of y when x is 2.4

b) the value of x when y is -5.6

POINTS TO REMEMBER WHEN DRAWING GRAPHS OF CURVES

1. Do not take too few points. About ten are usually necessary.

2. To decide where you need to draw the y-axis, look at the range of x-values.

3. To decide where to draw the x-axis, look at the range of y-values.

4. In some questions you will be given most of the y-values but you may have to calculate a few more for yourself. In this case always plot first those points that you were given and, from these, get an idea of the shape of the curve. Then you can plot the points you calculated and see if they fit on to the curve you have in mind. If they do not, go back and check your calculations.

5. When you draw a smooth curve to pass through the points, always turn the page into a position where your wrist is on the inside of the curve.

QUADRATIC GRAPHS

The most important family of curves we consider give what are called *quadratic graphs*. The simplest of these is the graph of $y = x^2$.

We will draw the graph of $y = x^2$ for values of x ranging from -3 to $+3$ at half-unit intervals. The corresponding values for x and y are given in the table below:

x	-3	-2.5	-2	-1.5	-1	-0.5	0	0.5	1	1.5	2	2.5	3
$y(=x^2)$	9	6.25	4	2.25	1	0.25	0	0.25	1	2.25	4	6.25	9

A suitable scale for you to take is 2 cm to represent 1 unit on both axes but we have taken 1 cm to represent 1 unit on both axes.

We draw the y-axis vertically in the centre of the page, since the x-values range from -3 to $+3$, i.e. they are symmetrical about O.

The x-axis is drawn along the bottom of the page since all the y-values are positive.

The table gives us thirteen points. In drawing any quadratic graph, aim for at least ten points. It is especially important to have plenty of points where the graph is changing direction most quickly. For quadratic graphs this means around the lowest (or highest) point.

From the graph we can find the value of y that corresponds to any value of x within the range -3 to $+3$. For any value of y between 0 and 9 we can find the two corresponding values of x.

For example: a) if $x = 1.8$, $y = 3.2$
b) if $x = -2.3$, $y = 5.3$
c) if $y = 2$, $x = 1.4$ or -1.4
(There are two values of x for which $y = 2$)

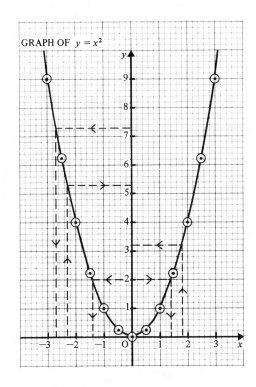

GRAPH OF $y = x^2$

EXERCISE 14c **1.** Draw on the same axes the graphs of $y = x^2$, $y = 2x^2$ and $y = 3x^2$, taking half-unit intervals for x in the range -2 to $+2$. Take 4 cm as the unit on the x-axis and 2 cm as the unit on the y-axis for values of y from 0 to 12
What can you deduce about the graph of $y = ax^2$ for any positive value of a?

2. Draw the graph of $y = x^2 + 3$ for values of x in the range -3 to 3. Take 4 cm as the unit for x and 1 cm as the unit for y.
a) Use your graph to find the values of x when $y = 6$
b) Are there any values of x for which y has the value 0?

3. Draw the graph of $y = -x^2 + 4$ for values of x in the range
-3 to 3. Take 2 cm as the unit for both x and y. Use your
graph to find the value of x when   a) $y = 0$ b) $y = 3$
Is the graph upside down compared with the graphs you drew in
questions 1 and 2?

4. Draw on the same axes the graphs of $y = x^2$, $y = x^2 + 4$ and
$y = x^2 - 4$. Use values of x from -3 to $+3$ at unit intervals,
taking 2 cm as 1 unit on the x-axis and 1 cm as 1 unit on the
y-axis. Let the scale on the y-axis range from -5 to $+14$
What can you say about the shapes of the three graphs?
What can you deduce about the graph of $y = x^2 + C$ for
different positive or negative values of C?

5. Complete the following table which gives values of $x(x - 3)$ for
values of x in the range -1 to 4 at half-unit intervals.

x	-1	-0.5	0	0.5	1	1.5	2	2.5	3	3.5	4
$x - 3$	-4			-2.5		-1.5					1
$x(x - 3)$	4			-1.25		-2.25					4

Hence draw the graph of $y = x(x - 3)$ within the given range
taking 2 cm as the unit on both axes.
Use your graph to write down
a) the values of x where the graph crosses the x-axis
b) the values of x when $x(x - 3) = 3$ (i.e. when $y = 3$).

6. Draw the graph of $y = x(2x - 3)$ for values of x in the range
-2 to 3 taking values of x at half-unit intervals. Use a scale of
2 cm for 1 unit on the x-axis and 1 cm for 1 unit on the y-axis.
Use your graph to find
a) the values of x where the graph crosses the x-axis
b) the lowest value of $x(2x - 3)$, i.e. the lowest value of y and
the corresponding value of x.

7. Draw the graph of $y = 2x(2 + x)$ for values of x in the range
-5 to 3. Take values of x at unit intervals, with extra values
where you think they are needed. Let the scale on your y-axis
range from -4 to $+32$. Let 1 cm represent 2 units. Use your
graph to find
a) the smallest value of $2x(2 + x)$ and the value of x for which
it occurs.
b) the value of $2x(2 + x)$ when $x = -3.5$
c) the values of x when $2x(2 + x) = 0$

Draw the graph of $y = x^2 + x - 6$ for whole number values of x from -4 to $+4$. Take 1 cm as 1 unit on the x-axis and 1 cm as 2 units on the y-axis. Use your graph to find

a) the lowest value of $x^2 + x - 6$ and the corresponding value of x

b) the values of x when $x^2 + x - 6$ is 4

x	-4	-3	-2	-1	0	1	2	3	4
x^2	16	9	4	1	0	1	4	9	16
x	-4	-3	-2	-1	0	1	2	3	4
-6	-6	-6	-6	-6	-6	-6	-6	-6	-6
$x^2 + x - 6$	6	0	-4	-6	-6	-4	0	6	14

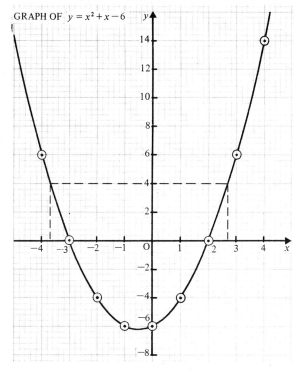

GRAPH OF $y = x^2 + x - 6$

a) From the graph, the lowest value of $x^2 + x - 6$ is $-6\frac{1}{4}$. This occurs when $x = -\frac{1}{2}$

b) The values of x when $x^2 + x - 6$ is 4 are -3.70 and 2.70

8. Draw the graph of $y = x^2 - 2x - 3$ for whole number values of x in the range -3 to 5. Take 2 cm as 1 unit for x and 1 cm as 1 unit for y. Use your graph to find

a) the lowest value of $x^2 - 2x - 3$ and the corresponding value of x

b) the values of x when $x^2 - 2x - 3$ has a value of

    i) 1   ii) 8

9. Draw the graph of $y = 6 + x - x^2$ for whole number values of x from -3 to 4. Take 2 cm as 1 unit on both axes. Use your graph to find

a) the highest value of $6 + x - x^2$ and the corresponding value of x

b) the values of x when $6 + x - x^2$ has a value of

    i) -2 ii) 4

15 POLYGONS

A polygon is a plane (flat) figure bounded by straight lines.

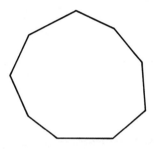

This is a nine-sided polygon.

Some polygons have names which you already know:

a three-sided polygon is a triangle

a four-sided polygon is a quadrilateral

a five-sided polygon is a pentagon

a six-sided polygon is a hexagon

an eight-sided polygon is an octagon

253

REGULAR POLYGONS

A polygon is called regular when all its sides are the same length *and* all its angles are the same size. The polygons below are all regular:

EXERCISE 15a State which of the following figures are regular polygons. Give a brief reason for your answer:

1. Rhombus
2. Square
3. Rectangle
4. Parallelogram

5. Isosceles triangle
6. Right-angled triangle
7. Equilateral triangle
8. Circle

Make a rough sketch of each of the following polygons. (Unless you are told that a polygon is regular, you must assume that it is *not* regular.)

9. A regular quadrilateral
10. A hexagon
11. A triangle
12. A regular triangle

13. A regular hexagon
14. A pentagon
15. A quadrilateral
16. A ten-sided polygon

When the vertices of a polygon all point outwards, the polygon is convex.

Sometimes one or more of the vertices point inwards, in which case the polygon is concave.

convex polygon

concave polygon

In this chapter we consider only convex polygons.

INTERIOR ANGLES

The angles enclosed by the sides of a polygon are the interior angles. For example,

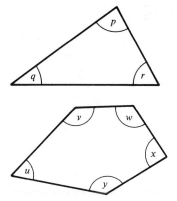

p, *q* and *r* are the interior angles of the triangle,

u, *v*, *w*, *x* and *y* are the interior angles of the pentagon.

THE EXTERIOR ANGLES

If we produce (extend) one side of a polygon, an angle is formed outside the polygon. It is called an *exterior angle*.

For example, *s* is an exterior angle of the quadrilateral.

If we produce all the sides in order we have all the exterior angles.

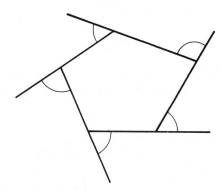

EXERCISE 15b　**1.**　What is the sum of the interior angles of any triangle?

2.　What is the sum of the interior angles of any quadrilateral?

3.　In triangle ABC, find
　　a) the size of each marked
　　　　angle
　　b) the sum of the exterior
　　　　angles.

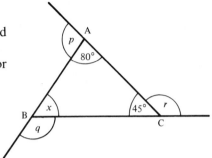

4.

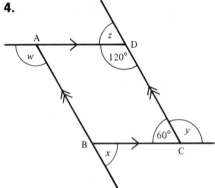

ABCD is a parallelogram.
Find
a) the size of each marked
　　angle
b) the sum of the exterior
　　angles.

5.　In triangle ABC, write down the
　　value of
　　a) $x + q$
　　b) the sum of all six marked angles
　　c) the sum of the interior angles
　　d) the sum of the exterior angles.

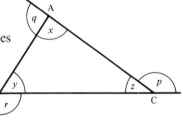

6.　Draw a pentagon. Produce the sides in order to form the five
　　exterior angles. Measure each exterior angle and then find their
　　sum.

7.　Construct a regular hexagon of side 5 cm. (Start with a circle of
　　radius 5 cm and then with your compasses still open to a radius
　　of 5 cm, mark off points on the circumference in turn.) Produce
　　each side of the hexagon in turn to form the six exterior angles.

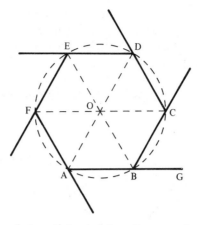

If O is the centre of the circle, joining O to each vertex forms
six triangles:

a) What kind of triangle is each of these triangles?
b) What is the size of each interior angle in these triangles?
c) Write down the value of $A\hat{B}C$.
d) Write down the value of $C\hat{B}G$.
e) Write down the value of the sum of the six exterior angles of
the hexagon.

THE SUM OF THE EXTERIOR ANGLES OF A POLYGON

In the last exercise, we found that the sum of the exterior angles is
360° in each case. This is true of any polygon, whatever its shape or
size.

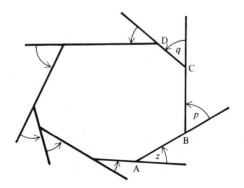

Consider walking round this polygon. Start at A and walk along AB.
When you get to B you have to turn through angle p to walk along
BC. When you get to C you have to turn through angle q to walk

along CD, . . . and so on until you return to A. If you then turn through angle z you are facing in the direction AB again. You have now turned through each exterior angle and have made just one complete turn, i.e.

EXERCISE 15c

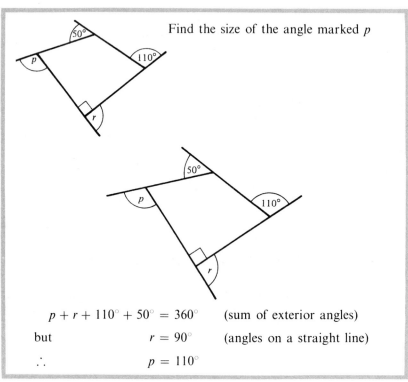

Find the size of the angle marked p

$$p + r + 110° + 50° = 360° \quad \text{(sum of exterior angles)}$$

but $\qquad\qquad\qquad r = 90° \qquad$ (angles on a straight line)

$\therefore \qquad\qquad\qquad p = 110°$

In each case find the size of the angle marked p:

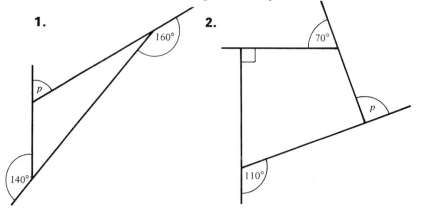

1.

2.

3.

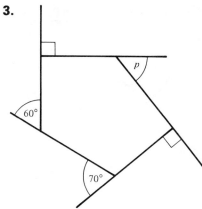

6.

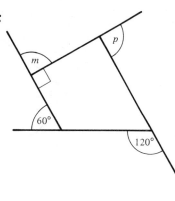

4.

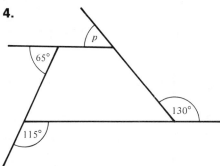

7.

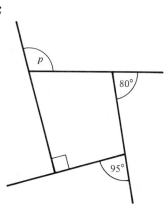

5.

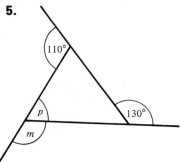

8.

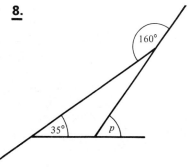

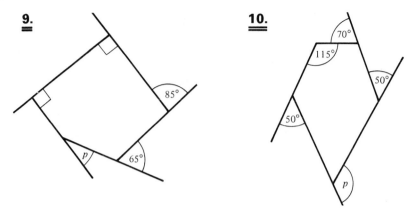

In questions 11 and 12 find the value of *x*.

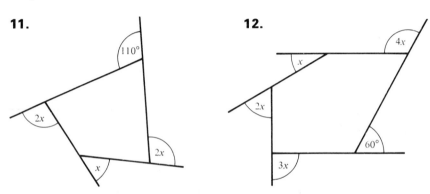

13. The exterior angles of a hexagon are *x*, 2*x*, 3*x*, 4*x*, 3*x* and 2*x*. Find the value of *x*.

14. Find the number of sides of a polygon if each exterior angle is   a) 72°   b) 45°.

THE EXTERIOR ANGLE OF A REGULAR POLYGON

If a polygon is regular, all its exterior angles are the same size. We know that the sum of the exterior angles is 360°, so the size of one exterior angle is easily found; we just divide 360° by the number of sides of the polygon, i.e.

> in a *regular* polygon with *n* sides, the size of an exterior angle is
>
> $$\frac{360°}{n}$$

EXERCISE 15d

Find the size of each exterior angle of a 24-sided regular polygon.

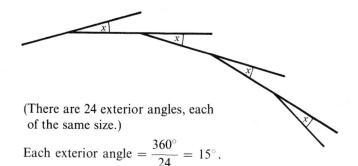

(There are 24 exterior angles, each of the same size.)

Each exterior angle $= \dfrac{360°}{24} = 15°$.

Find the size of each exterior angle of a regular polygon with:

1. 10 sides   **4.** 6 sides   **7.** 9 sides

2. 8 sides   **5.** 15 sides   **8.** 16 sides

3. 12 sides   **6.** 18 sides   **9.** 20 sides

THE SUM OF THE INTERIOR ANGLES OF A POLYGON

Consider an octagon:

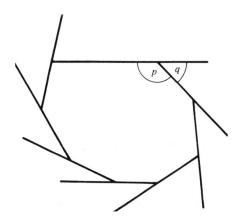

At each vertex there is an interior angle and an exterior angle and the sum of these two angles is $180°$ (angles on a straight line), i.e. $p + q = 180°$ at each one of the eight vertices.

Therefore, the sum of the interior angles and exterior angles together is

$$8 \times 180° = 1440°$$

The sum of the eight exterior angles is 360°.

Therefore, the sum of the interior angles is

$$1440° - 360° = 1080°$$

EXERCISE 15e

Find the sum of the interior angles of a 14-sided polygon.

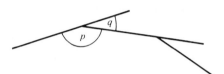

At each vertex $\qquad p + q = 180°$

$\therefore$ sum of interior angles and exterior angles is

$$14 \times 180° = 2520°$$

$\therefore$ sum of interior angles $\quad = 2520° - 360°$

$$= 2160°$$

Find the sum of the interior angles of a polygon with:

1. 6 sides   **4.** 4 sides   **7.** 18 sides

2. 5 sides   **5.** 7 sides   **8.** 9 sides

3. 10 sides   **6.** 12 sides   **9.** 15 sides

FORMULA FOR THE SUM OF THE INTERIOR ANGLES

If a polygon has n sides, then

the sum of the interior and exterior angles together is $n \times 180°$

$$= 180n°$$

so the sum of the interior angles only is $180n° - 360°$

which, as $360° = 180° \times 2$, can be written as $180°(n-2)$, i.e.

> in a polygon with *n* sides, the sum of the interior angles is
>
> $(180n - 360)°$ or $(n-2)180°$

EXERCISE 15f **1.** Find the sum of the interior angles of a polygon with
a) 20 sides b) 16 sides c) 11 sides.

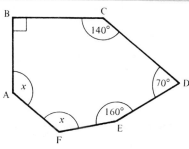

In the hexagon ABCDEF, the angles marked *x* are equal. Find the value of *x*.

The sum of the interior angles is $180° × 6 - 360° = 720°$

$$\therefore \quad 90° + 140° + 70° + 160° + 2x = 720°$$
$$460° + 2x = 720°$$
$$2x = 260°$$
$$x = 130°$$

In each of the following questions find the size of the angle(s) marked *x*:

2.

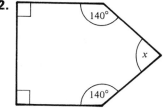

4.

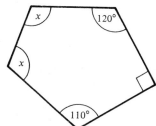

3.

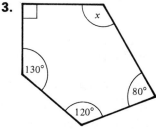

5.

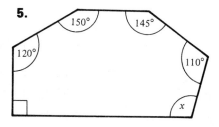

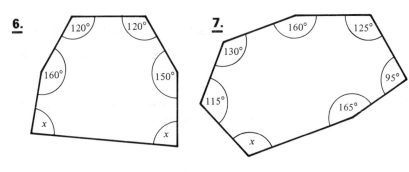

Find the size of each interior angle of a regular nine-sided polygon.

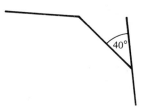

(As the polygon is regular, all the exterior angles are equal and all the interior angles are equal.)

Method 1 Sum of exterior angles $= 360°$
∴ each exterior angle $= 360° \div 9 = 40°$
∴ each interior angle $= 180° - 40° = 140°$

Method 2 Sum of interior angles $= 180° \times 9 - 360° = 1260°$
∴ each interior angle $= 1260° \div 9 = 140°$

Find the size of each interior angle of:

8. A regular pentagon.   **11.** A regular ten-sided polygon.

9. A regular hexagon.   **12.** A regular 12-sided polygon.

10. A regular octagon.   **13.** A regular 20-sided polygon.

14. How many sides has a regular polygon if each exterior angle is
a) 20°   b) 15°?

15. How many sides has a regular polygon if each interior angle is
a) 150°   b) 162°?   (Find the exterior angle first.)

16. Is it possible for each exterior angle of a regular polygon to be
a) 30°   b) 40°   c) 50°   d) 60°   e) 70°   f) 90°?
In those cases where it is possible, give the number of sides.

17. Is it possible for each interior angle of a regular polygon to be
a) 90°   b) 120°   c) 180°   d) 175°   e) 170°   f) 135°?
In those cases where it is possible, give the number of sides.

18. Construct a regular pentagon with sides 5 cm long. (Find the size of each interior angle and use your protractor.)

19. Construct a regular octagon of side 5 cm.

MIXED PROBLEMS

EXERCISE 15g

ABCDE is a pentagon, in which the interior angles at A and D are each $3x°$, the interior angles at B, C and E are each $4x°$. AB and DC are produced until they meet at F. Find $B\hat{F}C$.

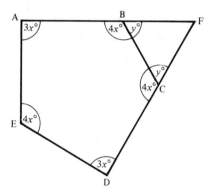

Sum of the interior angles of a pentagon $= 180° \times 5 - 360°$

$$= 540°$$

$\therefore \qquad 3x + 4x + 3x + 4x + 4x = 540$

$$18x = 540$$

$$x = 30°$$

$\therefore \qquad A\hat{B}C = 120° \qquad \text{and} \qquad B\hat{C}D = 120°$

$\therefore \qquad y = 60°$

$\therefore \qquad B\hat{F}C = 180° - 2 \times 60° \qquad \text{(angle sum of } \triangle BFC)$

$$= 60°$$

In questions 1 to 10 find the value of x:

1.

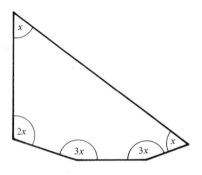

5.

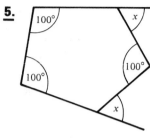

2.

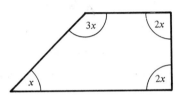

6.

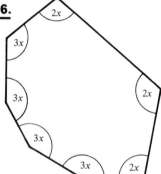

3.

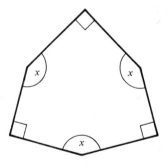

7.

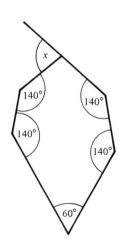

4.

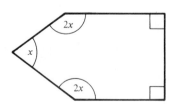

8.

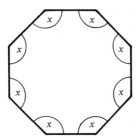

10.

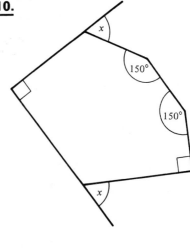

9.

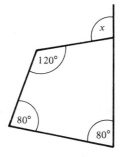

11. ABCDE is a regular pentagon.

OA = OB = OC = OD = OE.

Find the size of each angle at O.

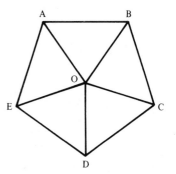

12. ABCDEFGH is a regular octagon. O is a point in the middle of the octagon such that O is the same distance from each vertex. Find AÔB.

13. ABCDEF is a regular hexagon. AB and DC are produced until they meet at G. Find BĜC.

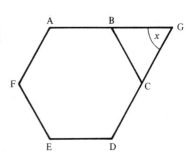

14. ABCDE is a regular pentagon. AB and DC are produced until they meet at F. Find BF̂C.

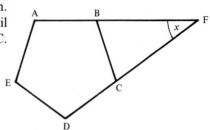

ABCDEF is a regular hexagon. Find AD̂B.

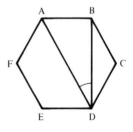

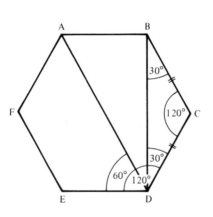

As ABCDEF is regular, the exterior angles are all equal.

$$\text{Each exterior angle} = 360° \div 6$$
$$= 60°$$
∴   each interior ∠ $= 180° - 60°$
$$= 120°$$

△BCD is isosceles (BC = DC).

∴   CB̂D = BD̂C = 30°   (angle sum of △BCD)

AD is a line of symmetry.

∴   ED̂A = CD̂A = 60°
∴   AD̂B = 60° − 30°
$$= 30°$$

In questions 15 to 20, each polygon is regular. Give answers correct to one decimal place where necessary:

15.

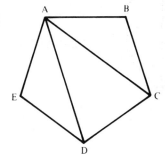

Find   a) A$\hat{C}$B   b) D$\hat{A}$C.

16.

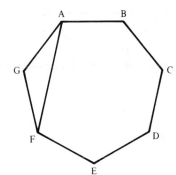

Find   a) A$\hat{G}$F   b) G$\hat{A}$F.

17.

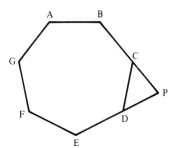

Find C$\hat{P}$D.

18.

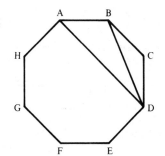

Find   a) C$\hat{B}$D   b) B$\hat{D}$A.

19.

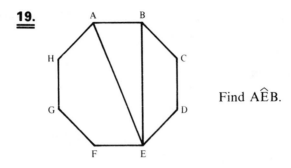

Find $A\widehat{E}B$.

20.

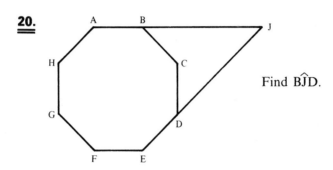

Find $B\widehat{J}D$.

PATTERN MAKING WITH REGULAR POLYHEDRA

Regular hexagons fit together without leaving gaps, to form a flat surface. We say that they *tessellate*.

Find $B\widehat{J}D$.

The hexagons tessellate because each interior angle of a regular hexagon is 120°, so three vertices fit together to make 360°.

EXERCISE 15h   1.

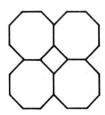

This is a pattern using regular octagons. They do not tessellate:

a) Explain why they do not tessellate.
b) What shape is left between the four octagons?
c) Continue the pattern. (Trace one of the shapes above, cut it out and use it as a template.)

2. Trace this regular pentagon and use it to cut out a template:

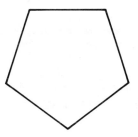

a) Will pentagons tessellate?

b)

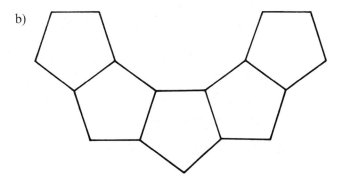

Use your template to copy and continue this pattern until you have a complete circle of pentagons. What shape is left in the middle?

c) Make up a pattern using pentagons.

3.

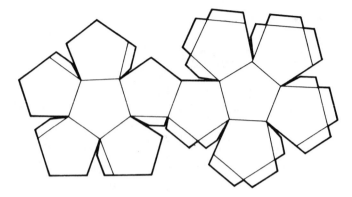

Use your template from question 2 to copy this net on to thick paper. Cut it out and fold along the lines. Stick the edges together using the flaps. You have made a regular dodecahedron.

4. Apart from the hexagon, there are two other regular polygons that tessellate. Which are they, and why?

5. Regular hexagons, squares and equilateral triangles can be combined to make interesting patterns. Some examples are given below:

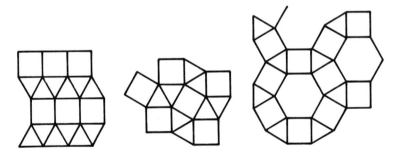

Copy these patterns and extend them. (If you make templates to help you, make each shape of side 2 cm.)

6. Make some patterns of your own using the shapes in question 5.

16 PROBABILITY

SINGLE EVENTS

If we toss an unbiased dice, each of the six numbers is equally likely to appear. We say that there are six equally likely *outcomes* to this experiment. The probability of scoring 2 is $\frac{1}{6}$, since out of six equally likely outcomes, only one is "successful", i.e. is a 2

We write
$$P(2) = \frac{1}{6}$$

Similarly
$$P(\text{even number}) = \frac{3}{6} = \frac{1}{2}$$

$$P(10) = 0$$

$$P(1 \text{ or } 2 \text{ or } 3 \text{ or } 4 \text{ or } 5 \text{ or } 6) = \frac{6}{6} = 1$$

$$P(\text{not } 2) = 1 - \frac{1}{6}$$

A successful event can be denoted by A and the probability of it happening is then $P(A)$. If A is "throwing a 2" then $P(A) = \frac{1}{6}$.

Summary

$$P(\text{successful event}) = \frac{\text{number of successful outcomes}}{\text{total number of possible outcomes}}$$

$P(\text{certainty}) = 1$
$P(\text{impossibility}) = 0$
$0 \leqslant P(A) \leqslant 1$
$P(\text{not } A) = 1 - P(A)$

EXERCISE 16a

On a bus, there are 48 passengers. 12 are men, 16 are women and 6 are girls. The rest are boys.
A passenger is chosen at random. What is the probability that the passenger is   a) a girl   b) male?

a) $P(\text{girl}) = \frac{6}{48} = \frac{1}{8}$

b) There are 16 women and 6 girls, i.e. 22 females, so there are $(48 - 22)$ males, i.e. 26 males.
$$P(\text{male}) = \frac{26}{48} = \frac{13}{24}$$

273

1. A bag contains four pink beads, three blue ones and two green ones. If one bead is drawn at random from the bag, find the probability that the bead is

a) pink   c) pink or blue or green

b) pink or blue   d) yellow.

2. If a card is drawn at random from a pack of 52, find the probability that it is

a) a king   b) a heart

c) a card which is neither a king nor a heart.

3. A two figure number is written down at random. Find the probability that

a) the number is greater than 68

b) the number is a multiple of 9.

4. A letter is picked at random from the English alphabet. Find the probability that the letter

a) is a consonant

b) comes from the part of the alphabet from G to N inclusive

c) appears in the word GEOMETRY.

5. A two figure number is written down at random. Find the probability that

a) it is less than 100

b) it contains at least one 5 (i.e. a number such as 15 or 55).

A board is marked as shown. A pin is stuck into the board at random. Find the probability that the pin is stuck into the shaded triangle.

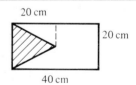

Area of board $= 20 \times 40 \, \text{cm}^2 = 800 \, \text{cm}^2$

Area of triangle $= \frac{1}{2}$ base $\times$ height

$= \frac{1}{2} \times 20 \times 20 \, \text{cm}^2 = 200 \, \text{cm}^2$

P(pin in shaded area) $= \frac{200}{800} = \frac{1}{4}$

6.

A square of side 4 cm is marked on a square board of side 6 cm. When a coin is thrown on to the board its centre is equally likely to fall anywhere on the board.

Find the probability that the centre of the coin falls

a) on the inner square
b) between the inner and outer squares.

7.

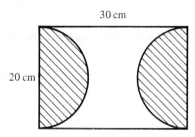

A rectangular board measuring 20 cm by 30 cm is marked as shown in the diagram. A pin is stuck into the board at random. Find the probability that the pin is stuck into the shaded region. (Give your answer as a decimal correct to 2 decimal places.)

8.

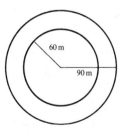

Two concentric circles, radii 90 m and 60 m, are drawn on the ground. A parachutist falls into the larger circle and is equally likely to land anywhere within it.

Find the probability that the parachutist lands

a) within the inner circle
b) between the inner and outer circles.

A box contains red and blue scarves. If a scarf is taken at random the probability that it is red is $\frac{4}{9}$. What is the probability of choosing a blue scarf?

$$P(\text{non-red scarf}) = 1 - \frac{4}{9}$$
$$= \frac{5}{9}$$

9. The probability of drawing a pearl bead out of a bag of mixed beads is $\frac{2}{3}$. What is the probability of drawing a bead which is not a pearl?

10. In a raffle the probability of buying a winning ticket is $\frac{2}{153}$. What is the probability of not winning a prize?

11. The probability that Jane wins a race is $\frac{3}{7}$. What is the probability that she loses?

ADDITION OF PROBABILITIES

If we select a card at random from a pack of 52, the probability of drawing a jack is $\frac{4}{52}$ and the probability of drawing a red four is $\frac{2}{52}$.

There are 4 jacks and 2 red fours so if we want to find the probability of drawing either a jack or a red four there are 6 cards which we would count as "successful".

The probability of drawing a jack or a red four is therefore $\frac{6}{52}$, which is the same as $\frac{4}{52} + \frac{2}{52}$ i.e. the *sum* of the separate probabilities.

We *add* the probabilities if there are several *separate* events we would count as "successful".

Notice though that the probability of drawing a jack or a heart is not $\frac{4}{52} + \frac{13}{52}$ because the jack of hearts is in both groups. The two events are not completely separate.

Sometimes it is useful to draw a diagram to represent the information about the probabilities.

If a box contains 3 red, 4 blue and 2 white beads and we draw one bead at random then $P(\text{red}) = \frac{3}{9}$, $P(\text{blue}) = \frac{4}{9}$ and $P(\text{white}) = \frac{2}{9}$; this information can be shown on a diagram.

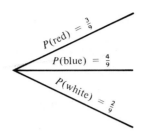

If we want either a red or a blue bead then we *add* the probabilities,

i.e. $$P(\text{red or blue}) = \frac{3}{9} + \frac{4}{9}$$
$$= \frac{7}{9}$$

Notice that all three probabilities add up to 1

i.e. $$P(\text{red or blue or white}) = \frac{3}{9} + \frac{4}{9} + \frac{2}{9}$$
$$= \frac{9}{9} = 1$$

We cannot draw a diagram to show the probabilities of picking a jack or a heart because they are not completely separate events.

This type of diagram is called a *tree diagram*; later we use it to show more information and it grows more branches.

FINDING THE NUMBER OF OUTCOMES

If we know that there are 24 beads in a bag and that the probability of drawing a red bead is $\frac{3}{8}$ then we know that the number of red beads is $\frac{3}{8}$ of 24, i.e. there are 9 red beads.

EXERCISE 16b

In a prize draw, the prizes are grocery hampers and garden centre vouchers. The possibility that the first ticket drawn is for a hamper is $\frac{1}{200}$ and the probability that it is for a voucher is $\frac{3}{200}$.

What is the probability that the first ticket is for

a) either a hamper or a voucher,
b) neither of these?

a) $P(\text{hamper or voucher}) = \frac{1}{200} + \frac{3}{200} = \frac{4}{200} = \frac{1}{50}$

b) $P(\text{neither a hamper nor a voucher}) = 1 - \frac{1}{50} = \frac{49}{50}$

1. A card is drawn at random from the 12 court cards (jacks, queens and kings).

What is the probability that the card is

a) a red queen
b) a black king
c) either a red queen or a black king?

2. Sophy is looking for her keys. The probability that she has put them in a pocket is $\frac{2}{9}$ and the probability that she has put them in her bag is $\frac{1}{3}$. Find the probability that

a) she has put them in a pocket or her bag
b) she has put them somewhere else.

3. The probability that Sean wins a race is $\frac{1}{10}$ and that Ewan wins is $\frac{1}{5}$. There are no dead heats.

What is the probability that

a) either Sean or Ewan wins the race
b) some other person wins the race?

The probability that the Blackwells' newspaper will be delivered before 8 a.m. is $\frac{3}{16}$, between 8 and 9 a.m. is $\frac{1}{2}$ and after 9 a.m. is $\frac{5}{16}$.

a) Draw a tree diagram to show this information.
b) Find the probability that the paper will be delivered at or after 8 a.m.

a)

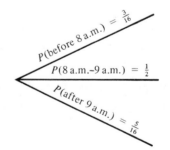

$P(\text{before 8 a.m.}) = \frac{3}{16}$

$P(\text{8 a.m.–9 a.m.}) = \frac{1}{2}$

$P(\text{after 9 a.m.}) = \frac{5}{16}$

b) $P(\text{at or after 8 a.m.}) = \frac{1}{2} + \frac{5}{16}$

$= \frac{13}{16}$

In questions 4 to 6, draw a tree diagram to show the given information.

4. Tony is waiting at a bus-stop, where only buses numbered 4, 25 or 72 stop. The probability is $\frac{2}{7}$ that the next bus is a 4, $\frac{3}{7}$ that it is a 25 and $\frac{2}{7}$ that it is a 72.

Find the probability that the next bus is

a) either a 4 or a 25   b) either a 4 or a 72.

5. When I draw the curtains in the morning the probability is $\frac{1}{8}$ that the first bird I see is a blackbird, $\frac{1}{4}$ that it is a blue-tit and $\frac{5}{8}$ that it is a sparrow.

Find the probability that the first bird I see is

a) either a blackbird or a sparrow

b) either a blue-tit or a blackbird.

6. Danny goes to work by bus or cycle or on foot. The probability that he chooses to go by bus is $\frac{5}{12}$, on his cycle is $\frac{1}{4}$ and on foot is $\frac{1}{3}$.

Find the probability that on Monday he chooses to go

a) by bus or cycle

b) by bus or on foot.

A board is divided into 16 small squares and some of these are coloured grey.
If a small square is picked at random, the probability that it is grey is $\frac{3}{4}$.
How many squares are grey?

$$\text{Number of grey squares} = \frac{3}{4} \times 16$$
$$= 12$$

---

7. The probability of drawing a king from a handful of 12 cards is $\frac{1}{6}$. How many kings are there in the hand?

8. In a car park there is a probability of $\frac{2}{7}$ that a car picked at random is red. There are 112 cars in the car park. How many of them are not red?

9. From a handful of cards the probability of drawing a club at random is $\frac{2}{3}$. There are 6 clubs. How many other cards are there?

10. In a cupboard there are only history and geography textbooks. There is a probability of $\frac{5}{8}$ that a book taken at random is on history. There are 30 geography books. How many history books are there?

TWO EVENTS

When two coins are tossed, the set of possible outcomes is

$$\{HH, \ HT, \ TH, \ TT\}$$

Each outcome is equally likely so we can see that $P(2 \text{ heads}) = \frac{1}{4}$.

By listing the possible equally likely outcomes the required probability can be found. However, if two dice are tossed the list of possibilities is long and confusing. In Book 2A we introduced an organised way of setting out the list so that there is no risk of missing any outcome. This is revised in the next section.

POSSIBILITY SPACES

We list the outcomes in the following table using crosses.

		First dice					
		1	2	3	4	5	6
Second dice	1	×	×	×	⊗	×	×
	2	×	×	⊗	×	×	☒
	3	×	⊗	×	×	☒	×
	4	⊗	×	×	☒	×	×
	5	×	×	☒	×	×	×
	6	×	☒	×	×	×	×

The table shows that there are 36 possible outcomes.

To find the probability of getting a total score of 5 we ring the crosses that mark the outcomes such as $1 + 4$, $2 + 3$ and so on.

Hence $$P(5) = \frac{4}{36} = \frac{1}{9}$$

Similarly, we can find the possibility of scoring, say, 8. We put squares around the outcomes that total 8.

$$P(8) = \frac{5}{36}$$

EXERCISE 16c In each question from 1 to 4 draw a possibility space to show the outcomes when two dice are tossed and use it to find the required probabilities. Use different marks such as a ring and a square, or different colours, for the first two parts of each question.

1. Find the probability that
   a) the sum of the two numbers is 8
   b) the difference between the two numbers is 3
   c) the sum is 8 and the difference is 3.

2. Find the probability that
   a) prime numbers appear on both dice
   b) at least one prime number appears
   c) only one prime number appears.

3. Find the probability that
   a) the sum of the two numbers is 7 or more
   b) the difference between the two numbers is 2 or less
   c) the sum of the two numbers is 7 or more and their difference is 2 or less.

4. Find the probability that
   a) an even number appears on both dice
   b) an odd number appears on both dice
   c) an even number less than 6 appears on both dice.

5. A four-sided spinner has the numbers 1 to 4 marked on it. It is spun twice and the two scores are noted.
   Find the probability that
   a) the total score is odd
   b) the two separate scores are both odd
   c) the product of the scores is odd.

6. I have two bags each containing four tulip bulbs and I know that each contains a pink, a red, a yellow and a white bulb. If I take one bulb at random from each bag, find the probability that
   a) the tulips will be the same colour
   b) the tulips will be of different colours.

7. One bag contains two 5p coins and one 10p coin. Another bag contains one 5p coin and three 10p coins. One coin is removed at random from each bag.
   Find the probability that the total value of the two coins is 15p.

PROBABILITY TREES

Suppose we have two discs, a red one marked A on one side and B on the other, and a blue one marked E on one side and F on the other.

Tossing the red disc, the probability that we get A is $\frac{1}{2}$ and the probability that we get B is also $\frac{1}{2}$.

This information can be shown in the following tree diagram.

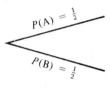

Red disc

Suppose that the red disc showed A and we go on to toss the blue disc. The probability of getting E is $\frac{1}{2}$ and the probability of getting F is $\frac{1}{2}$. We put this information on the diagram.

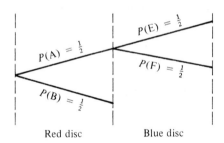

Red disc   Blue disc

We complete the diagram by considering what the probabilities are, supposing that the red disc shows a B before we toss the blue disc.

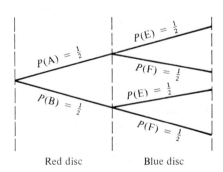

Red disc   Blue disc

To use the tree diagram to find the probability that we get an A and then an E, follow the path from left to right for an A on the first branch and an E on the second. The two probabilities we find there are $\frac{1}{2}$ and $\frac{1}{2}$. Multiply them together to get $\frac{1}{4}$.

To find the probability that we get a B on the red disc and an F on the blue one, follow the B and F path and multiply the probabilities, i.e.

$$P(\text{B and F}) = \frac{1}{2} \times \frac{1}{2}$$

$$= \frac{1}{4}$$

EXERCISE 16d

Two coins are tossed, one after the other. Find the probability that

a) both coins show heads

b) the first shows a tail and the second a head.

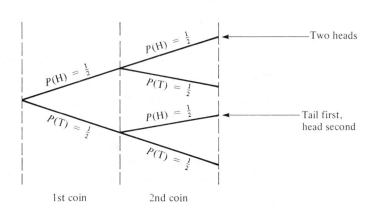

a) $P(\text{two heads}) = \frac{1}{2} \times \frac{1}{2}$

$$= \frac{1}{4}$$

b) $P(\text{first tail, second head}) = \frac{1}{2} \times \frac{1}{2}$

$$= \frac{1}{4}$$

In each question, draw a probability tree to show the given information.

1. The probability that Mark gets to work on time is $\frac{7}{8}$ and the probability that he leaves work on time is $\frac{3}{5}$.

a) Find the probability that he does not leave work on time.

b) Copy and extend the given probability tree.

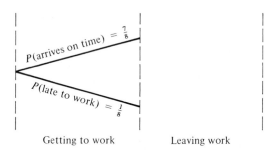

Getting to work   Leaving work

What is the probability that

c) Mark gets to work on time but does not leave on time

d) Mark is late for work but leaves on time?

2. When a drawing pin falls to the ground the probability that it lands point up is $\frac{1}{4}$.

a) Find the probability that a pin does not land point up.

Two drawing pins fall one after the other. Find the probability that

b) both drawing pins land point up

c) both drawing pins do not land point up.

3. The first of two boxes of tennis balls contains one white and two yellow; the second box contains three yellow and two lime green. A ball is taken at random from each box.

Find the probability that

a) both balls are yellow

b) one is white and one is lime green.

4. a) If a dice is rolled what is the probability of getting
   (i) a six   (ii) a number other than six?

b) Two dice, a red and a blue, are rolled. Find the probability that
   (i) both dice show sixes
   (ii) the red dice gives a six but the blue dice does not
   (iii) the blue dice gives a six but the red dice does not.

c) What is the total probability that just one six appears?

COMBINING PROBABILITIES

In the worked example on p 283, we see that there are two ways in which a tail and a head can appear. We have to follow two paths along the branches.

We *add* the probabilities resulting from these paths. Each new possible way of achieving the event increases the probability.

> We *multiply* the probabilities when we follow a path along the branches of the probability tree and *add* the results of following several paths.

EXERCISE 16e

The first of two parcels contains 3 French books and 2 German books. The second parcel contains 1 French book and 3 German books. Two books are taken at random, one from each parcel.

What is the probability that one book is French and one German?

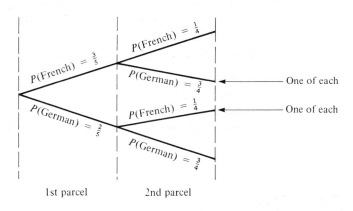

$P(\text{French}) = \frac{3}{5}$ $P(\text{French}) = \frac{1}{4}$

$P(\text{German}) = \frac{3}{4}$ ←———— One of each

$P(\text{German}) = \frac{2}{5}$ $P(\text{French}) = \frac{1}{4}$ ←———— One of each

$P(\text{German}) = \frac{3}{4}$

1st parcel   2nd parcel

(As the order of choosing the books does not matter, there are two paths on the tree that give a French and a German book.)

$$P(\text{one of each}) = \frac{3}{5} \times \frac{3}{4} + \frac{2}{5} \times \frac{1}{4}$$
$$= \frac{9}{20} + \frac{2}{20}$$
$$= \frac{11}{20}$$

For each question, draw a probability tree to illustrate the given information.

1. In a group of 6 girls, four are fair and two are dark. Of 5 boys, two are fair and three are dark. One boy and one girl are picked at random.

What is the probability, that of the two pupils picked, one is fair and one is dark?

2. In a class of 20, four are left-handed. In a second class of 24, six are left-handed. One pupil is chosen at random from each class.

What is the probability that one of the pupils is left-handed and one is not?

3. Derek and Alexis keep changing their minds about whether to send Christmas cards to each other. In any one year, the probability that Derek sends a card is $\frac{3}{4}$ and that Alexis sends one is $\frac{5}{6}$.

Find the probability that next year

a) they both send cards
b) only one of them sends a card
c) neither sends a card.

What should the three answers add up to?

4. Copy the probability tree in the worked example in Exercise 16d and, by adding more branches to the right, extend it to show the following information.

Three unbiased coins are tossed, one after the other.

Find the probability that

a) three heads appear
b) three tails appear
c) two heads and one tail appear in any order.

MIXED EXERCISE

EXERCISE 16f Unless the method is specified, choose the one you think is best for answering each question.

1. A letter is picked at random from the word CATASTROPHE

Find the probability that

a) the letter is a vowel
b) the letter is A or T.

2. A knitting wool sample card has 3 green, 1 black, 4 blue and 2 red samples.

    If one sample is picked at random, what is the probability that it is
    a) yellow
    b) black, green, red or blue?

3. The scores on a four-sided spinner are 1, 2, 3 or 4. On a second spinner the scores are 5, 6, 7 or 8.

    If the two are spun, find the probability that
    a) the sum of the two numbers is 9
    b) the product of the two numbers is a prime number
    c) the difference between the two numbers is 2.

4.

    A sector is chosen at random from each circle.

    What is the probability that
    a) both sectors picked are shaded
    b) one is shaded and the other not?

5. There are two bags. The first contains 2 white and 3 black marbles and the second contains 1 red and 2 blue marbles. Two marbles are drawn, one from each bag.
    a) Draw a possibility space to show this information.
    b) Hence find the probability that a white and a blue marble are drawn.
    c) Now draw a probability tree to show the information.
    d) Use the tree to find the answer to (b). Do your answers agree?

17 RATIO AND PROPORTION

The first few exercises are to remind you about ratios. Remember that ratios compare the sizes of related quantities.

SIMPLIFYING RATIOS

A ratio can be divided or multiplied throughout by the same number.

EXERCISE 17a

> Give the ratios
>
> a) $225 : 105$ b) $\frac{4}{7} : \frac{7}{8} : \frac{1}{2}$ in their simplest forms
>
> a) $225 : 105 = 45 : 21$ (dividing by 5)
>
> $= 15 : 7$ (dividing by 3)
>
> b) $\frac{4}{5} : \frac{7}{8} : \frac{1}{2} = \overset{8}{\cancel{40}} \times \frac{4}{\cancel{5}_1} : \overset{5}{\cancel{40}} \times \frac{7}{\cancel{8}_1} : \overset{20}{\cancel{40}} \times \frac{1}{\cancel{2}_1}$
>
> $= 32 : 35 : 20$

Give the following ratios in their simplest form:

1. $12 : 18$ **4.** $320 : 480$ **7.** $3.2 : 7.2$

2. $3 : 6 : 9$ **5.** $288 : 128 : 144$ **8.** $\frac{1}{2} : \frac{5}{6} : \frac{2}{3}$

3. $3.5 : 2.5$ **6.** $\frac{1}{2} : \frac{3}{4} : \frac{1}{4}$ **9.** $36 : 54 : 18$

> Which ratio is larger, $3 : 2$ or $14 : 9$?
>
> (To compare the sizes of the ratios, we write them as fractions and then convert them to equivalent fractions with a common denominator.)
>
> $\frac{3}{2} = \frac{27}{18}$ and $\frac{14}{9} = \frac{28}{18}$
>
> The second ratio is the larger.

288

Which ratio is the larger?

10. 6 : 11 or 2 : 5

11. 15 : 4 or 11 : 3

12. 20 : 3 or 31 : 4

13. 2 : 7 or 5 : 16

Express the ratio 7 : 5 in the form $n : 1$

$$7 : 5 = \frac{7}{5} : 1$$
$$= 1.4 : 1$$

Express the following ratios in the form $n : 1$, giving n correct to three significant figures where necessary:

14. 3 : 2

15. 12 : 5

16. 6 : 7

17. 30 : 11

18. 3 : 5

19. 21 : 8

20. 4 : 3

21. 3 : 4

22. 10 : 7

MIXED UNITS AND PROBLEMS

If we are asked to compare two quantities expressed in different units, we need to change one or both so that the two quantities are in the same unit. It is easier to change to smaller units (where multiplication is required) rather than to larger units (where division is required).

EXERCISE 17b

Simplify the ratio 63 cm : 0.72 m

$$63 \,\text{cm} : 0.72 \,\text{m} = 63 \,\text{cm} : 72 \,\text{cm}$$
$$= 7 : 8$$

Simplify the following ratios:

1. 45 cm : 0.1 m

2. 42 p : £1.05

3. 340 m : 1.2 km

4. 32 g : 2 kg

5. 450 mg : 1 g

6. 2.2 t : 132 kg

Find the ratio of 14 p per gram to £120 per kilogram

(In order to compare we will use both prices in £ per kg.)

$$14\text{p per g} = 14\,000\text{p per kg}$$
$$= £140 \text{ per kg}$$
$$14\text{p per g} : £120 \text{ per kg} = £140 \text{ per kg} : £120 \text{ per kg}$$
$$= 140 : 120$$
$$= 7 : 6$$

Find the ratios of the following prices:

7. 4 p per kilogram to £38 per tonne.
(First express 4 p per kg as a price per tonne.)

8. 6 p each to 70 p per dozen. (First find the cost per dozen.)

9. £16.20 per metre to 15 p per centimetre.

10. 72 p for twenty to 4 p each.

Give the ratio of the cost of 6 m of material at £2.40 per metre to the cost of 8 m at £2.20 per metre

$$\text{First cost} = £6 \times 2.40$$
$$= £14.40$$
$$\text{Second cost} = £8 \times 2.20$$
$$= £17.60$$

$$\text{Ratio of costs} = 14.40 : 17.60$$
$$= 144 : 176$$
$$= 18 : 22$$
$$= 9 : 11$$

11. In a school of 1029 pupils, 504 are girls. What is the ratio of the number of boys to the number of girls?

12. I spend £3.60 on groceries and £2.40 on vegetables. What is the ratio of the cost of   a) groceries to vegetables
b) vegetables to groceries   c) groceries to the total?

13. One rectangle has a length of 6 cm and a width of $4\frac{1}{2}$ cm. A second rectangle has a length of 9 cm and a width of $2\frac{1}{2}$ cm. Find the ratios of   a) their lengths   b) their widths
c) their perimeters   d) their areas.

14. Find the ratio of the cost of $12 \, \text{m}^2$ of carpet at £7.20 per m^2 to the cost of 50 carpet tiles at £2.40 per tile.

15.

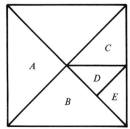

Find the ratios of the following areas

a) $B : A$ d) $E : D$

b) $C : B$ e) $E : C + D$

c) $E : A + B$ f) $C :$ whole square

16.

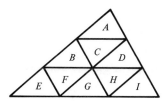

The areas of the small triangles are equal. Find the ratios of the following areas

a) $A :$ whole figure

b) $A : A + B + C + D$

c) $B + E + F + G :$ whole figure

FINDING MISSING QUANTITIES

If we are given the ratio $x : 6 = 3 : 5$ we may write this as $\dfrac{x}{6} = \dfrac{3}{5}$ and solve the equation for x.

It does not matter in which order we compare the two quantities as long as we are consistent. If the ratio is $4 : x = 5 : 3$ we may rewrite it as $x : 4 = 3 : 5$.

EXERCISE 17c

Find x if   a) $x : 6 = 13 : 15$ b) $4 : x = 3 : 2$

a)
$$\frac{x}{6} = \frac{13}{15}$$

$$\overset{1}{\cancel{6}} \times \frac{x}{\cancel{6}_1} = \overset{2}{\cancel{6}} \times \frac{13}{\cancel{15}_{5}}$$

$$x = \frac{26}{5}$$

$$= 5.2$$

b) $x : 4 = 2 : 3$ (rearrange the ratio so that x comes first)

$$\frac{x}{4} = \frac{2}{3}$$

$$4 \times \frac{x}{4} = 4 \times \frac{2}{3}$$

$$x = \frac{8}{3}$$

$$= 2\frac{2}{3}$$

Find x in questions 1 to 6:

1. $x : 5 = 2 : 9$ **3.** $x : 6 = 5 : 4$ **5.** $3 : 8 = 9 : x$

2. $x : 3 = 1 : 7$ **4.** $5 : x = 7 : 2$ **6.** $15 : 2 = x : 3$

Complete the following ratios:

7. $4 :\ \ = 3 : 7$ **9.** $3 :\ \ = 5 : 2$ **11.** $6 : 5 = 4 :$

8. $\ \ : 5 = 6 : 11$ **10.** $9 : 5 =\ \ : 4$ **12.** $\ \ : 12 = 5 : 7$

In a town, the ratio of the number of males to the number of females is $80 : 81$. There are 9680 males. How many females are there? What is the total population?

Suppose there are x females,

then $x : 9680 = 81 : 80$ (Notice the change of order.)

$$\frac{x}{9680} = \frac{81}{80}$$

$$\overset{1}{\cancel{9680}} \times \frac{x}{\underset{1}{\cancel{9680}}} = \overset{121}{\cancel{9680}} \times \frac{81}{\underset{1}{\cancel{80}}}$$

$$x = 9801$$

There are 9801 females and the total population is 19 481

13. The numbers of Mr and Mrs James' grandsons and grand-daughters are in the ratio 4 : 3. There are nine granddaughters. How many grandsons are there? What is the ratio of the number of granddaughters to the number of grandchildren?

14. The ratio of the number of cats to the number of dogs owned by the children in one year group in a school is 5 : 3. There are 95 cats. How many cats and dogs are there altogether?

15. The ratio of the lengths of two rectangles is 6 : 5. The length of the second is 8.4 cm. What is the length of the first?

16. The ratio of the numbers of orange flowers to white flowers in a garden is 6 : 11. There are 144 orange flowers. How many white flowers are there?

DIVISION IN A GIVEN RATIO

EXERCISE 17d

Share 72 p amongst three people so that their shares are in the ratio 3 : 4 : 5

There are 12 portions (i.e. 3 + 4 + 5).

$$\text{First share} = \frac{3}{12} \times 72\,\text{p}$$

$$= 18\,\text{p}$$

$$\text{Second share} = \frac{4}{12} \times 72\,\text{p}$$

$$= 24\,\text{p}$$

$$\text{Third share} = \frac{5}{12} \times 72\,\text{p}$$

$$= 30\,\text{p}$$

Check: 18 p + 24 p + 30 p = 72 p

1. Divide £45 into two parts in the ratio 4 : 5

2. Divide 96 m into two parts in the ratio 9 : 7

3. Divide 5 kg into three parts in the ratio 1 : 2 : 5

4. Divide seven hours into three parts in the ratio $1:5:8$

5. There are 32 children in a class. The ratio of the number of boys to the number of girls is $9:7$. How many boys and how many girls are there?

6. The angles of a triangle are in the ratio $6:5:7$. Find the sizes of the three angles.

7. Share the contents of a box containing 30 chocolates amongst Anne, Mary and Sue in the ratio $3:4:3$. How many chocolates will each get?

8. A marksman fires at a target and the ratio of hits to misses is $11:4$. He fires 90 times. How many hits does he score and how many times does he miss?

MIXED QUESTIONS

EXERCISE 17e **1.** Simplify the ratio $324:252$

2. Divide $72\,\text{m}$ into two parts in the ratio $5:7$

3. Complete the ratio $4:7 = 3:$

4. Find x if $4:x = 9:5$

5. The ratio of two lengths is $4:5$. If the first length is $22\,\text{cm}$ what is the second?

6. If $p:q = 2:3$, what is $5p:2q$?

7. Find the ratio of $2\,\text{p}$ per gram to £2.12 per kilogram

8. Simplify $4\frac{2}{3}:3\frac{1}{2}$

9. In a block of flats, 24 have two bedrooms and 32 have one bedroom. Give the ratio of the number of two-bedroomed flats to the number of one-bedroomed flats.

10. Simplify the ratio $3.2:4.8$

SIMPLE DIRECT PROPORTION

If we know the cost of *one* article, we can easily find the cost of ten similar articles, or if we know what someone is paid for *one* hour's work, we can find what the pay is for five hours.

EXERCISE 17f

> If $1\,cm^3$ of lead weighs $11.3\,g$, what is the weight of
> a) $6\,cm^3$ b) $0.8\,cm^3$?
>
> $$1\,cm^3 \text{ weighs } 11.3\,g$$
>
> a) $6\,cm^3$ weigh $11.3 \times 6\,g = 67.8\,g$
>
> b) $0.8\,cm^3$ weigh $11.3 \times 0.8\,g = 9.04\,g$

1. The cost of $1\,kg$ of sugar is $90\,p$. What is the cost of
a) $3\,kg$ b) $12\,kg$?

2. In one hour an electric fire uses $1\frac{1}{2}$ units. Find how much it uses
in   a) four hours   b) $\frac{1}{2}$ hour.

3. One litre of petrol takes a car $18\,km$. At the same rate, how far
does it travel on   a) four litres   b) 6.6 litres ?

4. A knitting pattern states that, at the correct tension, five rows
measure $1\,cm$. How many rows must be knitted to measure
a) $7\,cm$ b) $8.4\,cm$?

5. The cost of $1\,kg$ of mushrooms is £3.30. Find the cost of
a) $\frac{1}{2}\,kg$ b) $2.4\,kg$.

We can reverse the process and, for instance, find the cost of *one*
article if we know the cost of three similar articles.

EXERCISE 17g

> $18\,cm^3$ of copper weigh $162\,g$. What is the weight of $1\,cm^3$?
>
> $$18\,cm^3 \text{ weigh } 162\,g$$
>
> $$1\,cm^3 \text{ weighs } \frac{162}{18}\,g = 9\,g$$

1. Six pens cost £7.20. What is the cost of one pen?

2. A car uses eight litres of petrol to travel $124\,km$. At the same
rate, how far can it travel on one litre?

3. A man walks steadily for three hours and covers $13\,km$. How
far does he walk in one hour?

4. Dress material cost £14.60 for $4\,m$. What is the cost of $1\,m$?

5. A carpet costs £117.60. Its area is 12 m². What is the cost of 1 m²?

We can use the same process even if the quantities are not a whole number of units.

The mass of 0.6 cm³ of a metal is 3 g. What is the mass of 1 cm³?

The mass of 0.6 cm³ is 3 g

The mass of 1 cm³ is $\dfrac{3}{0.6}$ g = 5 g

---

6. The cost of 2.8 m of material is £11.76. What is the cost of 1 m?

7. 8.6 m² of carpet cost £71.38. What is the cost of 1 m²?

8. The cost of running a refrigerator for 3.2 hours is 4.8 p. What is the cost of running the refrigerator for one hour?

9. A bricklayer takes 0.8 hours to build a wall 1.2 m high. How high a wall (of the same length) could he build in one hour?

10. A piece of webbing is 12.4 cm long and its area is 68.2 cm². What is the area of a piece of this webbing that is 1 cm long?

DIRECT PROPORTION

If two varying quantities are always in the same ratio, they are said to be *directly proportional* to one another (or sometimes simply *proportional*).

For example, when buying pens which each cost the same amount, the total cost is proportional to the number of pens. The ratio of the cost of 11 pens to the cost of 14 pens is 11 : 14, and if we know the cost of 11 pens, we can find the cost of 14 pens.

One method for solving problems involving direct proportion uses ratio, another uses the ideas in the last two exercises. This method is called the *unitary* method because it makes use of the cost of *one* article or the time taken by *one* man to complete a piece of work.

EXERCISE 17h

If the mass of $16\,\text{cm}^3$ of a metal is $24\,\text{g}$, what is the mass of $20\,\text{cm}^3$?

First Method (using ratios)

Let the mass of $20\,\text{cm}^3$ be x grams

Then $\qquad x : 24 = 20 : 16 \qquad$ (The ratio of the masses
$\qquad\qquad\qquad\qquad\qquad\qquad\qquad$ = the ratio of the volumes)

$$\frac{x}{24} = \frac{20}{16}$$

$$\overset{1}{\cancel{24}} \times \frac{x}{\cancel{24}_1} = \overset{3}{\cancel{24}} \times \frac{\overset{10}{\cancel{20}}}{\cancel{16}_{\cancel{2}_1}}$$

$$x = 30$$

The mass of $20\,\text{cm}^3$ is $30\,\text{g}$.

Second Method (unitary method)

(Write the first sentence so that it ends with the quantity you want, i.e. the mass.)

$16\,\text{cm}^3$ has a mass of $24\,\text{g}$.

$1\,\text{cm}^3$ has a mass of $\dfrac{24}{16}\,\text{g}$. $\qquad$ (There is no need to work out
$\qquad\qquad\qquad\qquad\qquad\qquad\qquad\qquad$ the value of $\dfrac{24}{16}$ yet.)

$20\,\text{cm}^3$ has a mass of $\overset{10}{\cancel{20}} \times \dfrac{\overset{3}{\cancel{24}}}{\cancel{16}_{\cancel{2}_1}}\,\text{g} = 30\,\text{g}$

1. At a steady speed a car uses four litres of petrol to travel $75\,\text{km}$. At the same speed how much petrol is needed to travel $60\,\text{km}$?

2. A hiker walked steadily for four hours, covering $16\,\text{km}$. How long did he take to cover $12\,\text{km}$?

3. An electric fire uses $7\frac{1}{2}$ units in three hours. How many units does it use in five hours?

4. How long does the same electric fire take to use 9 units?

5. A rail journey of 300 miles costs £54. At the same rate per mile
a) what would be the cost of travelling 250 miles?
b) how far could you travel for £63?

6. It costs £162 to turf a lawn of area 63 m². How much would it cost to turf a lawn of area 56 m²?

7. A machine in a soft drinks factory fills 840 bottles in six hours. How many could it fill in five hours?

8. A 6 kg bag of sprouts costs 198 p. At the same rate, what would an 8 kg bag cost?

9. A knitting pattern states that the correct tension is such that 55 rows measure 10 cm. How many rows should be knitted to give 12 cm?

10. A scale model of a ship is such that the mast is 9 cm high and the mast of the original ship is 12 m high. The length of the original ship is 27 m. How long is the model ship?

Either method will work, whether the numbers are complicated or simple. Even if the question is about something unfamiliar, it is sufficient to know that the quantities are proportional.

In a spring balance, the extension in the spring is proportional to the load. If the extension is 2.5 cm when the load is 8 newtons, what is the extension when the load is 3.6 newtons?

Ratio Method

Let the extension be x cm.

$$x : 2.5 = 3.6 : 8$$

$$\frac{x}{2.5} = \frac{3.6}{8}$$

$$2.5 \times \frac{x}{2.5} = 2.5 \times \frac{3.6}{8}$$

$$x = 1.125$$

The extension is 1.125 cm.

> *Unitary Method*
>
> If a load of 8 newtons gives an extension of 2.5 cm,
> then a load of 1 newton gives an extension of $\frac{2.5}{8}$ cm.
> ∴ a load of 3.6 newtons gives an extension of
>
> $$3.6 \times \frac{2.5}{8} \text{ cm} = 1.125 \text{ cm}$$

11. It costs £392 to hire scaffolding for 42 days. How much would it cost to hire the same scaffolding for 36 days at the same rate per day?

12. The rates of currency exchange published in the newspapers on a certain day showed that 14 kroner could be exchanged for 210 pesos. How many pesos could be obtained for 32 kroner?

13. At a steady speed, a car uses 15 litres of petrol to travel 164 km. At the same speed, what distance could be travelled if six litres were used?

14. If a 2 kg bag of sugar contains 9×10^6 crystals, how many crystals are there in a) 5 kg   b) 1.8 kg   c) 0.03 kg?

15. The current flowing through a lamp is proportional to the voltage across the lamp. If the voltage across the lamp is ten volts the current is 0.6 amps. What voltage is required to make a current of 0.9 amps flow?

16. The amount of energy carried by an electric current is proportional to the number of coulombs. If five coulombs carry 19 joules of energy, how many joules are carried by 6.5 coulombs?

17. A recipe for date squares uses the following quantities:

Ingredients	Costs
125 g of brown sugar	500 g cost 76 p
75 g of oats	750 g cost 102 p
75 g of flour	$1\frac{1}{2}$ kg cost 88p
100 g of margarine	250 g cost 36 p
100 g of dates	250 g cost 84 p
Pinch of bicarbonate of soda	—
Squeeze of lemon juice	1 p

Find the cost of making these date squares as accurately as possible, then give your answer correct to the nearest penny.

18. A do-it-yourself enthusiast makes a base for a table.

Materials	Costs
4 legs each 30 cm long	2 m cost £3.20
4 stretchers each 70 cm long	3 m cost £2.10
4 stretchers each 35 cm long	2 m cost £1.50
3 pieces each 80 cm long	3 m cost £6.30
$\frac{3}{4}$ litre of varnish	1 litre costs £4.80
12 screws	20 screws cost 80 p

What is the total cost of the materials that are actually used?

INVERSE PROPORTION

Some quantities are not directly proportional to one another, although there is a connection between them. As one increases in size, the other may decrease, so that the reciprocal, or inverse, of the second is proportional to the first.

Suppose, for example, that a certain amount of food is available for several days. If each person eats the same amount each day, the more people there are, the shorter is the time that the food will last. The number of days the food will last is *inversely proportional* to the number of people eating it.

EXERCISE 17i   In this exercise, assume that the rates are constant.

Four bricklayers can build a certain wall in ten days. How long would it take five bricklayers to build it?

Ratio Method

Suppose it takes five bricklayers x days to build it.

(Five bricklayers will take a shorter time so we use the inverse ratio.)

$$\frac{x}{10} = \frac{4}{5}$$

$$\overset{1}{\cancel{10}} \times \frac{x}{\cancel{10}_1} = \overset{2}{\cancel{10}} \times \frac{4}{\cancel{5}_1}$$

$$x = 8$$

It would take them 8 days.

Unitary Method

Four bricklayers take 10 days.
One bricklayer would take 40 days.

Five bricklayers would take $\frac{40}{5}$ days

$$= 8 \text{ days}$$

---

1. Eleven taps fill a tank in three hours. How long would it take to fill the tank if only six taps are working?

2. Nine children share out equally the chocolates in a large tin and get eight each. If there were only six children, how many would each get?

3. The length of an essay is 174 lines with an average of 14 words per line. If it is rewritten with an average of 12 words per line, how many lines will be needed?

4. A field of grass feeds 24 cows for six days. How long would the same field feed 18 cows?

5. The dimensions of a block of stamps are 30 cm wide by 20 cm high. The same number of stamps could also have been arranged in a block 24 cm wide. How high would this second block be?

6. A batch of bottles were packed in 25 boxes taking 12 bottles each. If the same batch had been packed in boxes taking 15 each, how many boxes would be filled?

7. When knitting a scarf 48 stitches wide, one ball of wool will give a length of 18 cm. If there had been 54 stitches instead, how long a piece would the same ball give?

8. In a school, 33 classrooms are required if each class has 32 pupils. How many classrooms would be required if the class size was reduced to 22?

9. A factory requires 42 machines to produce a given number of articles in 63 days. How many machines would be required to produce the same number of articles in 54 days?

MIXED QUESTIONS

EXERCISE 17j   Some of the following questions cannot be answered because the quantities are neither in direct nor in inverse proportion. In these cases write "There is no answer". For those questions that can be solved, give answers correct to three significant figures where necessary:

1.   The list of exchange rates states that £1 = 12 French francs and £1 = 2300 lira, so that 12 francs = 2300 lira:
a) How many lira can 54 francs be exchanged for?
b) How many francs are 1000 lira worth?

2.   A man earned £30.60 for an eight-hour day. How much would he earn at the same rate for a 38-hour week?

3.   A typist typed 3690 words in $4\frac{1}{2}$ hours. How long would it take to type 2870 words at the same rate?

4.   At the age of twelve, a boy is 1.6 m tall. How tall will he be at the age of eighteen?

5.   A ream of paper (500 sheets) is 6.2 cm thick. How thick is a pile of 360 sheets of the same paper?

6.   If I buy balloons at 14 p each, I can buy 63 of them. If the price of a balloon increases to 18 p, how many can I buy for the same amount of money?

7.   A boy's mark for a test is 18 out of a total of 30 marks. If the test had been marked out of 40 what would the boy's mark have been?

8.   Twenty-four identical mathematics text books occupy 60 cm of shelf space. How many books will fit into 85 cm?

9.   A lamp post 4 m high has a shadow 3.2 m long cast by the sun. A man 1.8 m high is standing by the lamp post. At the same moment, what is the length of his shadow?

10.   A contractor decides that he can build a barn in nine weeks using four men. If he employs two more men, how long will the job take? Assume that all the men work at the same rate.

11.   A girl twelve years old gained 27 marks in a competition. How many marks did her six-year-old sister gain?

12.   For a given voltage, the current flowing is inversely proportional to the resistance. When the current flowing is 2.5 amps the resistance is 0.9 ohms. What is the current when the resistance is 1.5 ohms?

MIXED EXERCISES

EXERCISE 17k

1. Simplify the ratio $7.35 : 2.45$

2. Complete the ratio $: 9 = 2 : 5$

3. Divide 56 m into three parts in the ratio $1 : 2 : 4$

4. A car uses seven litres of petrol for a 100 km journey. At the same rate, how far could it go on eight litres?

5. Eight typists together could complete a task in five hours. If all the typists work at the same rate, how long would six typists take?

6. Simplify the ratio $7\frac{1}{2} : 2\frac{1}{2} : 1\frac{1}{4}$

7. The ratio of the numbers of eleven-year-olds to twelve-year-olds in a class is $8 : 3$. There are 24 eleven-year-olds. How many twelve-year-olds are there?

8. Give the ratio $6 : 5$ in the form $n : 1$

EXERCISE 17l

1. Simplify the ratio $1024 : 768$

2. Share 40 sweets amongst three people so that their shares are in the ratio $3 : 2 : 5$

3. Find x if $3 : 5 = x : 11$

4. Simplify the ratio $9\frac{1}{7} : 8$

5. Express the ratio $3 : 5$ in the form $n : 1$

6. The ratio of two sums of money is $4 : 5$. The first sum is £6. What is the second?

7. Mrs Jones buys 19 apples and five oranges. Mrs Brown buys five apples and three oranges. Find the ratio of the total numbers of oranges to apples.

8. A typist charges £25 for work which took her six hours. How much would she charge for nine hours' work at the same rate?

18 TRIGONOMETRY

TANGENT OF AN ANGLE

In Book 2A, we worked with right-angled triangles. This chapter is either a reminder of the work you did then or, if it is new to you, an introduction.

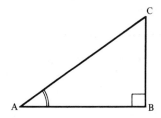

In triangle ABC, AC is the *hypotenuse*, opposite to the right angle.

BC is the *opposite side* to angle A.

AB is the *adjacent side* (or neighbouring side) to angle A.

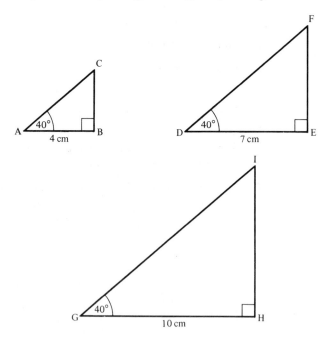

The three triangles above are similar, so their sides are proportional.

304

In particular, the opposite side and the adjacent side are in the same ratio in all three triangles and also in any other right-angled triangle containing an angle of 40°.

This ratio is called the *tangent* of 40° or, in shortened form, tan 40°. Its size is stored, together with the tangents of other angles, in some calculators.

$$\tan \widehat{A} = \frac{\text{opposite side}}{\text{adjacent side}}$$

EXERCISE 18a

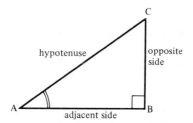

In questions 1 to 6, copy the diagram. Identify the hypotenuse and the sides opposite and adjacent to the marked angle:

1.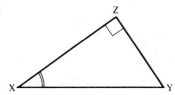

2.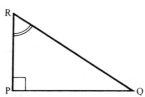

3.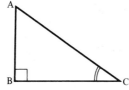

4.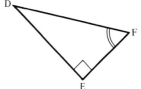

5.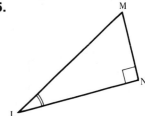

6.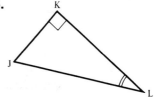

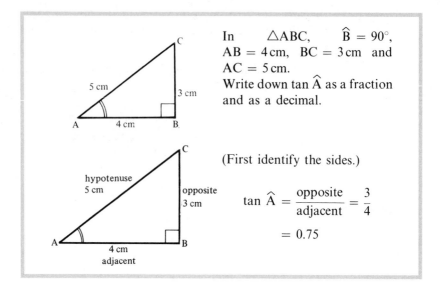

In   △ABC, $\widehat{B} = 90°$,
AB = 4 cm, BC = 3 cm and
AC = 5 cm.
Write down tan $\widehat{A}$ as a fraction
and as a decimal.

(First identify the sides.)

$$\tan \widehat{A} = \frac{\text{opposite}}{\text{adjacent}} = \frac{3}{4}$$

$$= 0.75$$

In each of the following questions write down the tangent of the marked angle as a fraction and as a decimal (correct to four decimal places where necessary):

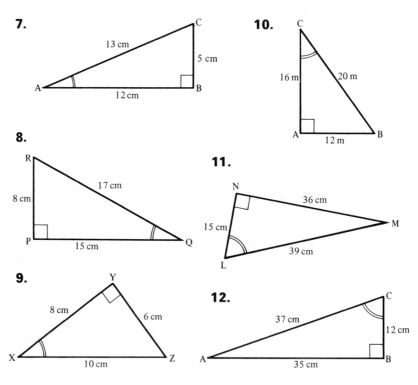

7.

8.

9.

10.

11.

12.

USING A CALCULATOR

To find the tangent of an angle, enter the size of the angle, then press the "tan" button. Write the answer correct to four decimal places.

$$\tan 42.4° = 0.9131$$

To find an angle given its tangent, enter the value of the tangent and then press the inverse button followed by the tangent button. Write down the size of the angle correct to one decimal place.

EXERCISE 18b

Find tan 38.2° and tan 80°

$$\tan 38.2° = 0.7869$$
$$\tan 80° = 5.671$$

Find the tangents of the following angles:

1. 62° **4.** 16.8° **7.** 78.4° **10.** 48.2°

2. 14° **5.** 4.6° **8.** 45° **11.** 3°

3. 30.5° **6.** 72° **9.** 30° **12.** 29.4°

Find the angle whose tangent is 0.628

$$\tan \widehat{A} = 0.628$$
$$\widehat{A} = 32.1°$$

Find the angles whose tangents are given in questions 13 to 24:

13. 0.179 **16.** 0.4326 **19.** 0.9213 **22.** 2.683

14. 0.356 **17.** 1.362 **20.** 0.8 **23.** 0.924

15. 1.43 **18.** 0.632 **21.** 0.3214 **24.** 0.0024

FINDING AN ANGLE

EXERCISE 18c

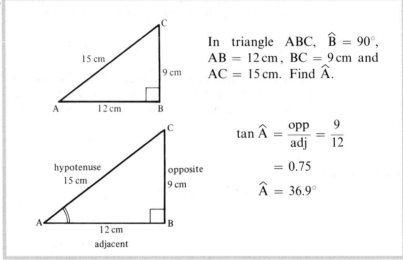

In triangle ABC, $\widehat{B} = 90°$, AB $= 12$ cm, BC $= 9$ cm and AC $= 15$ cm. Find $\widehat{A}$.

$$\tan \widehat{A} = \frac{\text{opp}}{\text{adj}} = \frac{9}{12}$$

$$= 0.75$$

$$\widehat{A} = 36.9°$$

Use the information given on the diagrams to find $\widehat{A}$:

1.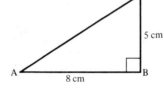

4.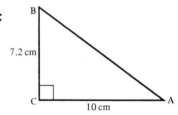

2.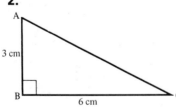

5.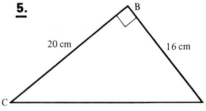

3.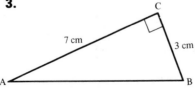

6.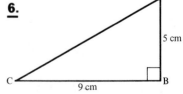

7. In triangle PQR, $\hat{P} = 90°$, QP = 6 cm and PR = 10 cm. Find $\hat{R}$.

8. In triangle XYZ, $\hat{Y} = 90°$, XY = 4 cm and YZ = 5 cm. Find $\hat{X}$.

9. In triangle LMN, $\hat{L} = 90°$, LM = 7.2 cm and LN = 6.4 cm. Find $\hat{N}$.

10. In triangle DEF, $\hat{D} = 90°$, DE = 210 cm and DF = 231 cm. Find $\hat{E}$.

11. In △ABC, $\hat{C} = 90°$, AC = 3.2 m and BC = 4.7 m. Find $\hat{B}$.

FINDING A SIDE

EXERCISE 18d

In triangle ABC, $\hat{B} = 90°$, AB = 4 cm and $\hat{A} = 32°$. Find BC.

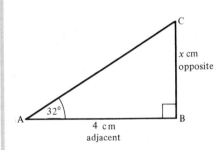

$$\frac{x}{4} = \frac{\text{opp}}{\text{adj}} = \tan 32°$$

$$\frac{x}{4} = 0.6249$$

$$\not{4} \times \frac{x}{\not{4}} = 4 \times 0.6249$$

$$x = 2.4996$$

∴ BC = 2.50 cm (to 3 s.f.)

Use the information given in the diagram to find the required side:

1. Find RQ.

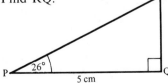

2. Find BC.

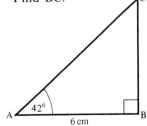

3. 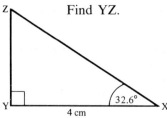 Find YZ.

5. 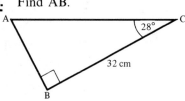 Find AB.

4. 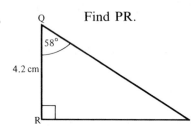 Find PR.

6. 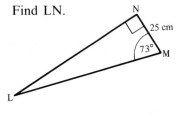 Find LN.

7. In triangle ABC, $\widehat{B} = 90°$, $\widehat{A} = 32°$ and $AB = 9\,cm$. Find BC.

8. In triangle DEF, $\widehat{D} = 90°$, $\widehat{E} = 48°$ and $DE = 20\,cm$. Find DF.

9. In triangle PQR, $\widehat{R} = 90°$, $\widehat{Q} = 10°$ and $RQ = 16\,cm$. Find PR.

10. In triangle XYZ, $\widehat{Z} = 90°$, $\widehat{Y} = 67°$ and $ZY = 3.2\,cm$. Find XZ.

In $\triangle ABC$, $\widehat{B} = 90°$, $\widehat{A} = 24°$ and $BC = 6\,cm$. Find AB.

(AB is opposite to $\widehat{C}$, so find $\widehat{C}$ first.)

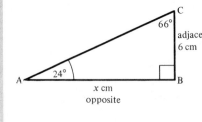

$\widehat{C} = 66°$ ($\angle$s of a triangle)

$$\frac{x}{6} = \frac{\text{opp}}{\text{adj}} = \tan 66°$$

$$\frac{x}{6} = 2.246$$

$$\cancel{6} \times \frac{x}{\cancel{6}} = 6 \times 2.246$$

$$x = 13.476$$

∴   AB is 13.5 cm   (to 3 s.f.)

Use the information given in the diagram to find the required side. It may be necessary to find the third angle of the triangle first.

11. Find AB.

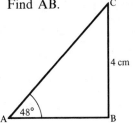

14. Find ZY.

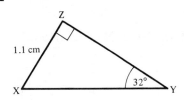

12. Find PQ.

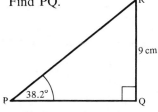

15. Find DE.

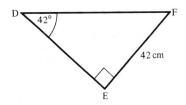

13. Find NL.

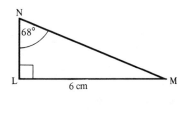

16. Find AC.

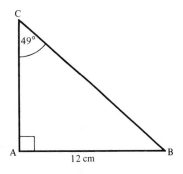

17. In triangle PQR, $\hat{P} = 90°$, $\hat{Q} = 52°$ and PR = 6 cm. Find QP.

18. In triangle ABC, $\hat{A} = 90°$, $\hat{B} = 31°$ and AC = 220 cm. Find AB.

19. In triangle XYZ, $\hat{Z} = 90°$, $\hat{X} = 67°$ and YZ = 2.3 cm. Find XZ.

20. In triangle LMN, $\hat{L} = 90°$, $\hat{M} = 9°$ and LN = 11 m. Find LM.

USING THE HYPOTENUSE

So far, we have used only the opposite and adjacent sides. If we wish to use the hypotenuse we need different ratios.

THE SINE OF AN ANGLE

For an angle in a right-angled triangle, the name given to the ratio $\dfrac{\text{opposite side}}{\text{hypotenuse}}$ is the *sine* of the angle.

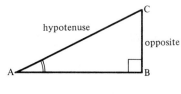

In triangle ABC $\dfrac{\text{BC}}{\text{AC}} = \sin \widehat{\text{A}}$

The use of sines is similar to the use of tangents.

EXERCISE 18e

Find the sine of 32.6°

$$\sin 32.6° = 0.5388$$

Find the sines of the following angles:

1.	62.4°	**5.**	15.2°	**9.**	82°
2.	70°	**6.**	37.5°	**10.**	27.8°
3.	14.3°	**7.**	59.6°	**11.**	15.8°
4.	9°	**8.**	30°	**12.**	87.2°

Find the angle whose sine is 0.6284

Let the angle be $\widehat{\text{A}}$ $\sin \widehat{\text{A}} = 0.6284$

$$\widehat{\text{A}} = 38.9°$$

Find the angles whose sines are given:

13. 0.271	**16.** 0.909	**19.** 0.614
14. 0.442	**17.** 0.6664	**20.** 0.7283
15. 0.524	**18.** 0.3720	**21.** 0.1232

EXERCISE 18f

In triangle ABC, $\widehat{B} = 90°$, BC = 3 cm and AC = 7 cm.
Find $\widehat{A}$.

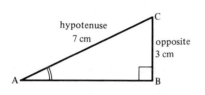

$$\sin \widehat{A} = \frac{\text{opp}}{\text{hyp}} = \frac{3}{7}$$

$$= 0.4286$$

$$\widehat{A} = 25.4°$$

Use the information given in the diagram to find the marked angle:

1.

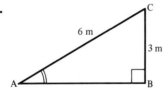

4.

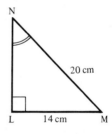

2.

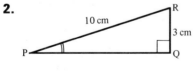

5.

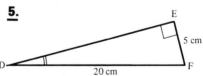

3.

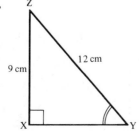

6.

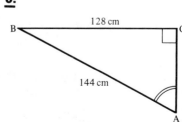

7. In triangle ABC, $\hat{C} = 90°$, BC = 7 cm and AB = 10 cm. Find $\hat{A}$.

8. In triangle PQR, $\hat{Q} = 90°$, PQ = 30 cm and PR = 45 cm. Find $\hat{R}$.

9. In triangle LMN, $\hat{M} = 90°$, MN = 3.2 cm and LN = 8 cm. Find $\hat{L}$.

10. In triangle DEF, $\hat{E} = 90°$, EF = 36 cm and DF = 108 cm. Find $\hat{D}$.

In triangle PQR, $\hat{P} = 90°$, $\hat{Q} = 32.4°$ and RQ = 4 cm. Find PR.

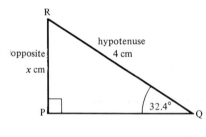

$$\frac{x}{4} = \frac{\text{opp}}{\text{hyp}} = \sin 32.4°$$

$$4 \times \frac{x}{4} = 4 \times 0.5358$$

$$x = 2.1432$$

∴   PR = 2.14 cm   (to 3 s.f.)

Use the information given in the diagram to find the required length:

11. Find AC.

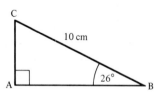

13. Find EF.

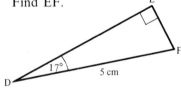

12. Find XY.

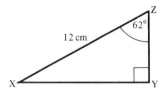

14. Find PR.

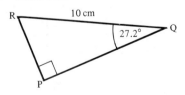

15. Find BC.

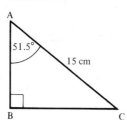

16. Find PR.

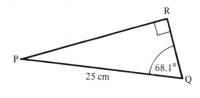

17. In triangle ABC, $\hat{A} = 90°$, BC = 11 cm and $\hat{C} = 35°$. Find AB.

18. In triangle PQR, $\hat{P} = 90°$, QR = 120 m and $\hat{Q} = 10.5°$. Find PR.

19. In triangle XYZ, $\hat{X} = 90°$, YZ = 3.6 cm and $\hat{Y} = 68°$. Find XZ.

20. In triangle DEF, $\hat{F} = 90°$, DE = 48 m and $\hat{D} = 72°$. Find EF.

THE COSINE OF AN ANGLE

For an angle in a right-angled triangle, the name given to the ratio $\dfrac{\text{adjacent side}}{\text{hypotenuse}}$ is the *cosine* of the angle.

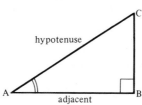

In triangle ABC $\qquad \dfrac{AB}{AC} = \cos \hat{A}$

EXERCISE 18g

Find the cosine of 51.6°

$$\cos 51.6° = 0.6211$$

Find the cosines of the following angles:

1. 32°

2. 41.8°

3. 82°

4. 47.8°

5. 60°

6. 15.6°

7. 52.1°

8. 49°

9. 72.3°

Find the angle whose cosine is 0.862

$$\cos \widehat{A} = 0.862$$
$$\widehat{A} = 30.5°$$

Find the angles whose cosines are given:

10. 0.347   **13.** 0.349   **<u>16.</u>** 0.865

11. 0.936   **14.** 0.6281   **<u>17.</u>** 0.014

12. 0.719   **15.** 0.3149   **<u>18.</u>** 0.0732

EXERCISE 18h

In triangle ABC, $\widehat{B} = 90°$, AC = 20 cm and AB = 15 cm. Find $\widehat{A}$.

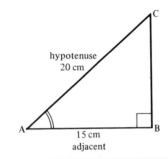

$$\cos \widehat{A} = \frac{\text{adj}}{\text{hyp}} = \frac{15}{20}$$
$$= 0.75$$
$$\widehat{A} = 41.4°$$

Find the marked angles in the following triangles:

1.

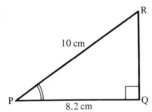

3.

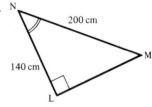

2.

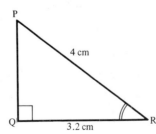

4.

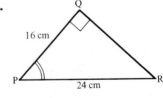

5.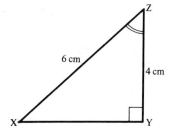

6.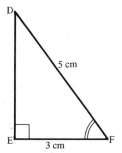

7. In triangle ABC, $\hat{B} = 90°$, $AB = 3.2\,cm$ and $AC = 5\,cm$. Find $\hat{A}$.

8. In triangle PQR, $\hat{P} = 90°$, $QR = 12\,cm$ and $PQ = 4.8\,cm$. Find $\hat{Q}$.

9. In triangle LMN, $\hat{L} = 90°$, $MN = 20\,cm$ and $ML = 3\,cm$. Find $\hat{M}$.

10. In triangle DEF, $\hat{F} = 90°$, $DE = 18\,cm$ and $DF = 16.2\,cm$. Find $\hat{D}$.

11. In triangle XYZ, $\hat{Z} = 90°$, $XY = 14\,m$ and $YZ = 11.6\,m$. Find $\hat{Y}$.

In triangle XYZ, $\hat{X} = 90°$, $\hat{Y} = 27°$, and $ZY = 3.2\,cm$. Find XY.

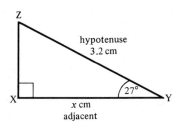

$$\frac{x}{3.2} = \frac{adj}{hyp} = \cos 27°$$

$$\frac{x}{3.2} = 0.8910$$

$$\cancel{3.2} \times \frac{x}{\cancel{3.2}} = 0.8910 \times 3.2$$

$$x = 2.8512$$

$\therefore$ $XY = 2.85\,cm$ (to 3 s.f.)

Use the information given in the diagrams to find the required lengths:

12. Find AB.

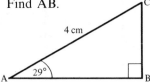

15. Find MN.

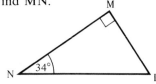

13. Find XY.

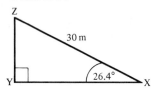

16. Find PQ.

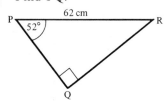

14. Find ED.

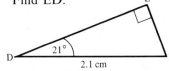

17. Find YZ.

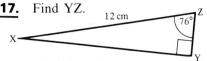

18. In triangle PQR, $\widehat{Q} = 90°$, $\widehat{P} = 31°$ and PR = 20 cm. Find PQ.

19. In triangle LMN, $\widehat{N} = 90°$, $\widehat{L} = 42°$ and LM = 3 cm. Find LN.

20. In triangle DEF, $\widehat{D} = 90°$, $\widehat{E} = 68°$ and EF = 11 cm. Find DE.

21. In triangle XYZ, $\widehat{Z} = 90°$, $\widehat{Y} = 15°$ and YX = 14 cm. Find ZY.

SUMMARY

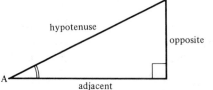

$$\text{Sin } \widehat{A} = \frac{\textbf{Opposite}}{\textbf{Hypotenuse}} \quad (\text{S O H})$$

$$\text{Cos } \widehat{A} = \frac{\textbf{Adjacent}}{\textbf{Hypotenuse}} \quad (\text{C A H})$$

$$\text{Tan } \widehat{A} = \frac{\textbf{Opposite}}{\textbf{Adjacent}} \quad (\text{T O A})$$

Some people remember by using the word "SOHCAHTOA" or a sentence like "Some Old Hangars Can Almost Hold Two Old Aeroplanes".

SINES, COSINES AND TANGENTS

EXERCISE 18i In questions 1 to 8, find the marked angles. Remember to label the given sides first and then decide which ratio you will have to use:

1.

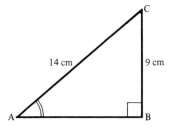

5.

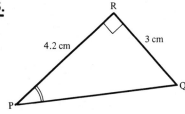

2.

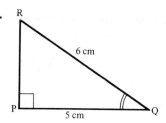

6.

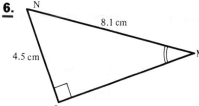

3.

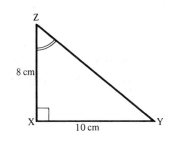

7.

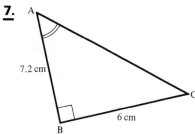

4.

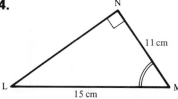

8.

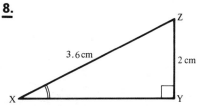

9.

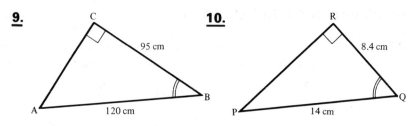

10.

11. In triangle ABC, $\hat{B} = 90°$, AC = 60 cm and BC = 22 cm. Find $\hat{C}$.

12. In triangle PQR, $\hat{R} = 90°$, PQ = 24 cm and QR = 6 cm. Find $\hat{P}$.

13. In triangle ABC, $\hat{B} = 90°$, AC = 1.5 cm and BC = 0.82 cm. Find $\hat{C}$.

14. In triangle PQR, $\hat{R} = 90°$, RQ = 8 cm and RP = 6.2 cm. Find $\hat{Q}$.

15. In triangle DEF, $\hat{F} = 90°$, DF = 16.2 cm and EF = 19.8 cm. Find $\hat{E}$.

16. In triangle XYZ, $\hat{X} = 90°$, YZ = 1.6 m and XY = 1.32 m. Find $\hat{Z}$.

17. In triangle DEF, $\hat{E} = 90°$, DE = 1.9 cm and EF = 2.1 cm. Find $\hat{F}$.

18. In triangle GHI, $\hat{H} = 90°$, GI = 52 cm and IH = 21 cm. Find $\hat{I}$.

Use the information given in the diagram to find the required length:

19. Find BC.

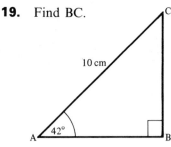

21. Find ZY.

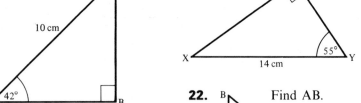

20. Find PQ.

22. Find AB.

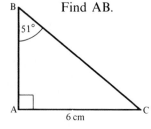

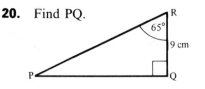

23. Find MN.

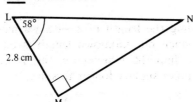

26. Find **AB**.

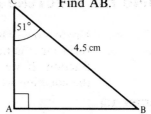

24. Find DE.

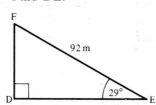

27. Find BC.

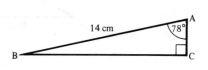

25. Find YZ.

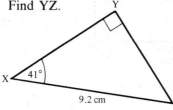

28. Find PQ.

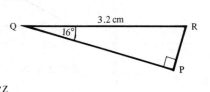

29. In triangle ABC, $\hat{C} = 90°$, $\hat{A} = 78°$ and $AC = 24$ cm. Find BC.

30. In triangle PQR, $\hat{P} = 90°$, $\hat{Q} = 36°$ and $QR = 3.2$ cm. Find PQ.

31. In triangle XYZ, $\hat{X} = 90°$, $\hat{Y} = 36°$ and $YZ = 17$ cm. Find XZ.

32. In triangle DEF, $\hat{F} = 90°$, $\hat{E} = 51°$ and $DF = 9.2$ cm. Find EF.

33. In triangle LMN, $\hat{M} = 90°$, $\hat{N} = 25°$ and $LN = 16$ cm. Find MN.

34. In triangle LMN, $\hat{L} = 90°$, $\hat{M} = 56.2°$ and $LN = 32$ cm. Find ML.

35. In triangle ABC, $\hat{C} = 90°$, $\hat{B} = 72.8°$ and $AB = 78$ cm. Find AC.

36. In triangle PQR, $\hat{R} = 90°$, $\hat{P} = 31.2°$ and $PQ = 117$ cm. Find QR.

FINDING THE HYPOTENUSE

Up to now, when finding the length of a side, we have been able to form an equation in which our unknown length is on the top of the fraction. If we wish to find the hypotenuse, this is not possible and the equation we form takes slightly longer to solve.

EXERCISE 18j

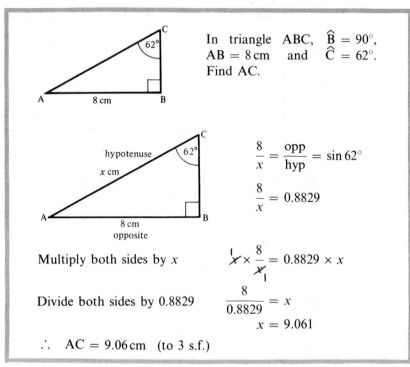

In triangle ABC, $\hat{B} = 90°$, AB = 8 cm and $\hat{C} = 62°$. Find AC.

$$\frac{8}{x} = \frac{\text{opp}}{\text{hyp}} = \sin 62°$$

$$\frac{8}{x} = 0.8829$$

Multiply both sides by x

$$x \times \frac{8}{x} = 0.8829 \times x$$

Divide both sides by 0.8829

$$\frac{8}{0.8829} = x$$

$$x = 9.061$$

∴   AC = 9.06 cm   (to 3 s.f.)

Use the information given in the diagram to find the hypotenuse:

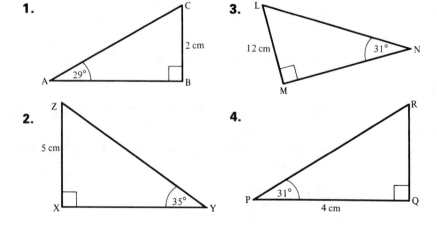

1.

3.

2.

4.

5. 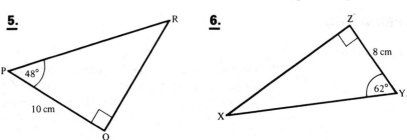   **6.**

7. In triangle ABC, $\hat{B} = 90°$, $\hat{A} = 43°$ and BC = 3 cm. Find AC.

8. In triangle PQR, $\hat{P} = 90°$, $\hat{Q} = 28°$ and PR = 7 cm. Find QR.

9. In triangle LMN, $\hat{L} = 90°$, $\hat{M} = 14°$ and LN = 8 cm. Find MN.

10. In triangle XYZ, $\hat{Z} = 90°$, $\hat{Y} = 62°$ and ZY = 20 cm. Find XY.

ANGLES OF ELEVATION AND DEPRESSION

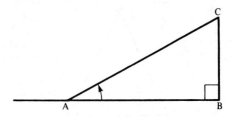

Imagine that you are at A, a point on level ground some distance from a cliff BC. You look horizontally at B, then *elevate* your line of view until you are looking at C.

$B\hat{A}C$ is the *angle of elevation* of C from A.

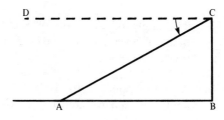

Now imagine you are on top of the cliff, at C. You look out horizontally to D, then *depress* your line of view until you are looking at A.

$D\hat{C}A$ is the *angle of depression* of A from C.

PROBLEMS

EXERCISE 18k

A flagpole stands on level ground. From a point on the ground 30 m away from its foot, the angle of elevation of the top of the pole is 22°. Find the height of the pole.

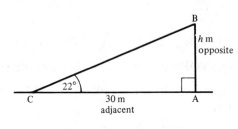

$$\frac{h}{30} = \frac{\text{opp}}{\text{adj}} = \tan 22°$$

$$\frac{h}{30} = 0.4040$$

$$\cancel{30} \times \frac{h}{\cancel{30}} = 0.4040 \times 30$$

$$h = 12.12$$

The pole is 12.1 m high   (to 3 s.f.)

1.

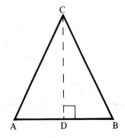

In triangle ABC, AC = CB = 10 m and $\hat{A}$ = 64°. Find the height of the triangle.

2. From a point on level ground 40 m from the base of a pine tree, the angle of elevation of the top of the tree is 50°. Find the height of the tree.

3. The angle of elevation of the top of a church tower, from a point of level ground 500 m away, is 16°. Find the height of the tower.

4. A is the point (2, 0), B is (8, 0) and C is (8, 5). Calculate the angle between AC and the *x*-axis.

5. ABCD is a rectangle with AB = 26 cm and BC = 48 cm. Find the angle between the diagonal AC and side AB.

6. A is the point (1, 2), B(3, 2) and C(1, 5). Find $A\hat{B}C$.

7. ABCD is a rhombus of side 15 cm. The diagonal AC is of length 20 cm. Find the angle between AC and the side CD.

8.

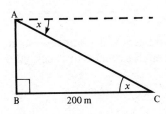

A boat C is 200 m from the foot B of a vertical cliff, which is 40 m high. What is the angle of depression of the boat from the top of the cliff?

9.

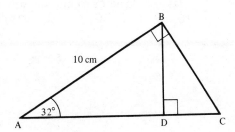

In the figure AB = 10 cm, $\hat{A}$ = 32° and A$\hat{B}$C = B$\hat{D}$C = 90°. Copy the figure and then mark in the sizes of the remaining angles.

Find   a) BD   b) BC.

10. Triangle ABC is an equilateral triangle of side 6 cm.
Find   a) its height   b) its area.

11. A lamp post stands on level ground. From a point which is 10 m from its foot, the angle of elevation of the top is 25°. How high is the lamp post?

12.

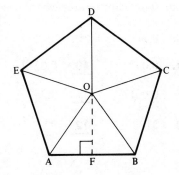

ABCDE is a regular pentagon of side 10 cm.

a) Find A$\hat{O}$B and O$\hat{A}$B.

b) Find OF.

c) Find the area of triangle AOB and hence find the area of the pentagon.

19 PYTHAGORAS' THEOREM

SQUARES AND SQUARE ROOTS

The following exercise revises the finding of squares and square roots. Remember to make a rough estimate to check your answer.

EXERCISE 19a

Use a calculator to find the squares of
a) 2.3   b) 23   c) 2300   d) 0.023

a)
$$2.3^2 \approx 2 \times 2 = 4$$
$$2.3^2 = 5.29$$

b)
$$23^2 \approx 20 \times 20 = 400$$
$$23^2 = 529$$

c)
$$2300^2 \approx 2000 \times 2000 = 4\,000\,000$$
$$2300^2 = 5\,290\,000$$

d)
$$0.023^2 \approx 0.02 \times 0.02 = 0.0004$$
$$0.023^2 = 0.000\,529$$

Use a calculator to find the squares of the following numbers, giving your answers correct to four significant figures where necessary:

1. 6.2	**5.** 0.71	**9.** 3.12	**13.** 5210
2. 13.7	**6.** 0.059	**10.** 0.0312	**14.** 52.1
3. 242	**7.** 0.0017	**11.** 9.2	**15.** 0.521
4. 2780	**8.** 312	**12.** 92	**16.** 0.0521

Use a calculator to find, correct to four significant figures, the square root of
a) 6.23   b) 62.3   c) 623 000   d) 0.0623

a) $\sqrt{6.23} = 2.496$ to 4 s.f.

b) $\sqrt{62.3} = 7.893$ to 4 s.f.

c) $\sqrt{623\,000} = 789.3$ to 4 s.f.

d) $\sqrt{0.0623} = 0.2496$ to 4 s.f.

Use a calculator to find the square roots of the following numbers, giving your answers correct to four significant figures:

17. 9.87	**21.** 0.0482	**25.** 2.62	**29.** 0.461
18. 19.9	**22.** 0.004 82	**26.** 0.062	**30.** 4.61
19. 124	**23.** 96	**27.** 0.000 78	**31.** 461
20. 96 800	**24.** 321	**28.** 0.5	**32.** 0.000 461

PYTHAGORAS' THEOREM

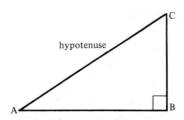

Pythagoras' theorem states that, in a right-angled triangle the square of the hypotenuse is equal to the sum of the squares of the other two sides,

i.e. $AC^2 = AB^2 + BC^2$

If we are given any two sides of a right-angled triangle we can find the third side.

EXERCISE 19b

In △ABC, $\widehat{B} = 90°$, AB = 62 cm and BC = 41 cm.
Find AC.

$$AC^2 = AB^2 + BC^2 \quad \text{(Pythagoras' theorem)}$$
$$= 62^2 + 41^2$$
$$= 3844 + 1681$$
$$= 5525$$

$$AC = \sqrt{5525}$$
$$= 74.33 \text{ cm}$$
$$= 74.3 \text{ cm} \qquad \text{(to 3 s.f.)}$$

Use the information given in the diagrams to find the required lengths, giving your answers correct to three significant figures.

1. Find AC.

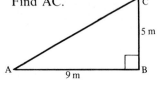

2. Find RQ.

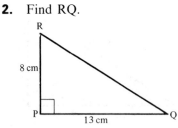

3. Find ZY.

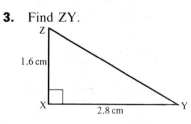

4. Find LN.

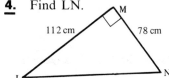

5. Find DF.

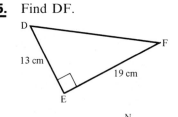

6. Find LN.

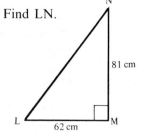

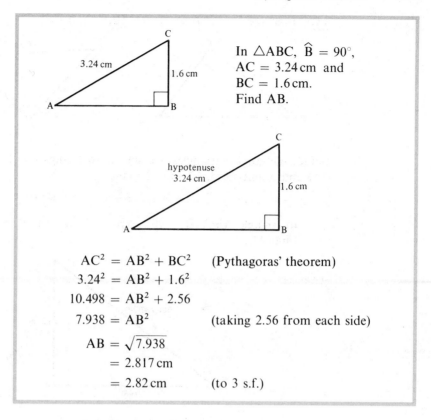

In △ABC, $\hat{B} = 90°$,
AC = 3.24 cm and
BC = 1.6 cm.
Find AB.

$AC^2 = AB^2 + BC^2$ (Pythagoras' theorem)
$3.24^2 = AB^2 + 1.6^2$
$10.498 = AB^2 + 2.56$
$7.938 = AB^2$ (taking 2.56 from each side)
$AB = \sqrt{7.938}$
$= 2.817$ cm
$= 2.82$ cm   (to 3 s.f.)

Use the information in the diagrams to find the required lengths, giving your answers correct to three significant figures where necessary:

7. Find AB.

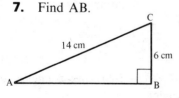

9. Find LM.

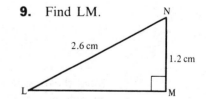

8. Find PQ.

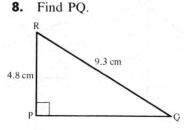

10. Find XY.

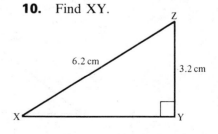

11. Find DE.

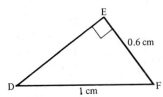

12. Find LN.

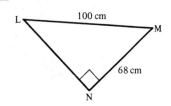

Some right-angled triangles are similar to triangles with sides of 3, 4 and 5 units or 5, 12 and 13 units.

In triangle ABC, $\widehat{B} = 90°$, AB = 16 cm and BC = 12 cm. Find AC.

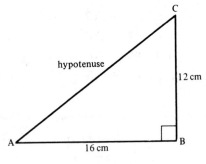

$$12 = 4 \times 3, \ 16 = 4 \times 4$$

∴   △ABC is similar to a 3, 4, 5 triangle

∴ $AC = 4 \times 5$ cm
                          $= 20$ cm

Find the missing lengths in the following triangles, using a 3, 4, 5 triangle or a 5, 12, 13 triangle:

13.

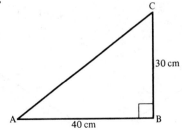

14.

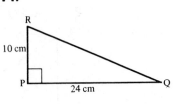

15.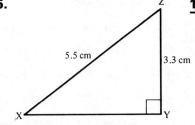

17.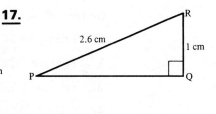

16.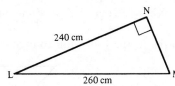

18.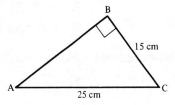

THE CONVERSE OF PYTHAGORAS' THEOREM

If we are given three sides of a triangle, we can tell whether or not the triangle contains a right angle. Bear in mind that, *if* there is a right angle the longest side will be the hypotenuse.

EXERCISE 19c

Are the following triangles right-angled?
a) △ABC: AB = 17 cm, BC = 8 cm, CA = 15 cm
b) △PQR: PQ = 15 cm, PR = 7 cm, RQ = 12 cm

a)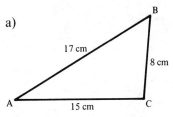

$$AB^2 = 17^2 = 289$$
$$AC^2 + BC^2 = 15^2 + 8^2$$
$$= 225 + 64$$
$$= 289$$
$$\therefore \quad AC^2 + BC^2 = AB^2$$

∴ by Pythagoras' theorem, $\widehat{C} = 90°$.

b)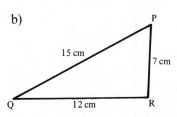

$$PQ^2 = 15^2 = 225$$
$$PR^2 + RQ^2 = 12^2 + 7^2$$
$$= 144 + 49$$
$$= 193$$
$$\therefore \quad PR^2 + RQ^2 \neq PQ^2$$

∴ the triangle is not right-angled.

Are the following triangles right-angled?

1. Triangle ABC: AB = 48 cm, BC = 64 cm and CA = 80 cm.

2. Triangle PQR: PQ = 2.1 cm, QR = 2.8 cm and RP = 3.5 cm.

3. Triangle LMN: LM = 6 cm, MN = 7.2 cm and NL = 9 cm.

4. Triangle ABC: AB = 9.2 cm, BC = 6.3 cm and CA = 4.6 cm.

5. Triangle DEF: DE = 6.4 cm, EF = 12 cm and DF = 13.6 cm.

6. Triangle XYZ: XY = 32 cm, YZ = 40 cm and ZX = 48 cm.

EXERCISE 19d Find the missing lengths in the following triangles:

1.
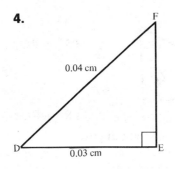

2.
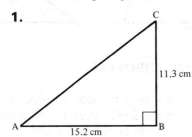

3.
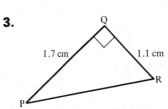

4.
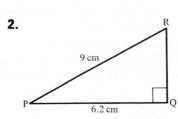

5.
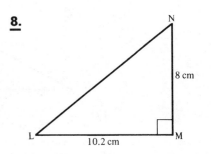

6.
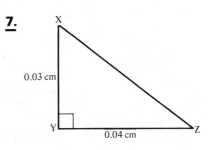

7.
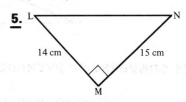

8.
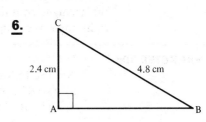

9. In triangle ABC, $\hat{A} = 90°$, AB = 3.2 cm and BC = 4.8 cm.
   Find AC.

10. In triangle PQR, $\hat{Q} = 90°$, PQ = 56 cm and QR = 32 cm.
    Find PR.

11. In triangle ABC, AB = 1 cm, BC = 2.4 cm and CA = 2.6 cm.
    Is $\hat{B}$ a right angle?

12. In triangle DEF, $\hat{F} = 90°$, DF = 2.8 cm and DE = 4.2 cm.
    Find EF.

13. In triangle XYZ, $\hat{Y} = 90°$, XY = 17 cm and YZ = 20 cm.
    Find XZ.

14. In triangle LMN, NL = 25 cm, LM = 24 cm and MN = 7 cm.
    Is the triangle right-angled? If it is, which angle is 90°?

PROBLEMS

If you find one length first and use it to find a second length or an angle, do
not use a corrected figure for the first length. Whenever you can, use at least
four significant figures. Your answer will be more accurate.

EXERCISE 19e

In a circle with centre O and radius 7 cm, there is a chord
11 cm long. How far is the chord from the centre?

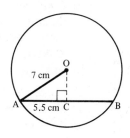

(We need the *perpendicular*
distance from O to the chord.)

From symmetry,
   AC = $\frac{1}{2}$ × 11 cm = 5.5 cm

$AO^2 = AC^2 + OC^2$ (Pythag. th.)

$7^2 = 5.5^2 + OC^2$

$49 = 30.25 + OC^2$

$18.75 = OC^2$

$OC = \sqrt{18.75}$

$= 4.330$

The distance of the chord from the centre is 4.33 cm (to 3 s.f.)

1. A is the point $(3, 1)$ and B is the point $(7, 9)$. Find the length of AB.

2. A ship sails 32 nautical miles due north then 22 nautical miles due east. How far is it from its starting point?

3. A pole 4.5 m high stands on level ground. It is supported in a vertical position by two wires attached to its top and to points on opposite sides of the pole each 3.2 m from the foot of the pole.
   a) How long is each wire?
   b) What is the angle between each wire and the pole?

4. The diagonal AC of a rectangle ABCD is 0.67 m long and side AB is 0.32 m long. How long is side BC?

5. Find the length of the diagonal of a square of side 15 cm.

6. ABCD is a kite and AC is its line of symmetry. $\hat{B} = \hat{D} = 90°$, AB = 36 cm and BC = 16 cm. Find   a) AC   b) BÂD.

7.

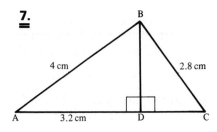

In the figure, $A\hat{D}C = 180°$, AB = 4 cm, AD = 3.2 cm and BC = 2.8 cm.
Find   a) BD   b) AC.
Is $A\hat{B}C$ a right angle?
Give a reason for your answer.

8.

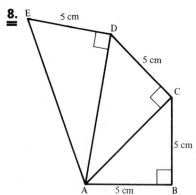

a) *Construct* the figure in the diagram, starting with △ABC then adding △ADC and △ADE.

b) Measure AC, AD and EA.

c) *Calculate* AC, AD and AE and check the accuracy of your drawing.

9. Construct a right-angled triangle, choosing whole numbers of centimetres for the lengths of the two shorter sides, such that the hypotenuse will be $\sqrt{65}$ cm long. Check the accuracy of your drawing by measuring the hypotenuse and by calculating $\sqrt{65}$.

THREE-DIMENSIONAL PROBLEMS

We can use Pythagoras' theorem and the trigonometry of right-angled triangles to find lengths and angles of various solids.

CUBOIDS

EXERCISE 19f   **1.**

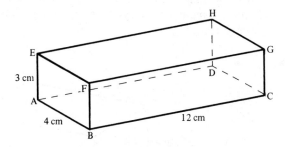

The solid above is a cuboid. Each face is a rectangle.

a) Copy the diagram, noticing that, in the drawing, some faces look like parallelograms.

Which edges are equal in length to EA?
Which edges are equal in length to AB?
Which edges are equal in length to BC?
How many right angles are there?

b) Join B to E on your diagram and draw triangle ABE (notice that EÂB = 90° and should now be drawn as a right angle).

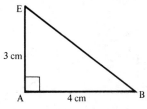

Find EB by using Pythagoras' theorem, and EB̂A by using its tangent.

c) Join F and C on your drawing of the cuboid. Draw triangle FBC, marking in all the information you know about it. Find FC and FĈB.

2.   a) Copy the picture of the cuboid again. Join A and C. Draw triangle ABC (notice AB̂C = 90°) and find AC.

b) Join E and C. What is the size of EÂC? Draw triangle EAC and mark in the sizes of any sides and angles you know. Find EC and EĈA.

3. Draw a picture of a cuboid like the first one but such that AB = 5 cm, EA = 2 cm and BC = 8 cm.

a) Draw the appropriate triangle and find FC.

b) Find AF and $F\hat{A}B$.

c) Find EG and the angle between FG and EG.

When dealing with solids like cuboids and the other types in this chapter, remember always to make a separate drawing of the triangle you are using.

EXERCISE 19g

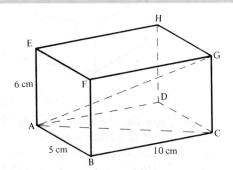

In the cuboid, EA = 6 cm, AB = 5 cm and BC = 10 cm. Find AG.

(To find AG, we use triangle AGC so we need to calculate AC^2 first.)

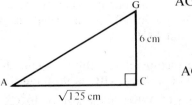

$AC^2 = AB^2 + BC^2$ (Pythag. th.)

$= 25 + 100$

$= 125$

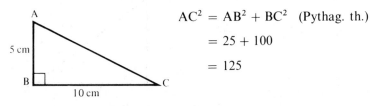

$AG^2 = AC^2 + GC^2$ (Pythag. th.)

$= 125 + 36$

$= 161$

$AG = \sqrt{161}$

$= 12.7 \text{ cm}$ (to 3 s.f.)

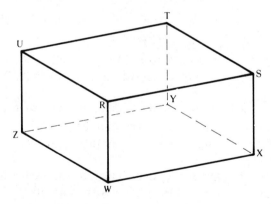

1. In the cuboid above, AB = 8 cm, BC = 12 cm and CG = 5 cm.
Find   a) AC   b) EC   c) $E\hat{C}A$.

2. In the cuboid above, EF = 3 cm, EA = 2 cm and AD = 6 cm.
Find   a) EB   b) the angle between AB and EB   c) EG.

3. In the cuboid above, HG = 6 cm, GC = 8 cm and BC = 12 cm.
Find   a) HC   b) HB   c) $H\hat{B}C$.

4. In the cuboid above, AD = 11 cm, AB = 10 cm and EA = 12 cm.
Find   a) AC   b) EC   c) BH   d) $G\hat{B}C$ e) $E\hat{G}H$.

5. In the cuboid above, BC = 20 cm, CG = 8 cm and DC = 12 cm.
Find EC.

6. In the cuboid above, RS = 12 cm, SX = 7 cm and XY = 9 cm.
Find   a) WY   b) WT   c) $T\hat{W}Y$.

7. In the cuboid above, TX = 10 cm, $T\hat{X}Y$ = 45° and UT = 12 cm.
Find   a) XY   b) TY   c) the volume of the cuboid.

8. In the cuboid above, SX = 6 cm, WX = 9 cm and XY = 4 cm.
Find   a) $S\hat{W}X$ b) $S\hat{Y}X$ c) $S\hat{Z}X$.

9. In the cuboid above, WX = 11 cm, XY = 5 cm and WT = 14 cm.
Find   a) TY   b) the surface area of the cuboid.

WEDGES

EXERCISE 19h   **1.**

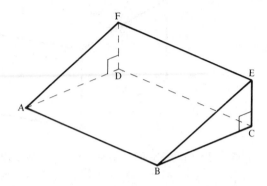

The solid above is a *wedge*. Two faces are right-angled triangles and three faces are rectangles.

a) Copy the diagram. (Notice that, in the drawing, each rectangular face looks like a parallelogram.)
AB = 10 cm, BC = 6 cm and CE = 4 cm.
Which edges are equal to AB? Which edges are equal to BC? Which edges are equal to EC?
How many right angles are there?

b) Draw triangle EBC. Find $E\hat{B}C$ and BE.

c) Join A to C. Draw triangle ABC. Find AC and $C\hat{A}B$. Is AC at right angles to EC?

d) Join A to E. Draw triangle AEC. Find AE. What other length (not yet drawn in) is equal to AE?

2. Draw a wedge similar to the wedge in question 1 but such that AB = BE = 8 m and $E\hat{B}C = 22°$.
Find a) EC   b) BC   c) AC   d) $E\hat{A}C$.

3. Draw a wedge similar to the wedge in question 1 but such that EF = 40 m, EC = 10 m and $B\hat{E}C = 70°$.
Find a) BC   b) AC   c) BE   d) AE   e) $E\hat{A}C$ f) $E\hat{A}F$.

4. Draw a wedge similar to the wedge in question 1, but such that BC = 4.4 m, AB = 6.8 m and EC = 2 m.
Find a) $E\hat{B}C$ b) $E\hat{A}C$.

5. Draw a wedge similar to the wedge in question 1, but such that AE = 7 cm, $E\hat{A}C = 22°$ and AD = 3 cm.
Find a) EC   b) BE   c) AB.

PYRAMIDS

1.

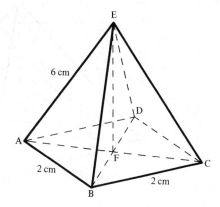

The solid above is called a square pyramid because its base is a square. Its vertex, E, is directly above the centre, F, of its base.

a) Copy the diagram. (Start by drawing the base as a parallelogram. Draw the diagonals of the base, to cut at F, and then mark E directly above F.) Name as many right angles as you can find. Which edges are equal to AE?

b) Draw the base ABCD as a square. Find AC and FC.

c) Draw triangle EFC. Find EF and $E\hat{C}F$.

2. Draw a square pyramid similar to the pyramid in question 1, but such that $AB = 4\,\text{cm}$ and $EF = 5\,\text{cm}$.

a) Find AC and AF.

b) Find AE and $E\hat{A}F$.

c) G is the midpoint of AB. Draw triangle EFG and find EG and $E\hat{G}F$.

3.

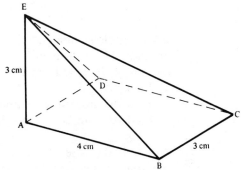

The base of the given pyramid is a rectangle and E is directly above A. $AB = 4\,\text{cm}$, $EA = BC = 3\,\text{cm}$.

Find   a) $E\hat{B}A$ and $E\hat{D}A$ b) AC   c) EC.

4.

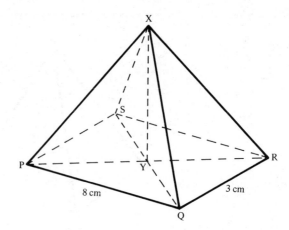

The base of the pyramid above is a rectangle. PQ = 8 cm and QR = 3 cm. Its vertex, X, is directly above the centre, Y, of the base. XY = 6 cm.

Find   a) PR   b) PY   c) $X\hat{P}Y$ d) XR.

MIXED QUESTIONS

EXERCISE 19j   1.

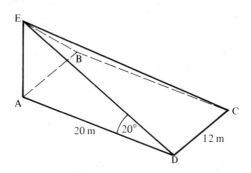

EA is a pole standing at the corner of a rectangular plot of level ground. AD = 20 m and DC = 12 m. The angle of elevation of the top of the pole from D is 20°.
a) Find EA.
b) Find the angle of elevation of E from B (i.e. $A\hat{B}E$).
c) Find AC and hence find the angle of elevation of E from C.

2.   ABCD is the base of a cube of side 4 cm. A′, B′, C′ and D′ are the four vertices of the cube directly above A, B, C and D respectively.
a) Find AC, CD′ and AD′. What sort of triangle is △ACD′?
b) What sort of quadrilateral is ACC′A′?
   Find AC′. What other lengths are equal to AC′?

3.

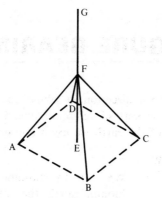

A pole, GE, is held upright by four equal guy ropes fastened to its midpoint, F. The other ends of the ropes are fastened to the four corners of a square, ABCD, on the level ground. AB = 6 m. Each rope is 6 m long.

a) Find BD and BE.

b) Find EF and the height of the pole.

c) Find the angle that each rope makes with the pole.

4.

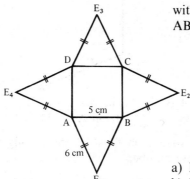

This is the net of a square pyramid with base ABCD and vertex E. AB = 5 cm. AE = 6 cm.

a) Find AC.

b) Draw a picture of the pyramid.

c) Find the height of the pyramid.

5.

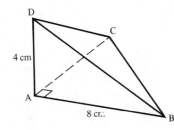

This is a triangular pyramid. The base is triangle ABC. $\widehat{CAB} = 90°$ and AB = AC = 8 cm. D is directly above A and AD = 4 cm.

Find   a) BD   b) $\widehat{DBA}$ c) BC   d) DC   e) $\widehat{DCA}$.

20 THREE-FIGURE BEARINGS

If we wish to describe a direction in which one point lies relative to another, we need some sort of reference line. On paper, we can use axes. On the surface of the earth we use a line pointing north.

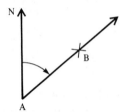

We start by standing at the first point A looking north, then turn clockwise until we are looking at the second point B.

The angle we have turned through gives the *bearing* of B from A.

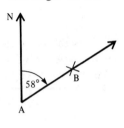

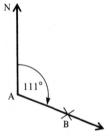

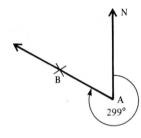

The bearing of B from A is 058°.

The bearing of B from A is 111°.

The bearing of B from A is 299°.

Notice that we write 0 in front of the two-figure number to make it into a three-figure number.

A bearing gives a *line*. If a point is on a given bearing, it lies on that line.

EXERCISE 20a

Mark a point P, then mark point Q, so that the bearing of Q from P is 212°

In questions 1 to 9, mark point A, then mark point B so that the bearing of B from A is:

1. 042°   **4.** 082°   **7.** 108°

2. 140°   **5.** 222°   **8.** 355°

3. 320°   **6.** 008°   **9.** 092°

Use the information in the diagram to find the bearing of Q from P.

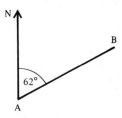

The reflex angle at P $= 360° - 72°$

$= 288°$

The bearing of Q from P is 288°.

Use the information given in each diagram to find the bearing of B from A:

10.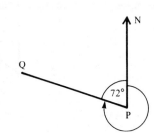

11.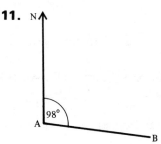

12.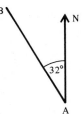

13.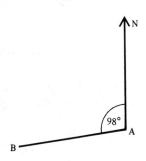

14.

16.

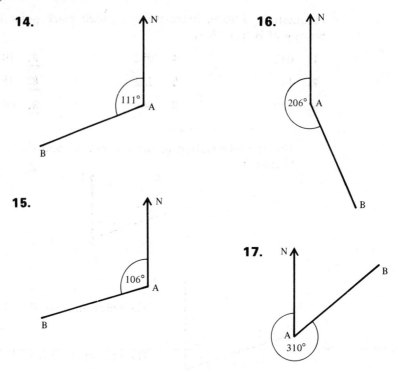

15.

17.

A man starts from point A and walks 4 km on a bearing of 036° to point B. Draw a diagram to represent this information.

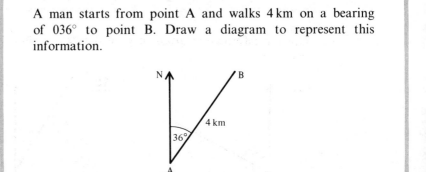

Draw diagrams to represent the information given in questions 18 to 27:

18. A car is driven 30 km on a bearing of 168°. It starts at point P and finishes at point Q.

19. A boy starts from C and cycles for 12 km, on a bearing of 213°, to D.

20. A town X is 260 km from town Y. The bearing of X from Y is 282°.

21. A man walks from A on a bearing of 027° to B. B is 11 km from A.

A car starts from A and is driven 50 km to B on a bearing of 072°. It is then driven 30 km to C on a bearing of 326°. Draw a diagram to represent this information.

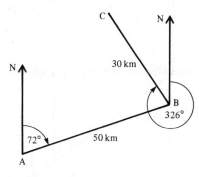

22. A ship sails 10 km from A, on a bearing of 100°, to point B. It then sails 12 km to C on a bearing of 070°.

23. Ship X is 5 km from ship Y on a bearing of 080°. Ship Z is 6 km from Y on a bearing of 300°.

24. A man walks 7 km from A, on a bearing of 285°, to B. Then he walks 3 km due east to C.

25. From town P, town Q is 45 km away on a bearing of 098°. Town R is 60 km from P on a bearing of 003°.

26. ABC is a triangular field. B is 100 m from A on a bearing of 127° and C is 130 m from B on a bearing of 330°.

27. Ship L is 10 km due east of ship M. The bearing of ship N from L is 343° and the bearing of N from M is 044°.

FINDING BEARINGS

If we are given the bearing of P from Q, we can draw a diagram in order to find the bearing of Q from P.

EXERCISE 20b

The bearing of B from A is 126°. What is the bearing of A from B?

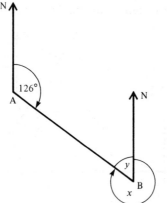

(Draw a line pointing north at B. The bearing of A from B is x.)

$y = 180° - 126°$ (interior angles)

$\quad = 54°$

$x = 360° - 54°$ (angles at a point)

$\quad = 306°$

The bearing of A from B is 306°.

1. The bearing of P from Q is 060°. What is the bearing of Q from P?

2. The bearing of C from D is 292°. What is the bearing of D from C?

3. The bearing of Y from X is 162°. What is the bearing of X from Y?

4. The bearing of A from B is 212°. What is the bearing of B from A?

<u>5.</u> The bearing of D from C is 352°. What is the bearing of C from D?

<u>6.</u> The bearing of Z from Y is 125°. What is the bearing of Y from Z?

ANGLE CALCULATIONS

EXERCISE 20c

The bearing of a ship S from a lighthouse A is 055°. A second lighthouse B is due east of A. The bearing of S from B is 302°. Find AŜB.

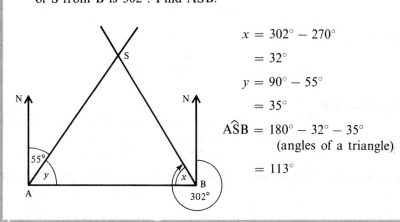

$$x = 302° - 270°$$
$$= 32°$$
$$y = 90° - 55°$$
$$= 35°$$
$$A\hat{S}B = 180° - 32° - 35°$$
$$\text{(angles of a triangle)}$$
$$= 113°$$

1. A car starts from C and is driven for 10 km on a bearing of 278°, to D. The car is then driven for 15 km due north to E. Find ED̂C.

2. The bearing of a town P from a town Q is 031°. The bearing of town R from town Q is 300°. Find PQ̂R.

3. A man starts from A and walks 5 km to B on a bearing of 112°. He then walks 3 km to C on a bearing of 260°. Find AB̂C.

4. A yacht starts from L and sails 12 km due east to M. It then sails 9 km on a bearing of 142° to K. Find LM̂K.

5. Point B is 60 km from A, on a bearing of 062°. C is 60 km from B on a bearing of 182°. Find AB̂C and BĈA.

6. The bearing of a ship X from a lighthouse Y is 101°. Ship Z is due west of X. The bearing of Z from Y is 230°. Find the angles of triangle XYZ.

7. A ship sails 10 km from point A, on a bearing of 005°, to B. Then it sails another 10 km from B, on a bearing of 095°, to C. Find $C\hat{A}B$.

8. From a church tower P, the bearing of a bridge Q is 340° and the bearing of a crossroads R is 111°. Find $Q\hat{P}R$.

PROBLEMS

These problems use Pythagoras' theorem and trigonometry.

EXERCISE 20d

P is 12 km due east of Q. R is 8 km due north of Q.
Find   a) the distance of R from P
      b) the bearing of P from R.

a)

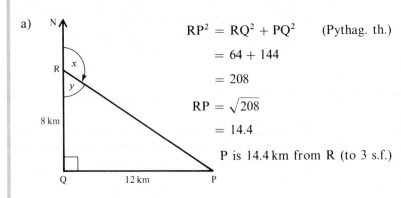

$$RP^2 = RQ^2 + PQ^2 \qquad \text{(Pythag. th.)}$$
$$= 64 + 144$$
$$= 208$$
$$RP = \sqrt{208}$$
$$= 14.4$$

P is 14.4 km from R (to 3 s.f.)

b) $\tan y = \dfrac{\text{opp}}{\text{adj}} = \dfrac{12}{8}$

$$= 1.5$$
$$y = 56.3°$$
$$x = 180° - 56.3° \qquad \text{(angles on a straight line)}$$
$$= 123.7°$$

The bearing of P from R is 124° (to the nearest degree).

Give bearings correct to the nearest degree:

1. From a point P, a man walks 9 km north to Q, then 5 km east to R. What is the bearing of R from P?

2. Three towns P, Q and R, lie in such positions that the bearing of P from R is 027°, the bearing of Q from R is 297° and PR = QR = 21 km. Find PR̂Q. How far is P from Q?

3. A ship sails 4 km on a bearing of 032° then 5 km on a bearing of 122°. How far is it from its starting point?

4. The bearing of a ship A from a ship B is 324°. Ship C is 8 km due north of B and is due east of A.

a) How far is C from A?

b) What is the bearing of B from A?

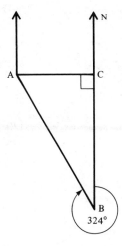

5. A is 20 km due west of B. The bearing of C from A is 044° and the bearing of C from B is 314°.

Find a) the angles of triangle ABC   b) the distance BC.

6. P is 16 km due north of Q. The bearing of R from P is 152°. The bearing of R from Q is 062°.

Find a) RQ   b) the bearing of P from R.

7. X is 25 km due west of Y. The bearing of Z from X is 052°. XYZ = 90°. Find the distances of Z from X and Y.

8. L is 5 km from M on a bearing of 241°. K is 7 km from M on a bearing of 331°.

Find a) ML̂K
      b) the bearing of L from K
      c) the bearing of K from L.

21 INEQUALITIES AND REGIONS

USING TWO-DIMENSIONAL SPACE

In a previous chapter, we discussed inequalities in a purely algebraic way. In this chapter, we will be looking at them in a more visual way, using graphs.

If we have the inequality $x \geqslant 2$, x can take any value greater than or equal to 2. This can be represented by the following diagram.

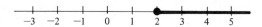

On this number line, x can take any value on the heavy part of the line including 2 itself, as indicated by the solid circle at 2.

If $x > 2$ then the diagram is as shown below.

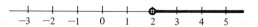

In this case, x cannot take the value 2 and this is shown by the open circle at 2.

It is sometimes more useful to use two-dimensional space with x and y axes, rather than a one-dimensional line. We represent $x \geqslant 2$ by the set of points whose x coordinates are greater than or equal to 2. (y is not mentioned in the inequality so y can take any value.)

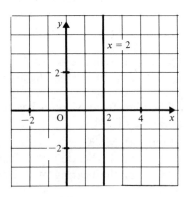

The boundary line represents all the points for which $x = 2$ and the region to the right contains all points with x coordinates greater than 2.

To indicate this, and to make future work easier, we use a continuous line for the boundary when it is included and we shade the region we do *not* want.

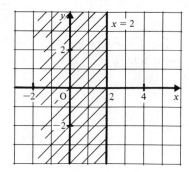

The inequality $x > 2$ tells us that x may not take the value 2. In this case we use a broken line for the boundary.

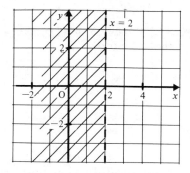

We can draw a similar diagram for $y > -1$

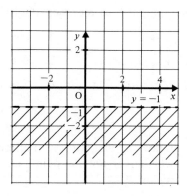

EXERCISE 21a

Draw diagrams to represent the inequalities
a) $x \leqslant 1$ b) $2 < y$

a) $x \leqslant 1$

The boundary line is $x = 1$
(included).

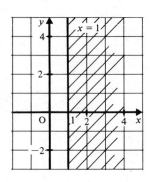

The unshaded region represents $x \leqslant 1$

b) $2 < y$

The boundary line is $y = 2$
(not included).

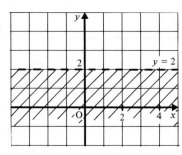

The unshaded region represents $2 < y$.

Draw diagrams to represent the following inequalities:

1. $x \geqslant 2$	**4.** $y < 4$	**7.** $x \leqslant -4$
2. $y \leqslant 3$	**5.** $x \geqslant 0$	**8.** $y > -3$
3. $x > -1$	**6.** $0 > y$	**9.** $2 < x$

Draw a diagram to represent $-3 < x < 2$ and state whether or not the points $(1, 1)$ and $(-4, 2)$ lie in the given region.

$(-3 < x < 2$ gives two inequalities, $-3 < x$ and $x < 2)$

Boundary lines are $x = -3$ and $x = 2$ (neither included)

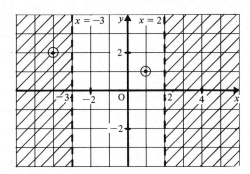

The unshaded region represents $-3 < x < 2$

$(-4, 2)$ does not lie in the given region.

$(1, 1)$ lies in the given region.

Draw diagrams to represent the following pairs of inequalities:

10. $2 \leqslant x \leqslant 4$ **13.** $4 < y < 5$ **16.** $-\frac{1}{2} \leqslant x \leqslant 1\frac{1}{2}$

11. $-3 < x < 1$ **14.** $0 \leqslant x < 4$ **17.** $-2 \leqslant y < -1$

12. $-1 \leqslant y \leqslant 2$ **15.** $-2 < y \leqslant 3$ **18.** $3 \leqslant x < 5$

19. In each of the questions 10 to 12, state whether or not the point $(1, 4)$ lies in the unshaded region.

Give the inequality that defines the unshaded region

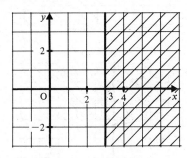

Boundary line $x = 3$ (included)

Inequality $x \leqslant 3$

Give the inequalities that define the unshaded region

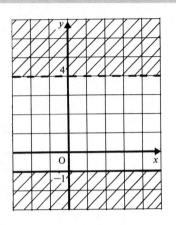

Boundary lines $y = 4$ (not included)
and $y = -1$ (included)
Inequalities $-1 \leqslant y < 4$

Give the inequalities that define the unshaded regions:

20.

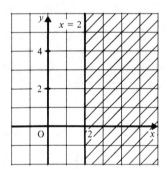

22.

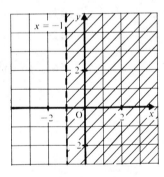

21.

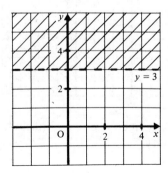

23.

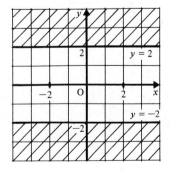

24.   **25.**

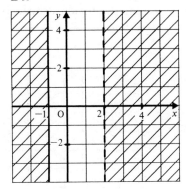   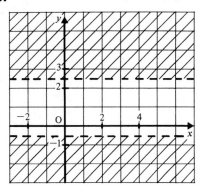

26. In each of the questions 20 to 25 state whether or not the point (2, −1) is in the unshaded region.

Give the inequalities that define the *shaded* regions:

27.   **29.**

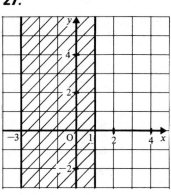   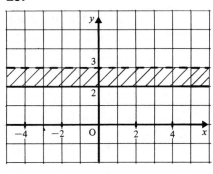

28.   **30.**

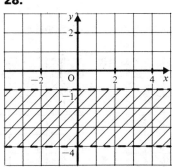   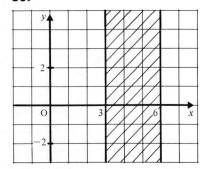

31. In each of the questions 27 to 30 state whether or not the point (0, 2) is in the shaded region.

EXERCISE 21b

Draw a diagram to represent the region defined by the set of inequalities $-1 \leqslant x \leqslant 2$ and $-5 \leqslant y \leqslant 0$

The boundary lines are

$$x = -1$$
$$x = 2$$
$$y = -5$$
$$y = 0$$

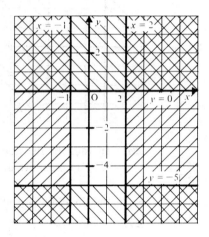

The unshaded region represents the inequalities.

Draw diagrams to represent the regions described by the following sets of inequalities. In each case, draw axes for values of x and y from -5 to 5.

1. $2 \leqslant x \leqslant 4$, $-1 \leqslant y \leqslant 3$

2. $-2 < x < 2$, $-2 < y < 2$

3. $-3 < x \leqslant 2$, $-1 \leqslant y$

4. $0 \leqslant x \leqslant 4$, $0 \leqslant y \leqslant 3$

5. $-4 < x < 0$, $-2 < y < 2$

6. $-1 < x < 1$, $-3 < y < 1$

7. $x \geqslant 0$, $y \geqslant 0$

8. $x \geqslant 1$, $-1 \leqslant y \leqslant 2$

Give the sets of inequalities that describe the unshaded regions:

9.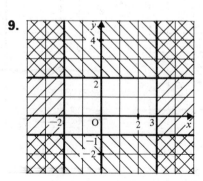

10.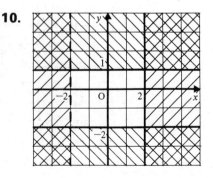

11. Is the point $(2\frac{1}{2}, 0)$ in either of the unshaded regions in questions 9 and 10?

Give the sets of inequalities that describe the unshaded regions:

12.
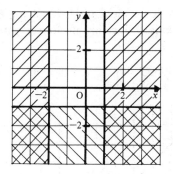

15.
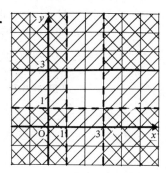

13.
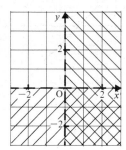

16.

14.
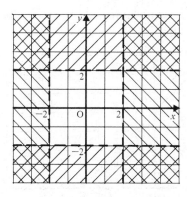

17.
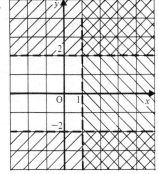

18. Is the point $(2, -1)$ in either of the unshaded regions in questions 16 and 17?

INEQUALITIES INVOLVING TWO VARIABLES ━━━━━━━━━━━━

The boundary lines for inequalities are parallel to the x or y axis when the equalities contain either x or y, but not both.

Now we will consider some inequalities involving *both x and y* and we will find that the boundary lines are no longer parallel to an axis.

Consider $x + y \geqslant 4$.
In this case the boundary line is $x + y = 4$; as it is included in the region it is drawn as a solid line.

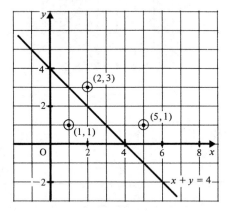

The boundary line divides the space into two regions, one on each side of the line. We need to decide which of the two regions is the one that we want.

Test a point such as $(2, 3)$.
When $x = 2$ and $y = 3$, $x + y = 5$ which is greater than 4, so the point $(2, 3)$ is in the required region.
The point $(5, 1)$ is also in the region, but the point $(1, 1)$ is not.
We can now see which region is required.

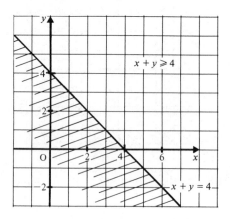

EXERCISE 21c

Leave unshaded the region defined by the inequality $2x + 3y < 12$

The boundary line is $2x + 3y = 12$ (not included in the inequality).

x	0	6	3
y	4	0	2

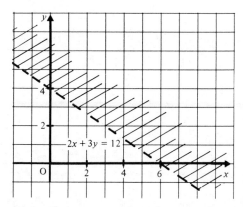

(To decide which of the two regions is wanted, we test an easy point such as $(0, 0)$.)

When $x = 0$ and $y = 0$, $2x + 3y = 0$ which is less than 12

Therefore $(0, 0)$ is in the required region. (Shade the region on the other side of the line.)

The unshaded region represents the inequality $2x + 3y < 12$

Find the regions defined by the following inequalities (draw axes for values of x and y from -6 to 6).

1. $x + y \leqslant 3$

2. $x + 4y \leqslant 8$

3. $x + y > 1$

4. $x + y \leqslant 2$

5. $2x + 5y \geqslant -6$

6. $3x + 4y \geqslant 12$

7. $4x + y < 4$

8. $2x + 5y > 10$

9. $2x + y \leqslant 6$

10. $3x + 2y > 5$

Sometimes the x and y terms are not on the same side of the inequality.

Leave unshaded the region defined by the inequality $y \geqslant 2x + 1$

The boundary line is $y = 2x + 1$

x	-2	0	2
y	-3	1	5

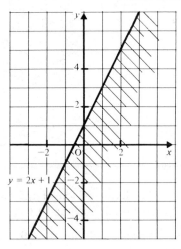

Test $(0, 0)$ $y = 0$

$2x + 1 = 1$

$0 < 1$ so $(0, 0)$ is not in the required region.

The unshaded region represents the inequality $y \geqslant 2x + 1$

Find the regions defined by the following inequalities:

11. $y \leqslant x + 1$

12. $y > 2x - 1$

13. $y \geqslant \frac{1}{2}x + 1$

14. $y < 2 - 2x$

15. $y < 4 - x$

16. $y \geqslant 2x - 2$

17. $y < 2x + 3$

18. $y < 5 + 3x$

19. $y > 3 + x$

20. $y \leqslant 5 - 2x$

21. $y \leqslant x - 4$

22. $y \geqslant 1 - x$

EXERCISE 21d

Find the inequality defining the unshaded region.

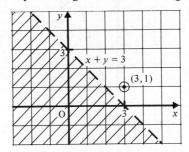

The boundary line is $x + y = 3$ and is not included.

Test the point $(3, 1)$, which is in the required region.

When $x = 3$ and $y = 1$, $x + y = 4$

$4 > 3$, so the inequality is $x + y > 3$

Find the inequalities that define the unshaded regions:

1.

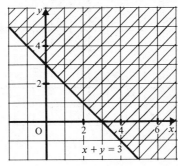

3.

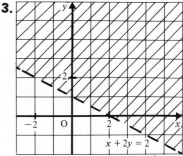

2.

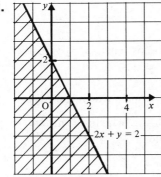

4.

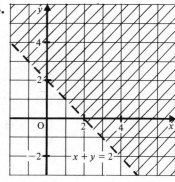

5.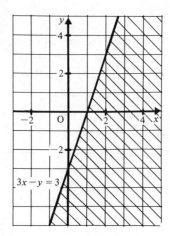

6.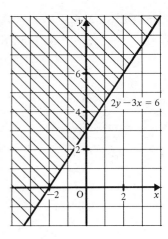

Find the inequality that defines the unshaded region

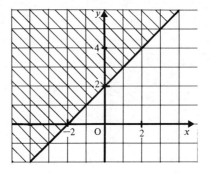

The gradient of the boundary line is

$$\frac{2-0}{0-(-2)} = \frac{2}{2} = 1$$

The line cuts the y-axis when $y = 2$ so the equation of the boundary line is $y = x + 2$

The boundary line is included in the region.

Test the point $(0, 0)$, which is in the required region.
At $(0, 0)$, $y = 0$ and $x + 2 = 2$
$0 < 2$ so the inequality is $y \leqslant x + 2$

Find the inequalities that define the unshaded regions:

7.
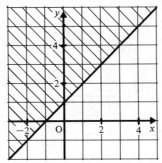

8.
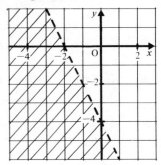

9.
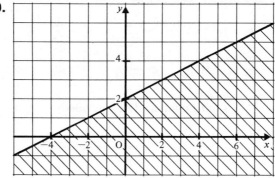

10.
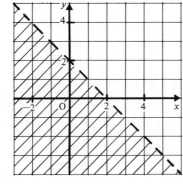

11.
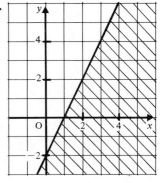

12. 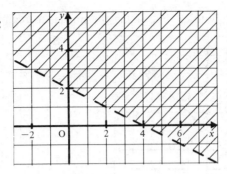   **13.**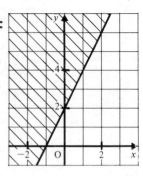

EXERCISE 21e

Leave unshaded the region defined by the set of inequalities
$x + y < 4$, $x \geqslant 0$ and $x + 2y \geqslant 2$

1st boundary line (not included) $x + y = 4$

x	4	0	2
y	0	4	2

2nd boundary line (included) $x = 0$

3rd boundary line (included) $x + 2y = 2$

x	0	2	4
y	1	0	−1

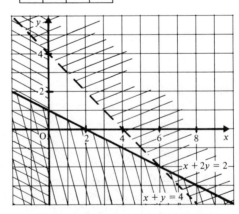

The unshaded region is defined by the given inequalities.

Leave unshaded the regions defined by the following sets of inequalities:

1. $x \geqslant -3$, $y \geqslant -2$, $x + y \leqslant 3$

2. $x > -1$, $-2 < y < 4$, $x + y < 4$

3. $y > 0$, $y \leqslant \frac{1}{2}x$, $x + y \geqslant 1$, $x + y \leqslant 5$

4. $y < 3$, $2x + 3y \geqslant 6$, $y > x - 2$

5. $y \leqslant 0$, $x \leqslant 0$, $x + y \geqslant -4$

6. $y > x$, $y < 4x$, $x + y < 5$

7. $y \geqslant 1$, $x \leqslant 0$, $y \leqslant x + 2$

8. $x + y \leqslant 6$, $3x + y \geqslant 3$, $y \geqslant -1$, $x \geqslant -1$

9. $x \geqslant 0$, $y \geqslant x - 1$, $2y + x < 4$

10. $x > 0$, $y \geqslant \frac{1}{2}x$, $x + y \geqslant 1$, $x + y \leqslant 5$

11. $y \geqslant 0$, $x \geqslant 0$, $x + y \leqslant 1$

12. What can you say about the region defined by $x + y > 4$, $x + y < 1$, $x > 0$ and $y > 0$?

13. Do the regions defined by the following sets of inequalities exist?
a) $x + y \geqslant 3$, $y \leqslant 2$, $y \geqslant 2x$ b) $x + y > 3$, $y < 2$, $y > 2x$

SHADING THE REQUIRED REGION

In some simple cases you might be asked to shade the region defined by the inequality, instead of leaving it unshaded.

Occasionally, you may be asked to shade the required region when it is defined by several inequalities. If you try to do it by shading the required side of each boundary line, you will find yourself with overlapping shadings, resulting in a confused diagram.

For instance if $y \geqslant 0$, $x \geqslant 0$ and $x + y \leqslant 3$, the diagram looks like this and the required region disappears in a muddle.

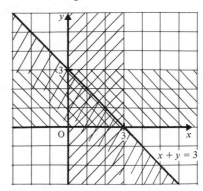

A better method is to do the shading as before so that the required region is left unshaded, then draw a second diagram on which you shade the required area.

1st diagram   2nd diagram

The required region is shaded.

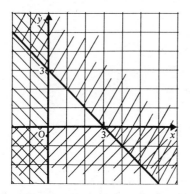
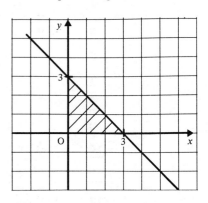

EXERCISE 21f

Shade the region defined by the set of inequalities
$x < 3$, $y < 4$, $x + y > 0$

Boundary lines are $x = 3$, $y = 4$ and $x + y = 0$

1st diagram

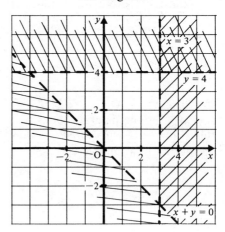

2nd diagram
The required region is shaded.

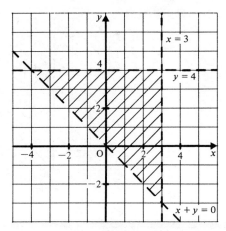

Shade the regions defined by the sets of inequalities:

1. $x \geqslant 0$, $y \geqslant 0$, $x + y \leqslant 3$

2. $x > -2$, $y \geqslant 2x$, $x + y < 4$

3. $x \leqslant 4$, $y \leqslant 3$, $x + y \geqslant 0$

4. $2x + y \geqslant 4$, $y \leqslant 0$, $x \leqslant 4$

5. $y > \frac{1}{2}x$, $0 < x < 2$, $x + y < 4$, $y < 3$

6. $\frac{x}{2} + \frac{y}{3} \geqslant 1$, $\frac{x}{2} - \frac{y}{3} \geqslant 1$, $x \leqslant 8$

7. $3x + 4y \leqslant 12$, $y \leqslant 2x + 1$, $y \geqslant -1$

8. $5x + 2y \leqslant -10$, $y \geqslant -1$, $x \geqslant -6$

9. $\frac{x}{5} + \frac{y}{4} \leqslant 1$, $y \leqslant 3x$, $y \geqslant \frac{1}{2}x$

10. $3x + 2y \geqslant -6$, $y \geqslant -3$, $x \geqslant -1$

If you wish to use diagrams for solving problems, it is best to leave the required regions unshaded.

EXERCISE 21g

Give the inequalities that define the unshaded region

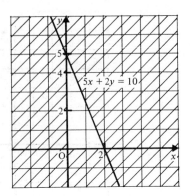

The first two inequalities are $x \geqslant 0$, $y \geqslant 0$

The 3rd boundary line is $5x + 2y = 10$

Test the point $(1, 1)$

When $x = 1$ and $y = 1$, $5x + 2y = 7$

$7 < 10$ so the 3rd inequality is $5x + 2y \leqslant 10$

Give the sets of inequalities that define the unshaded regions:

1.

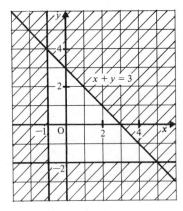

2.

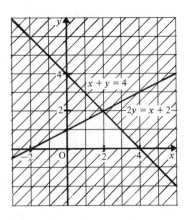

3.

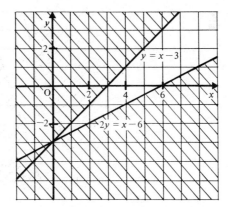

4.

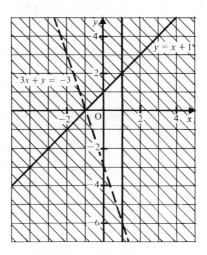

In the next four questions it is necessary to find the equations of the boundary lines first:

5.

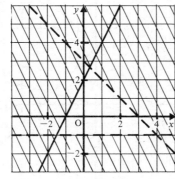

6.

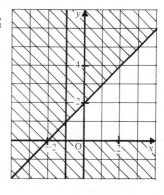

7.

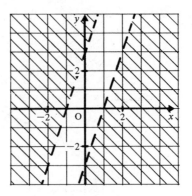

8.

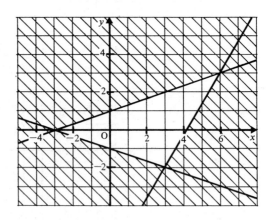

9. Use inequalities to describe the regions:

a) A   b) B   c) C   d) D   e) E   f) A + F

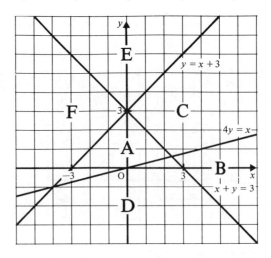

10.

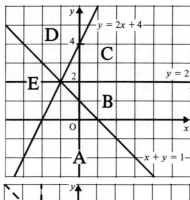

Use inequalities to describe the regions:

a) A   d) D

b) B   e) E

c) C   f) A + B

(The axes are not boundary lines.)

11.

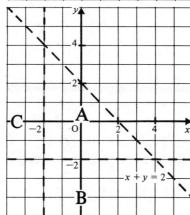

In the diagram, which region (A, B or C) does each of the following sets of inequalities refer to?

a) $x + y < 2$, $x < -2$, $y > -2$

b) $x + y < 2$, $x > -2$, $y > -2$

c) $x + y < 2$, $x > -2$, $y < -2$

COORDINATES OF POINTS IN A REGION

EXERCISE 21h

From the given diagram read off the coordinates of the vertices of the unshaded region

The vertices are $(0, 1)$, $(0, 4)$ and $(6, -2)$

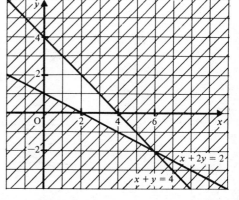

From the diagrams, give the coordinates of the vertices of the unshaded regions in questions 1 to 4:

1.

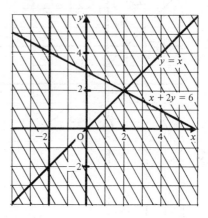

2.

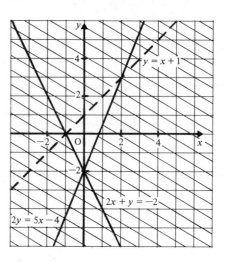

3.

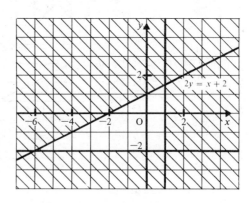

4.

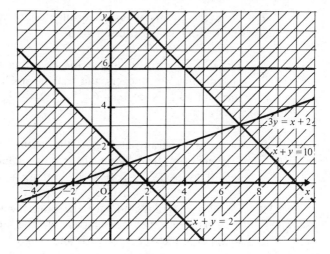

5. Draw a diagram and find the coordinates of the vertices of the region defined by the inequalities $x \geqslant -2$, $y \geqslant 2x$ and $y \leqslant x$

6. Draw a diagram and give the vertices of the region defined by the inequalities $y \geqslant 1$, $x \leqslant 0$ and $y \leqslant x + 2$

Give the points whose coordinates are integers and that lie in the region given in the worked example at the beginning of the exercise

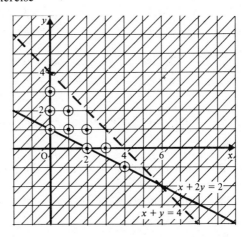

(Notice that points on the broken line are *not* in the region.)

Points are $(0, 1)$, $(0, 2)$, $(0, 3)$, $(1, 1)$, $(1, 2)$, $(2, 0)$, $(2, 1)$, $(3, 0)$ and $(4, -1)$.

7. Give the points whose coordinates are integers and that lie in the regions in questions 1 to 3

8. Draw a diagram and give the coordinates of the points whose coordinates are integers and that lie in the region defined by the inequalities $y \leqslant 3x + 6$, $y > x - 2$, $x + y > -2$ and $x + y \leqslant 3$

9. Draw a diagram and give the points whose coordinates are integers and that lie in the region defined by the inequalities $y > 0$, $x > 0$, $3x + 4y < 12$ and $4x + 3y < 12$

10. Draw a diagram and give the points with coordinates that are integers, on the boundaries of the region defined by the inequalities $x \geqslant 2$, $y \geqslant -1$ and $x + y \leqslant 4$

GREATEST AND LEAST VALUES

EXERCISE 21i

Find the value of $x - 2y$ at the points $(4, 1)$, $(3, 0)$ and $(2, -1)$. At which point is the value greatest?

At $(4, 1)$ $x - 2y = 4 - 2 = 2$
At $(3, 0)$ $x - 2y = 3 - 0 = 3$
At $(2, -1)$ $x - 2y = 2 - (-2) = 4$

$x - 2y$ has its greatest value at $(2, -1)$.

1. Find the value of $x + y$ at the points $(2, 3)$, $(4, -2)$ and $(-3, -1)$.

2. Find the value of $x - y$ at the points $(6, 2)$, $(1, 4)$ and $(4, -3)$.

3. Find the value of $2x + y$ at the points $(5, 1)$, $(-6, -2)$ and $(-1, 2)$.

4. Find the value of $3x - 2y$ at the points $(5, 5)$ and $(2, -8)$. At which point is the value greatest?

5. Find the value of $y - 3x$ at the points $(-6, -8)$ and $(3, 8)$. At which point is the value least?

From the diagram give the coordinates of the vertices of the unshaded region. At which vertex is the value of $x + 2y$ least?

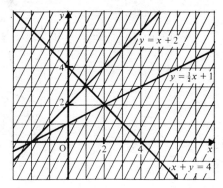

The three vertices are $(1, 3)$, $(2, 2)$ and $(-2, 0)$.

At $(1, 3)$	$x + 2y = 1 + 6 = 7$
At $(2, 2)$	$x + 2y = 2 + 4 = 6$
At $(-2, 0)$	$x + 2y = -2 + 0 = -2$

$\therefore$ $x + 2y$ is least at point $(-2, 0)$.

6. a) Find from the diagram the coordinates of the vertices of the unshaded region.

b) Find the vertex at which the value of $x + y$ is greatest.

c) Find the vertex at which the value of $x + y$ is least.

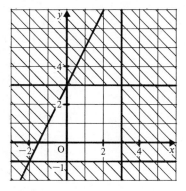

d) Mark the other points in the unshaded region whose coordinates are integers. How many are there?

e) Is there a point in the region where the value of $x + y$ is greater than its value at the vertex chosen in (b)?

7.

a) Find from the diagram the coordinates of the vertices of the region.

b) Find the vertex at which the value of $2x - y$ is greatest.

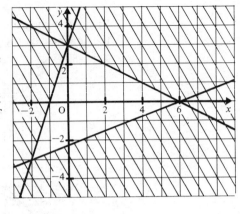

8.

a) From the diagram find the coordinates of the vertices of the unshaded region.

b) Find the vertex at which the value of $3x - 2y$ is
   i) greatest
   ii) least.

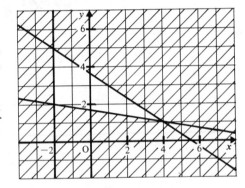

c) List the other points in the unshaded region whose coordinates are integers.

d) Is there a point in the region where the value of $3x - 2y$ is less than its value at the vertex chosen in (b) (ii)?

9. a) Draw axes for the values of x and y from -5 to 5 and leave unshaded the region defined by the set of inequalities $x \geqslant -2$, $y \geqslant -3$ and $x + y \leqslant 2$.

b) Find the coordinates of the vertices of the region.

c) Mark the other points in the region with integer coordinates. How many are there?

d) At which of these points in the region is the value of $x - y$ greatest and at which point is it least?

As we see from the answers to questions 7 to 9, the greatest and least values of any expression such as $2x + 3y$ occur at or near the vertices of a region and not at points between the vertices or in the middle of the region. Therefore to find the greatest or least values, we need test points at or near the vertices only.

EXERCISE 21j

For the region defined by the set of inequalities $x \geqslant -1$, $y \geqslant -2$ and $x + y < 3$, draw a diagram and find the points with coordinates that are integers, where the value of $2x - y$ is   a) greatest   b) least

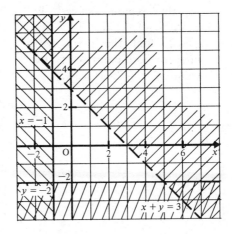

At $(-1, 3)$ $2x - y = -2 - 3 = -5$
At $(-1, -2)$ $2x - y = -2 + 2 = 0$
At $(4, -2)$ $2x - y = 8 + 2 = 10$

a) $(2x - y)$ is greatest at $(4, -2)$.
b) $(2x - y)$ is least at $(-1, 3)$.

1.

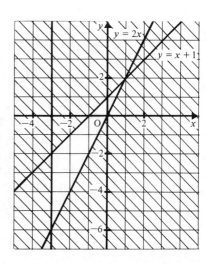

In the region defined by $x \geqslant -3$, $y \leqslant x + 1$ and $y \geqslant 2x$, find the point with integer coordinates where the value of $3x - y$ is greatest.

2.

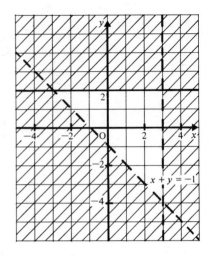

In the region defined by $x + y > -1$, $x < 3$ and $y \leqslant 2$, find the point with integer coordinates where the value of $x + 2y$ is least.

3.

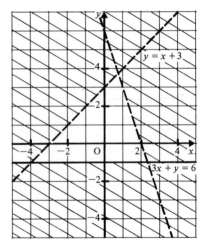

In the region defined by $y < x + 3$, $3x + y < 6$ and $y \geqslant -1$, find the point with integer coordinates where the value of $3x - 2y$ is greatest.

4. In the region defined by $y \leqslant 4 - 2x$, $y \geqslant -2$ and $x \geqslant 0$, by drawing a diagram, find the points or point with integer coordinates at which the value of $2x + y$ is greatest.

5. In the region defined by $x + y < 4$, $x > -1$ and $y > 1$, by drawing a diagram, find the point with integer coordinates at which the value of $2x - y$ is least.

6. In the region defined by $4x + y \leqslant 4$, $y \leqslant 3$ and $x \leqslant 1$, by drawing a diagram, find the point with integer coordinates at which the value of $x + y$ is greatest. Is there a point at which the value of $x + y$ is least?

22 COORDINATES IN THREE DIMENSIONS

TWO DIMENSIONS

In Book 1A we saw that we can locate a point in the x–y plane, i.e. in two dimensions (2-D for short), by starting with a fixed point, O, (the origin) and two perpendicular axes Ox and Oy. A point P is then located by associating with it an ordered pair of numbers (x, y) called its x and y coordinates. The first number gives the distance from the origin along or parallel to the x-axis, while the second number gives the distance along or parallel to the y-axis.

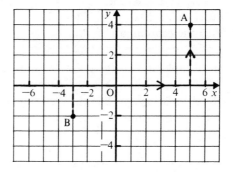

Thus, if A is the point $(5, 4)$ we start at O, move 5 units to the right along the x-axis, then 4 units up, parallel to the y-axis. Similarly, if B is the point $(-3, -2)$ we start at O, move 3 units to the left along the x-axis followed by 2 units down parallel to the y-axis.

THREE DIMENSIONS

To give the position of a point in space, i.e. in three dimensions (3-D for short), we use an ordered triple of numbers. Each number shows the distance of the point from the origin, O, along or parallel to three mutually perpendicular axes Ox, Oy and Oz. (Mutually perpendicular means that each axis is perpendicular to the other two.)

379

If we draw Ox and Oy as we did for two dimensions, then Oz comes "out of the page".

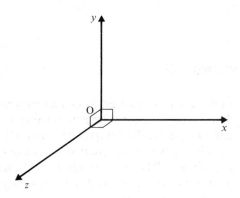

A point P is then located by giving its distance from O in the positive directions of these axes. These distances, given as an ordered triple, are called the *coordinates* of P.

For example, if P is the point $(4, 3, 2)$ then
the distance from O in the direction Ox is 4 units
the distance from O in the direction Oy is 3 units
the distance from O in the direction Oz is 2 units

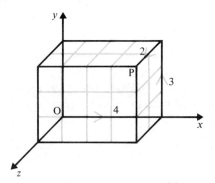

To find the position of P(4, 3, 2) we start at O, move 4 units along the x-axis, then 3 units up parallel to the y-axis, followed by 2 units parallel to the z-axis.

(You will find this much easier to follow if you build the model indicated above using small cubes. You are strongly advised to do this!)

EXERCISE 22a In this exercise the origin is at one vertex of the solid and the axes lie along three of its edges. Each axis is graduated in units. Where you are asked to draw diagrams you will find it an advantage to used squared paper.

1. For each diagram write down the coordinates of A.

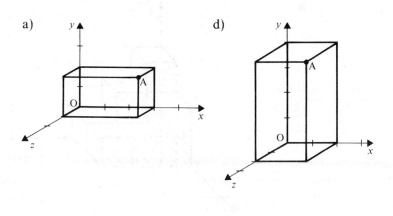

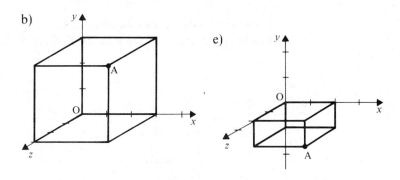

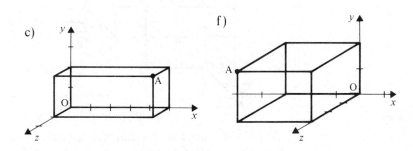

2. Draw diagrams, similar to those given in question 1, to show the position of A when A has each of the following sets of coordinates:

a) (2, 2, 4)   b) (5, 3, 2)   c) (1, 3, 4)   d) (4, 4, 4)

e) (2, 4, −2)

3.

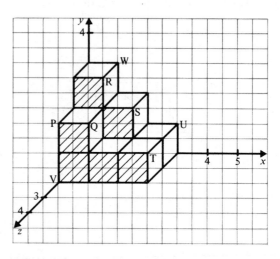

Each of the blocks in this diagram has an edge of one unit. Write down the coordinates of the vertices P, Q, R, S, T, U, V and W.

4.

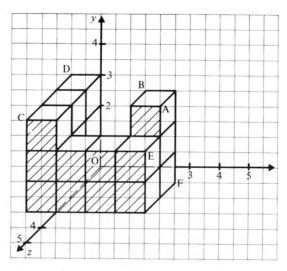

Each of the blocks in this diagram has an edge of one unit. Write down the coordinates of the points marked A, B, C, D, E and F.

5.

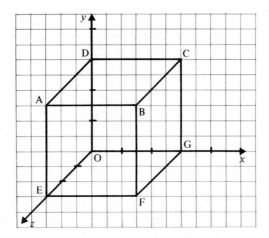

The diagram shows a cube of edge 3 units. Find

a) the coordinates of all eight vertices

b) the coordinates of the midpoints of the edges
   (i) BC   (ii) BF   (iii) DC   (iv) AD

c) the coordinates of the points of intersection of the diagonals of the square
   (i) ABFE   (ii) BCGF   (iii) ABCD.

Questions 6 to 8 refer to the diagram given below.

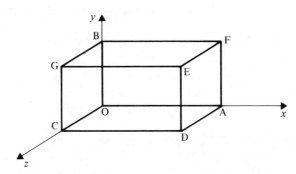

6. If OA = 2, OB = 6 and OC = 3 write down the coordinates of   a) D   b) E   c) A.

7. If the diagram represents a cube of edge 4 units write down the coordinates of each vertex.

8. If E is the point (4, 6, 2) write down the coordinates of each of the other vertices.

9.

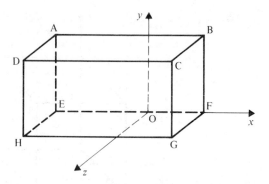

The diagram shows a cuboid with its vertices labelled A to H. The coordinates of B are $(3, 4, 0)$ and the coordinates of D are $(-5, 4, 2)$.

a) Write down the coordinates of the other vertices.
b) Write down the coordinates of the midpoint of
   (i) AB   (ii) CD.

In questions 10 and 11 the units are centimetres.

10. A, B, and C are the points $(6, 0, 0)$, $(6, 4, 0)$ and $(6, 4, 8)$ respectively.

a) Draw a diagram showing clearly the points A, B and C.
b) How far does a spider walk if it goes from A to C via B?

11. A, B, and C are the points $(3, 0, 0)$, $(3, 4, 0)$ and $(3, 4, 12)$ respectively.

a) Draw a diagram showing clearly the points A, B and C.
b) How far is it from O to C via A and B?
c) How far is it directly from   (i) O to B   (ii) O to C?

12. P, Q, and R are the points $(3, 0, 0)$, $(3, 3, 0)$ and $(3, 3, 3)$ respectively.

a) Draw a diagram showing clearly the points P, Q and R.
b) How far is it from O to R via P and Q?
c) What is the direct distance from   (i) O to Q   (ii) O to R?

13. D, E, and F are the points $(2, 0, 0)$, $(2, 6, 0)$ and $(2, 6, 2)$ respectively.

a) Draw a diagram showing clearly the points D, E and F.
b) How far is it from O to F via D and E?
c) What is the direct distance from   (i) O to E   (ii) O to F?

14. The cube of edge 3 units which is shown in the diagram is rotated through 90° about the axis OG in the sense C to B to F.

Find the new coordinates of   a) A   b) B   c) C.

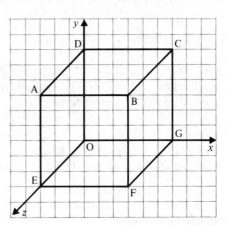

23 ALGEBRAIC FRACTIONS

SIMPLIFYING FRACTIONS

We simplify a fraction such as $\frac{10}{50}$ by recognising that 10 is a common factor of the numerator and denominator and then cancelling that common factor,

i.e. $$\frac{10}{50} = \frac{\cancel{10}^{1}}{5 \times \cancel{10}_{1}} = \frac{1}{5}$$

To simplify an algebraic fraction, we do exactly the same: we find and then cancel the common factors of the numerator and denominator.

Note that we do not have to write the number 50 as 5×10 but that when the factors are letters it helps at this stage to put in the multiplication sign.

For example xy can be written as $x \times y$

and $2(a + b)$ can be written as $2 \times (a + b)$

EXERCISE 23a

Simplify   a) $\dfrac{2xy}{6y}$ b) $\dfrac{2a}{a^2b}$

a)
$$\frac{2xy}{6y} = \frac{\cancel{2}^{1} \times x \times \cancel{y}^{1}}{\cancel{6}_{3} \times \cancel{y}_{1}}$$
$$= \frac{x}{3}$$

b)
$$\frac{2a}{a^2b} = \frac{2 \times \cancel{a}^{1}}{\cancel{a}_{1} \times a \times b}$$
$$= \frac{2}{ab}$$

Simplify:

1. $\dfrac{2x}{8}$

2. $\dfrac{ab}{2b}$

3. $\dfrac{p^2}{pq}$

4. $\dfrac{a^2}{ab}$

5. $\dfrac{xy}{y^2}$

6. $\dfrac{3}{6a}$

386

7. $\dfrac{2ab}{4bc}$ **11.** $\dfrac{7a}{14}$ **15.** $\dfrac{3pq}{6p}$

8. $\dfrac{6p}{3pq}$ **12.** $\dfrac{xz}{2x}$ **16.** $\dfrac{10x}{15xy}$

9. $\dfrac{5p^2q}{10p}$ **13.** $\dfrac{b^2}{bd}$ **17.** $\dfrac{m^2n}{kmn}$

10. $\dfrac{a^2b}{abc}$ **14.** $\dfrac{4}{12x}$ **18.** $\dfrac{5s^2}{20st}$

FACTORS

We know that $3 \times 2 = 6$ but neither $3 + 2$ nor $3 - 2$ is equal to 6. We can write a number as the product of its factors but, in general, we cannot write a number as the sum or difference of its factors.

Thus $\begin{cases} p \text{ and } q \text{ are factors of } pq \\ a \text{ and } (a - b) \text{ are factors of } a(a - b) \end{cases}$

but in general $\begin{cases} p \text{ is } not \text{ a factor of } p + q \\ b \text{ is } not \text{ a factor of } a - b \end{cases}$

This means that in the fraction $\dfrac{p + q}{pq}$ we *cannot* cancel q because q is not a factor of the numerator.

Sometimes the common factors in a fraction are not very obvious.

Consider $\dfrac{x - 2}{y(x - 2)}$

Placing the numerator in brackets and using the multiplication sign gives $\dfrac{(x - 2)}{y \times (x - 2)}$

Now we can see clearly that $(x - 2)$ is a common factor, so

$$\frac{\cancel{(x - 2)}}{y \times \cancel{(x - 2)}} = \frac{1}{y}$$

EXERCISE 23b

Simplify where possible   a) $\dfrac{2a(a-b)}{a-b}$ b) $\dfrac{pq}{p-q}$

a) $\dfrac{2a(a-b)}{a-b} = \dfrac{2 \times a \times (a - b)}{(a - b)}$

$= 2a$

b) $\dfrac{pq}{p-q} = \dfrac{p \times q}{(p-q)}$ which cannot be simplified.

Simplify where possible:

1. $\dfrac{x-y}{x(x-y)}$

4. $\dfrac{p+q}{2p}$

7. $\dfrac{(p-q)(p+q)}{p+q}$

2. $\dfrac{st}{s(s-t)}$

5. $\dfrac{4x}{8(x-y)}$

8. $\dfrac{(4+a)}{(4+a)(4-a)}$

3. $\dfrac{2a}{a-b}$

6. $\dfrac{3(a+b)}{6ab}$

9. $\dfrac{a-b}{3(a+b)}$

10. $\dfrac{u-v}{v(u-v)}$

13. $\dfrac{10a}{15(a-b)}$

16. $\dfrac{(u+v)(u-v)}{u+v}$

11. $\dfrac{xy}{x(x+y)}$

14. $\dfrac{8(x-y)}{12xy}$

17. $\dfrac{x+y}{2(x-y)}$

12. $\dfrac{s-t}{2(s-t)}$

15. $\dfrac{s-t}{3s}$

18. $\dfrac{s+6}{(s+6)(s-6)}$

Sometimes we have to factorise the numerator and/or the denominator before we look for common factors.

Consider for example $\dfrac{12a-4b}{3a-b}$

In the numerator there is a common factor of 4, so

$$12a - 4b = 4(3a - b)$$

The fraction becomes $\dfrac{4(3a - b)}{(3a - b)}$

It is now clear that 4 and $(3a - b)$ are factors of $12a - 4b$ so we

can cancel $(3a - b)$, i.e. $\dfrac{4(3a - b)^{1}}{(3a - b)_{1}} = 4$

We have been writing expressions such as $4(3a - b)$ in the form $4 \times (3a - b)$, but the multiplication sign is not necessary and we will no longer put it in. If, however, you find that the multiplication sign makes it easier to see individual factors then continue to use it.

EXERCISE 23c

> Simplify $\dfrac{xy - y^2}{3y}$
>
> $$\dfrac{xy - y^2}{3y} = \dfrac{{}^{1}y(x - y)}{3y_{1}}$$
>
> $$= \dfrac{x - y}{3}$$

Simplify:

1. $\dfrac{4a}{8a - 2b}$

2. $\dfrac{2pq}{p^2 - pq}$

3. $\dfrac{a - b}{a^2 - ab}$

4. $\dfrac{3a - 6b}{5a - 10b}$

5. $\dfrac{2x - x^2}{3xy}$

6. $\dfrac{a^2}{3a - ab}$

7. $\dfrac{2a - 3b}{6a^2 - 9ab}$

8. $\dfrac{2s^2 - st}{2s - t}$

9. $\dfrac{3a - 6b}{a^2 - 2ab}$

10. $\dfrac{6x}{9x - 3y}$

11. $\dfrac{3ab}{ab + b^2}$

12. $\dfrac{p^2 + pq}{5p}$

13. $\dfrac{2p - 4q}{6p - 12q}$

14. $\dfrac{3a + a^2}{4ab}$

15. $\dfrac{2x - xy}{x^2}$

16. $\dfrac{2x + y}{6xy + 3y^2}$

17. $\dfrac{a^2b - ac}{ab - c}$

18. $\dfrac{p^2 + 2pq}{2p + 4q}$

Simplify $\dfrac{pq - 3q}{p^2 - 6p + 9}$

$$\frac{pq - 3q}{p^2 - 6p + 9} = \frac{q(p-3)}{(p - 3)(p-3)}$$

$$= \frac{q}{p - 3}$$

Simplify:

19. $\dfrac{a - 4}{a^2 - 6a + 8}$

22. $\dfrac{2a - 8}{a^2 - a - 12}$

25. $\dfrac{xy - 2y}{x^2 - 4x + 4}$

20. $\dfrac{x - 2}{x^2 - 6x + 8}$

23. $\dfrac{3x + 12}{x^2 + 7x + 12}$

26. $\dfrac{pq + 5q}{p^2 + 7p + 10}$

21. $\dfrac{y + 3}{y^2 + 5y + 6}$

24. $\dfrac{9y - 36}{y^2 - 2y - 8}$

27. $\dfrac{st - t}{s^2 - 8s + 7}$

28. $\dfrac{p + 3}{p^2 + 6p + 9}$

30. $\dfrac{2x + 6}{x^2 - x - 12}$

32. $\dfrac{uv + 6v}{u^2 + 12u + 36}$

29. $\dfrac{x - 5}{x^2 + x - 30}$

31. $\dfrac{3x - 15}{x^2 - 9x + 20}$

33. $\dfrac{xy - 7y}{x^2 - 9x + 14}$

FURTHER SIMPLIFYING

The next exercise contains slightly more complicated factorising.

Remember that $\qquad a^2 - b^2 = (a - b)(a + b)$

It sometimes happens that the numerator has a factor $(a - b)$

and the denominator has a factor $(b - a)$

Now $\qquad (a - b) = (-1)(b - a)$

therefore a fraction such as $\dfrac{a - b}{b - a}$ can be simplified,

i.e. $\qquad \dfrac{a - b}{b - a} = \dfrac{(-1)(b-a)}{(b-a)} = -1$

EXERCISE 23d

Simplify   a) $\dfrac{2x^2+5x-3}{4x^2-1}$ b) $\dfrac{a^2-2ab+b^2}{b^2-ab}$

a)
$$\frac{2x^2+5x-3}{4x^2-1}=\frac{\cancel{(2x-1)}(x+3)}{\cancel{(2x-1)}(2x+1)}$$
$$=\frac{(x+3)}{(2x+1)}$$

b)
$$\frac{a^2-2ab+b^2}{b^2-ab}=\frac{(a-b)(a-b)}{b(b-a)}$$
$$=\frac{(-1)\cancel{(b-a)}(a-b)}{b\cancel{(b-a)}}$$
$$=\frac{(-1)(a-b)}{b}$$
$$=\frac{b-a}{b}$$

Simplify:

1. $\dfrac{x^2-9}{2x^2-7x+3}$

2. $\dfrac{4x-8}{x^2-4}$

3. $\dfrac{4x^2-1}{2x^2-3x-2}$

4. $\dfrac{2-x}{x^2-4x+4}$

5. $\dfrac{a^2-b^2}{b^2-2ab+a^2}$

6. $\dfrac{2a^2+ab-b^2}{4a^2-b^2}$

7. $\dfrac{3x^2-xy-2y^2}{9x^2-4y^2}$

8. $\dfrac{x^2-5x+6}{3y-xy}$

9. $\dfrac{a-a^2}{a-1}$

10. $\dfrac{2y^2+5y-3}{4y^2-1}$

11. $\dfrac{x^2-6xy+9y^2}{x^2-3xy}$

12. $\dfrac{4x^2-7x-2}{4x^2-8x}$

13. $\dfrac{2x^2-13x+15}{x^2-10x+25}$

14. $\dfrac{a-1}{1-a^2}$

15. $\dfrac{ac + bc - ad - bd}{c - d}$

16. $\dfrac{2x^2 + 3x - 35}{7 + 5x - 2x^2}$

17. $\dfrac{4x^2 + 6x - 4}{x^2 - x - 6}$

18. $\dfrac{x^2 - xy - 2y^2}{y^2 + xy}$

19. $\dfrac{1 - x^2}{3x^2 + 9x + 6}$

20. $\dfrac{y + x + xy + y^2}{x^2 + 2xy + y^2}$

MULTIPLYING AND DIVIDING FRACTIONS

Reminder: The product of two fractions is found by multiplying the numerators and multiplying the denominators,

e.g.
$$\frac{2}{3} \times \frac{4}{5} = \frac{2 \times 4}{3 \times 5} = \frac{8}{15}$$

and
$$\frac{p}{q} \times \frac{(a - b)}{(a + b)} = \frac{p(a - b)}{q(a + b)}$$

To divide by a fraction, we multiply by its reciprocal,

e.g.
$$\frac{2}{3} \div \frac{3}{4} = \frac{2}{3} \times \frac{4}{3} = \frac{8}{9}$$

$$\frac{p}{q} \div r = \frac{p}{q} \times \frac{1}{r} = \frac{p}{qr}$$

and
$$\frac{p}{q} \div \frac{a}{(a - b)} = \frac{p}{q} \times \frac{(a - b)}{a} = \frac{p(a - b)}{qa}$$

EXERCISE 23e Find:

1. $\dfrac{a}{b} \times \dfrac{c}{d}$

2. $\dfrac{a}{b} \div \dfrac{c}{d}$

3. $\dfrac{x - y}{2} \times \dfrac{5}{x}$

4. $\dfrac{x - y}{2} \div \dfrac{5}{x}$

5. $\dfrac{a}{b} \div c$

6. $\dfrac{a}{b} \times c$

7. $\dfrac{(a - b)}{4} \div \dfrac{(a + b)}{3}$

8. $\dfrac{(x - 2)}{3} \times (x + 3)$

9. $\dfrac{(x - 2)}{3} \div (x + 3)$

10. $\dfrac{p}{q} \div \dfrac{1}{r}$

As is the case in number fractions, it is sometimes possible to simplify before multiplying.

Simplify $\dfrac{ab}{4} \times \dfrac{8}{a^2}$

$$\frac{ab}{4} \times \frac{8}{a^2} = \frac{{}^1ab}{\cancel{4}_1} \times \frac{\cancel{8}^2}{\cancel{a^2}a}$$

$$= \frac{2b}{a}$$

Simplify:

11. $\dfrac{2a}{b} \div \dfrac{a^2}{3b^2}$

12. $\dfrac{pq}{6} \times \dfrac{3}{p^2}$

13. $\dfrac{4xy}{3} \times \dfrac{9}{x^2}$

14. $\dfrac{2ab}{5} \div \dfrac{a}{b}$

15. $\dfrac{2p^2}{3} \times \dfrac{q}{4p}$

16. $\dfrac{x^2}{4} \div \dfrac{xy}{2}$

17. $\dfrac{1}{b^2} \div \dfrac{2}{b}$

18. $\dfrac{7p}{5q} \times \dfrac{10q}{21p^2}$

19. $\dfrac{a^2}{2b} \div 2a$

20. $\dfrac{a}{b} \times \dfrac{2a}{3b} \div \dfrac{2b}{3a}$

Simplify $\dfrac{(a^2 - b^2)}{a} \times \dfrac{b}{(a + b)}$

$$\frac{(a^2 - b^2)}{a} \times \frac{b}{(a + b)} = \frac{(a - b)\cancel{(a + b)} \times b}{a\cancel{(a + b)}}$$

$$= \frac{b(a - b)}{a}$$

(Notice that we leave the answer in factor form.)

Simplify $(x^2 - 3x - 4) \div (x - 4)$

$\left(\begin{array}{l}\text{Remember that } 7 \div 4 \text{ can be written } \frac{7}{4}; \\ \text{similarly } (x^2 - 3x - 4) \div (x - 4) \text{ can be written } \dfrac{x^2 - 3x - 4}{x - 4}\end{array}\right)$

$$(x^2 - 3x - 4) \div (x - 4) = \frac{x^2 - 3x - 4}{x - 4}$$

$$= \frac{\cancel{(x - 4)}^1 (x + 1)}{\cancel{(x - 4)}_1}$$

$$= x + 1$$

Simplify:

21. $\dfrac{(b - 2)}{4} \times \dfrac{1}{(b^2 - 4b + 4)}$

22. $\dfrac{(x^2 - 4)}{3} \times \dfrac{6}{(x + 2)}$

23. $\dfrac{(a^2 - 9)}{2} \div \dfrac{(a - 3)}{4}$

24. $\dfrac{(3b - 6)}{5} \div \dfrac{(b - 2)}{10}$

25. $(x^2 - 6x + 9) \div (x - 3)$

26. $\dfrac{1}{(x - 2)} \times (x^2 - 5x + 6)$

27. $(x - 4) \div (x^2 - 6x + 8)$

28. $(2x - 4) \times \dfrac{1}{(x^2 + 2x - 8)}$

29. $\dfrac{(x^2 - 4)}{5} \times \dfrac{3}{(x^2 + 8x + 12)}$

30. $\dfrac{(4x^2 - 9)}{3} \div \dfrac{(6x + 9)}{2}$

31. $(3x - 6) \div (3x^2 - 4x - 4)$

32. $\dfrac{(2x^2 - 5x + 3)}{2} \times \dfrac{(x - 1)}{(x^2 - 2x + 1)}$

33. $(4x^2 - 1) \div (12x^2 + 8x + 1)$

34. $\dfrac{(ab + a^2)}{b} \times \dfrac{b}{a + b}$

35. $\dfrac{a^2 - b^2}{c} \div \dfrac{b^2 - ab}{c^2}$

36. $\dfrac{(x^2 - 5x + 4)}{(x + 2)} \times \dfrac{(x^2 - 4)}{(x - 1)}$

LOWEST COMMON MULTIPLE

Before we can simplify $\frac{2}{3}+\frac{1}{5}$ we must change both $\frac{2}{3}$ and $\frac{1}{5}$ into equivalent fractions with the same denominator. This common denominator must contain both 3 and 5 as factors; there are many numbers we could choose but 15 is the lowest such number, i.e. 15 is the *lowest common multiple* (LCM) of 3 and 5.

To simplify $\dfrac{3}{x}+\dfrac{2}{y}$ we follow the same pattern. We need a common denominator with both x and y as factors. Again there are many we could use, but the simplest is xy; this is the LCM of x and y.

EXERCISE 23f

Find the LCM of ab and c

The LCM is abc

Find the LCM of:

1. p, q **4.** a, b, c **7.** v, uw

2. r, st **5.** x, y, wz **8.** $3, 7, 8$

3. $2, 3, 5$ **6.** a, d **9.** p, q, r

Find the LCM of   a) $4, 10$ b) ab, a^2 c) $2x, 6x$

a) $4 = 2 \times 2$ and $10 = 2 \times 5$

(The LCM is the *lowest* number that 4 and 10 divide into exactly, so the factors it must include are

2×2 from 4 and 5 from 10

The factor of 2 from 10 is not needed as 2 is already included.)

$\therefore$ the LCM of 4 and 10 is $2 \times 2 \times 5 = 20$

b) $ab = a \times b$ and $a^2 = a \times a$

$\therefore$ the LCM is $a \times b \times a = a^2b$

c) $2x = 2 \times x$ and $6x = 2 \times 3 \times x$

$\therefore$ the LCM is $2 \times 3 \times x = 6x$

Find the LCM of:

10. x, xy **13.** $x^2, 2xy$ <u>**16.**</u> $3p, p^2$

11. $x^2, 2x$ **14.** ab, bc <u>**17.**</u> $5a, ab$

12. $pq, 3p$ **15.** s, st <u>**18.**</u> $3pq, q^2$

19. $2x, 3x$ **22.** $6, 4, 10$ <u>**25.**</u> $4x, 6x$

20. $4x, 8x$ **23.** a, ab, a^2 <u>**26.**</u> $3y, 5y$

21. $6a, 9a$ **24.** $10x, 15x$ <u>**27.**</u> $2x, 3x, 4x$

ADDITION AND SUBTRACTION OF FRACTIONS

To add or subtract fractions we first have to change them into equivalent fractions with a common denominator.

Thus to find $\frac{2}{3} + \frac{1}{5}$, we choose a common denominator of 15 which is the LCM of 3 and 5.

Now $\dfrac{2}{3} = \dfrac{2 \times 5}{3 \times 5} = \dfrac{10}{15}$ and $\dfrac{1}{5} = \dfrac{1 \times 3}{5 \times 3} = \dfrac{3}{15}$

Therefore $\dfrac{2}{3} + \dfrac{1}{5} = \dfrac{10 + 3}{15} = \dfrac{13}{15}$

To simplify $\dfrac{3}{x} + \dfrac{2}{y}$ we follow the same pattern:

xy is the LCM of x and y.

$$\dfrac{3}{x} = \dfrac{3 \times y}{x \times y} = \dfrac{3y}{xy} \quad \text{and} \quad \dfrac{2}{y} = \dfrac{2 \times x}{y \times x} = \dfrac{2x}{xy}$$

$\therefore$ $\dfrac{3}{x} + \dfrac{2}{y} = \dfrac{3y + 2x}{xy}$

EXERCISE 23g

Simplify $\dfrac{1}{2a} + \dfrac{1}{b}$

($2ab$ is the LCM of $2a$ and b)

$$\dfrac{1}{2a} + \dfrac{1}{b} = \dfrac{(1) \times (b) + (1) \times (2a)}{2ab}$$

$$= \dfrac{b + 2a}{2ab}$$

Simplify $\dfrac{3}{4x} - \dfrac{1}{6x}$

(12x is the LCM of 4x and 6x)

$$\frac{3}{4x} - \frac{1}{6x} = \frac{(3) \times (3) - (1) \times (2)}{12x}$$

$$= \frac{7}{12x}$$

Simplify:

1. $\dfrac{1}{x} + \dfrac{1}{y}$ **7.** $\dfrac{2}{x} - \dfrac{3}{y}$ **13.** $\dfrac{2}{y} - \dfrac{3}{4y}$

2. $\dfrac{3}{p} + \dfrac{2}{q}$ **8.** $\dfrac{4}{3p} + \dfrac{2}{q}$ **14.** $\dfrac{3}{8p} - \dfrac{1}{4p}$

3. $\dfrac{2}{s} - \dfrac{1}{t}$ **9.** $\dfrac{3}{x} - \dfrac{2}{y}$ **15.** $\dfrac{1}{a} + \dfrac{5}{8a}$

4. $\dfrac{3}{a} + \dfrac{1}{2b}$ **10.** $\dfrac{5}{7a} + \dfrac{3}{4b}$ **16.** $\dfrac{1}{3x} - \dfrac{1}{7x}$

5. $\dfrac{1}{3x} - \dfrac{2}{5y}$ **11.** $\dfrac{1}{2x} + \dfrac{1}{3x}$ **17.** $\dfrac{4}{7x} - \dfrac{2}{5x}$

6. $\dfrac{1}{a} + \dfrac{5}{2b}$ **12.** $\dfrac{2}{5x} - \dfrac{3}{7x}$ **18.** $\dfrac{1}{y} - \dfrac{2}{3y}$

Simplify $\dfrac{4a}{3b} - \dfrac{b}{6a}$

($3b = 3 \times b$ and $6a = 2 \times 3 \times a$, $\therefore$ LCM $= 6ab$)

$$\frac{4a}{3b} - \frac{b}{6a} = \frac{(4a) \times (2a) - (b) \times (b)}{6ab}$$

$$= \frac{8a^2 - b^2}{6ab}$$

Simplify:

19. $\dfrac{1}{2a} + \dfrac{3}{4b}$

25. $\dfrac{s}{t^2} + \dfrac{s^2}{2t}$

31. $\dfrac{5}{7x} - \dfrac{3}{14xy}$

20. $\dfrac{a}{2b} - \dfrac{a^2}{b^2}$

26. $\dfrac{5}{2a} + \dfrac{2}{3ab}$

32. $\dfrac{9}{a^2} - \dfrac{3}{2ab}$

21. $\dfrac{3}{x} - \dfrac{4}{xy}$

27. $\dfrac{1}{x^2} + \dfrac{2}{3x}$

33. $\dfrac{3x}{2y} - \dfrac{3y}{2x}$

22. $\dfrac{2}{p^2} - \dfrac{3}{2p}$

28. $\dfrac{2y}{3x} - \dfrac{3x}{2y}$

34. $\dfrac{7}{9p} - \dfrac{5}{6q}$

23. $\dfrac{3a}{4b} + \dfrac{b}{6a}$

29. $\dfrac{5}{8x} + \dfrac{2}{4y}$

35. $\dfrac{a^2}{b^2} + \dfrac{4a}{5b}$

24. $\dfrac{5}{2p} - \dfrac{3}{4q}$

30. $\dfrac{p}{3q} + \dfrac{p^2}{q^2}$

36. $\dfrac{7}{5pq} + \dfrac{8}{15q}$

EXERCISE 23h

Simplify $\dfrac{x-2}{3} - \dfrac{x-4}{2}$

$$\dfrac{(x-2)}{3} - \dfrac{(x-4)}{2} = \dfrac{2(x-2) - 3(x-4)}{6}$$

$$= \dfrac{2x - 4 - 3x + 12}{6}$$

$$= \dfrac{-x + 8}{6}$$

$$= \dfrac{8 - x}{6}$$

(Notice that we placed brackets round the numerators *before* putting the fractions over a common denominator. This ensured that each numerator was kept together and that the signs were not confused.)

Simplify:

1. $\dfrac{x+2}{5} + \dfrac{x-1}{4}$

5. $\dfrac{x+3}{7} - \dfrac{x+2}{5}$

2. $\dfrac{x+3}{4} - \dfrac{x+1}{3}$

6. $\dfrac{x+4}{5} + \dfrac{x-1}{2}$

3. $\dfrac{2x-1}{3} + \dfrac{x+2}{5}$

7. $\dfrac{2x-1}{7} - \dfrac{x-2}{5}$

4. $\dfrac{2x+3}{4} - \dfrac{x-2}{6}$

8. $\dfrac{3x+1}{14} - \dfrac{2x+3}{21}$

Remember that $\frac{1}{5}(x-2) = \dfrac{(x-2)}{5}$

9. $\frac{1}{7}(2x-3) - \frac{1}{3}(4x-2)$

14. $\dfrac{2x+3}{5} - \dfrac{3x-2}{4}$

10. $\frac{1}{4}(5x-1) - \frac{1}{3}(2x-3)$

15. $\dfrac{3-x}{2} + \dfrac{1-2x}{6}$

11. $\dfrac{5-2x}{3} + \dfrac{4-3x}{2}$

16. $\dfrac{2+5x}{8} - \dfrac{3-4x}{6}$

12. $\frac{1}{4}(3-x) + \frac{1}{6}(1-2x)$

17. $\frac{1}{5}(4-3x) + \frac{1}{10}(3-x)$

13. $\frac{1}{8}(5x+4) - \frac{1}{3}(4x-1)$

18. $\frac{1}{9}(4-x) - \frac{1}{6}(2+3x)$

Simplify $\dfrac{2(x+1)}{3} - \dfrac{3(x-2)}{5}$

$$\dfrac{2(x+1)}{3} - \dfrac{3(x-2)}{5} = \dfrac{5 \times 2(x+1) - 3 \times 3(x-2)}{15}$$

$$= \dfrac{10(x+1) - 9(x-2)}{15}$$

$$= \dfrac{10x+10 - 9x+18}{15}$$

$$= \dfrac{x+28}{15}$$

Simplify:

19. $\dfrac{4(x + 2)}{3} + \dfrac{2(x - 1)}{5}$

23. $\dfrac{3(x - 1)}{2} + \dfrac{3(x + 4)}{7}$

20. $\dfrac{3(x - 1)}{4} + \dfrac{2(x + 1)}{3}$

24. $\dfrac{7(x - 3)}{3} - \dfrac{2(x + 5)}{9}$

21. $\dfrac{2(x - 2)}{3} - \dfrac{3(x - 1)}{7}$

25. $\dfrac{2(3x - 1)}{5} + \dfrac{4(2x - 3)}{15}$

22. $\dfrac{5(2x - 1)}{2} - \dfrac{4(x + 3)}{5}$

26. $\dfrac{3(x - 2)}{5} - \dfrac{7(x - 4)}{6}$

Simplify $\dfrac{2}{x} - \dfrac{1}{x + 2}$

$$\frac{2}{x} - \frac{1}{(x + 2)} = \frac{(2)(x + 2) - (1)(x)}{x(x + 2)}$$

$$= \frac{2x + 4 - x}{x(x + 2)}$$

$$= \frac{x + 4}{x(x + 2)}$$

(Notice that we placed the two-term denominator in brackets. Notice also that we left the common denominator in factorised form.)

Simplify:

27. $\dfrac{2}{a} + \dfrac{1}{a + 3}$

31. $\dfrac{3}{a} + \dfrac{2}{a + 4}$

28. $\dfrac{4}{x + 2} + \dfrac{2}{x}$

32. $\dfrac{3}{x - 1} + \dfrac{4}{x}$

29. $\dfrac{3}{x - 4} + \dfrac{1}{2x}$

33. $\dfrac{3}{2x + 1} + \dfrac{1}{3x}$

30. $\dfrac{2}{2x + 1} - \dfrac{3}{4x}$

34. $\dfrac{5}{2x + 3} - \dfrac{2}{5x}$

MIXED QUESTIONS

EXERCISE 23i Simplify:

1. $\dfrac{2}{a} - \dfrac{b}{c}$

2. $\dfrac{pq}{r} \times \dfrac{r^3}{p^2}$

3. $\dfrac{x+2}{4} + \dfrac{x-5}{3}$

4. $\dfrac{a^2 + ab}{a^2 - b^2}$

5. $\dfrac{3}{4x} - \dfrac{2}{3x}$

6. $\dfrac{x+3}{x^2 + 5x + 6}$

7. $\dfrac{p^2 - pq}{q^2 - p^2}$

8. $\dfrac{4}{x^2} - \dfrac{2}{3x}$

9. $\dfrac{1}{x} - \dfrac{3}{x+1}$

10. $\dfrac{a^2}{bc} \div \dfrac{a}{b^2}$

11. $\dfrac{2}{5x} \div \dfrac{3}{4x}$

12. $\dfrac{2}{5x} + \dfrac{3}{4x}$

13. $\dfrac{2}{5x} \times \dfrac{3}{4x}$

14. $\dfrac{x+4}{5} + \dfrac{2x-1}{10}$

15. $\dfrac{x+4}{5} \times \dfrac{2x-1}{10}$

16. $\dfrac{5}{4x} + \dfrac{5}{6x}$

17. $\dfrac{5}{4x} \times \dfrac{5}{6x}$

18. $\dfrac{5}{4x} \div \dfrac{5}{6x}$

19. $\dfrac{1}{3x} + \dfrac{6}{x-1}$

20. $\dfrac{1}{3x} \times \dfrac{6}{x-1}$

21. $\dfrac{3}{2a} - \dfrac{2}{a-1}$

22. $\dfrac{3}{2a} \times \dfrac{2}{a-1}$

23. $\dfrac{3}{4y-3} \div \dfrac{y}{4y-3}$

24. $\dfrac{3}{4y-3} - \dfrac{4y}{4y-3}$

SOLVING EQUATIONS WITH FRACTIONS

Remember that when solving an equation we *must* keep the equality true. This means that if we alter the size of one side of the equation then we must alter the other side in the same way.

Consider the equation $\dfrac{1}{x} + \dfrac{1}{2x} = \dfrac{5}{6}$

If we choose to multiply each side by the LCM of the denominators, we can remove all fractions from the equation.

The LCM of x, $2x$ and 6 is $6x$. Multiplying each side by $6x$ gives

$$6x\left(\frac{1}{x} + \frac{1}{2x}\right) = 6x \times \frac{5}{6}$$

$$\therefore \qquad \frac{6x^1}{1} \times \frac{1}{x_1} + \frac{{}^3 6x}{1} \times \frac{1}{2x_1} = \frac{{}^1 6x}{1} \times \frac{5}{6_1}$$

$$\therefore \qquad\qquad\qquad 6 + 3 = 5x$$

$$9 = 5x$$

$$\frac{9}{5} = x \quad \text{i.e.} \quad x = 1\frac{4}{5}$$

EXERCISE 23j Solve the following equations:

1. $\dfrac{1}{2} + \dfrac{4}{x} = 1$

6. $\dfrac{3}{x} + \dfrac{3}{10} = \dfrac{9}{10}$

2. $\dfrac{2}{3} - \dfrac{1}{x} = \dfrac{13}{15}$

7. $\dfrac{3}{8} - \dfrac{2}{x} = \dfrac{1}{6}$

3. $\dfrac{3}{4} - \dfrac{2}{x} = \dfrac{5}{12}$

8. $\dfrac{3}{2x} + \dfrac{1}{4x} = \dfrac{1}{3}$

4. $\dfrac{1}{x} - \dfrac{1}{3x} = \dfrac{1}{2}$

9. $\dfrac{1}{x} - \dfrac{1}{2} = \dfrac{3}{2x}$

5. $\dfrac{1}{2x} + \dfrac{2}{x} = \dfrac{1}{4}$

10. $\dfrac{3}{2x} + \dfrac{2}{5} = \dfrac{5}{x}$

Solve the equation $\dfrac{x-2}{4} - \dfrac{x-3}{6} = 2$

$$\frac{(x-2)}{4} - \frac{(x-3)}{6} = 2$$

Multiply each side by 12

$$12\left[\frac{(x-2)}{4} - \frac{(x-3)}{6}\right] = 12 \times 2$$

$$\therefore \quad \frac{\overset{3}{\cancel{12}}}{1} \times \frac{(x-2)}{\underset{1}{\cancel{4}}} - \overset{2}{\cancel{12}} \times \frac{(x-3)}{\underset{1}{\cancel{6}}} = 24$$

$$\therefore \quad 3(x-2) - 2(x-3) = 24$$
$$3x - 6 - 2x + 6 = 24$$
$$x = 24$$

Solve the following equations:

11. $\dfrac{x+2}{4} + \dfrac{x-3}{2} = \dfrac{1}{2}$

12. $\dfrac{x}{4} - \dfrac{x+3}{3} = \dfrac{1}{2}$

13. $\dfrac{x}{5} + \dfrac{x+1}{4} = \dfrac{8}{5}$

14. $\dfrac{2x}{5} - \dfrac{x-3}{8} = \dfrac{1}{10}$

15. $\dfrac{x-4}{3} - \dfrac{x+1}{4} = \dfrac{1}{6}$

16. $\dfrac{x+3}{5} + \dfrac{x-2}{4} = \dfrac{3}{10}$

17. $\dfrac{2}{3} - \dfrac{x+1}{9} = \dfrac{5}{6}$

18. $\dfrac{x+3}{4} - \dfrac{x}{2} = 5$

19. $\dfrac{3x}{20} + \dfrac{x-2}{8} = \dfrac{3}{10}$

20. $\dfrac{x+3}{7} - \dfrac{x-4}{3} = 1$

21. $\dfrac{2x-1}{7} + \dfrac{3x-3}{4} = \dfrac{1}{7}$

22. $\dfrac{2x}{9} - \dfrac{3x+2}{4} = \dfrac{7}{12}$

Sometimes a fractional equation is a quadratic equation in disguise.

Solve the equation $\dfrac{x-4}{2} + \dfrac{4}{x} = 1$

$$\frac{(x-4)}{2} + \frac{4}{x} = 1$$

Multiply each side by $2x$ $2x\left[\dfrac{(x-4)}{2} + \dfrac{4}{x}\right] = 2x \times 1$

$$\frac{2x}{1} \times \frac{(x-4)}{2} + \frac{2x}{1} \times \frac{4}{x} = 2x$$

$$x(x-4) + 8 = 2x$$
$$x^2 - 4x + 8 = 2x$$

(This contains an x^2 term so it is a quadratic equation. Therefore we collect all terms on one side.)

Take $2x$ from each side $x^2 - 6x + 8 = 0$

∴ $(x-2)(x-4) = 0$

∴   either $x - 2 = 0$ or $x - 4 = 0$

∴ $x = 2$ or $x = 4$

Solve the equations in questions 23 to 42:

23. $\dfrac{x+5}{2} + \dfrac{1}{x} = 1$ **26.** $\dfrac{x+12}{3} + \dfrac{3}{x} = 2$

24. $\dfrac{2}{x} + \dfrac{x+1}{3} = 2$ **27.** $\dfrac{x+9}{2} - \dfrac{2}{x} = 3$

25. $\dfrac{x+2}{2} + \dfrac{2}{x} + 1 = 0$ **28.** $x + 8 + \dfrac{9}{x} = 2$

29. $\dfrac{2-x}{4} + \dfrac{1}{2x} = \dfrac{3}{4x}$ **31.** $\dfrac{1}{3x} - \dfrac{x-2}{4} = \dfrac{1}{6}$

30. $\dfrac{1}{x} - \dfrac{2}{x-2} = 3$ **32.** $\dfrac{2}{x+3} + x = 0$

Not all of the following questions give quadratic equations:

33. $\dfrac{1}{2x} + \dfrac{1}{4x} = \dfrac{1}{6}$

36. $\dfrac{2x-1}{3} - \dfrac{3x}{2} = 2$

34. $\dfrac{x+2}{3} - \dfrac{3x}{2} = \dfrac{1}{5}$

37. $\dfrac{x+4}{3} + 2 = \dfrac{x}{4}$

35. $\dfrac{x-1}{2} + \dfrac{1}{x} = 1$

38. $\dfrac{x+1}{x} + \dfrac{1}{2} = 4$

39. $\dfrac{x}{5} + \dfrac{1}{x+1} = 1$

41. $\dfrac{4x+1}{2} - \dfrac{1}{2x} = \dfrac{1}{2}$

40. $\dfrac{1}{2}(x-1) + \dfrac{1}{3}(2x-3) = 2$

42. $\dfrac{x-3}{x} = 0$

MIXED EXERCISES

EXERCISE 23k

1. Simplify:

a) $\dfrac{ab^2}{2ab}$

b) $\dfrac{a(a+b)}{a+b}$

c) $\dfrac{a^2-b^2}{a+b}$

2. Simplify:

a) $\dfrac{1}{x} + \dfrac{1}{3x}$

b) $\dfrac{1}{x} \times \dfrac{1}{3x}$

c) $\dfrac{1}{x} \div \dfrac{1}{3x}$

3. Solve the equations:

a) $\dfrac{x+1}{3} - \dfrac{x-1}{2} = 3$

b) $\dfrac{x+1}{3} - \dfrac{1}{x} = 1$

4. a) Simplify $\dfrac{1}{2}(x-1) + \dfrac{1}{3}(x-2)$

b) Solve the equation $\dfrac{1}{2}(x-1) + \dfrac{1}{3}(x-2) = \dfrac{1}{4}$

EXERCISE 23l   **1.** Simplify:

$$\text{a) } \frac{6xy}{3y^2} \qquad \text{b) } \frac{2x(x-y)}{4x^2} \qquad \text{c) } \frac{x^2+4x+3}{x+1}$$

2. Simplify:

$$\text{a) } \frac{1}{2p} - \frac{1}{3p} \qquad \text{b) } (x^2 - 2x) \div x \qquad \text{c) } \frac{2y^2}{3} \times \frac{9}{4xy}$$

3. Solve the equations:

$$\text{a) } \frac{x}{4} + \frac{3x-2}{6} = \frac{1}{3} \qquad \text{b) } \frac{x-5}{4} + \frac{1}{2x} = \frac{4}{x}$$

4. a) Simplify $\dfrac{x-2}{4} + \dfrac{3}{x}$

b) Solve the equation $\dfrac{x}{2} - \dfrac{2x-4}{3} = \dfrac{1}{4}$

EXERCISE 23m   **1.** Simplify:

$$\text{a) } \frac{uv^2}{wu^2v} \qquad \text{b) } \frac{a-b}{(b-a)(b-2a)} \qquad \text{c) } \frac{x^2}{3x-x^2}$$

2. Simplify:

$$\text{a) } \frac{3s}{t} \div \frac{1}{6st} \qquad \text{b) } \frac{x^2-4}{3} \times \frac{6}{x+2} \qquad \text{c) } \frac{3}{4x-1} - \frac{2}{x}$$

3. Solve the equations:

$$\text{a) } \frac{x-2}{4} - \frac{x-3}{5} = \frac{3}{10} \qquad \text{b) } \frac{x-2}{4} + \frac{1}{2x} = \frac{1}{4}$$

4. a) Simplify $\frac{1}{2}(x-2) - \frac{1}{3}(x-3)$
b) Solve the equation $\frac{1}{2}(x-2) - \frac{1}{3}(x-3) = 5$

24 LOCI

LOCI IN TWO DIMENSIONS

A *locus* is the set of all the points whose positions satisfy a given rule.

When the locus is a straight or curved line it is convenient to think of the locus as the path that is traced out by a single moving point.

Remember that every point on a locus must obey the given conditions or law, and that every point which obeys the law must lie on the locus.

The plural of locus is *loci*.

EXERCISE 24a

Describe, and illustrate with a sketch, the locus of the tip of the minute hand of a clock as it moves between 12 noon and 12.30 p.m.

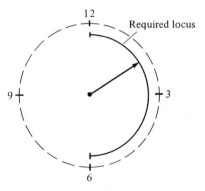

The required locus is a semicircle, centre at the centre of the clockface, radius the length of the minute hand.

In questions 1 to 10 describe, and illustrate with a sketch, the given locus.

1. The tip of the minute hand of a clock as it moves between 1 a.m. and 2 a.m.

2. The tip of the hour hand of a clock as it moves between 1 a.m. and 2 a.m.

407

3. A cricket ball when bowled at the wicket.

4. A cricket ball when hit along the ground for four.

5. A cricket ball when hit for six.

6. The centre of the wheel of a bicycle as the bicycle
a) travels in a straight line
b) travels around a bend.

7. The number at the top of this page as you turn the page over.

8. A satellite circling the earth.

9. The earth moving around the sun.

10. A goat on the end of a rope winding it around a tree.

A rod, OA, turns through a complete revolution about one end, O, which is fixed. Describe the locus of A.

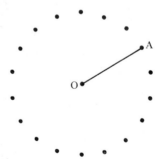

(Mark several possible positions for A until the overall shape of the path of A becomes clear.)

The locus of A is the circumference of a circle, centre O radius OA.

---

11. The minute hand of a clock is 80 cm long. Describe the locus of its tip
a) from 5 a.m. to 6 a.m.   b) from 2.15 p.m. to 2.45 p.m.

12. Describe the locus of a point on this page which moves so that it is always 3 cm from the top edge of this page.

13. Draw a straight line AB on a page of your exercise book. Describe the locus of a point X on the page which moves so that it is always 3 cm away from AB.

14.

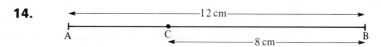

The rod AB is rotated about C. Describe the locus of
a) the point A   b) the point B.

15.

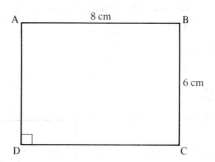

ABCD is a rectangle in which AB = 8 cm and BC = 6 cm. Describe the locus of points that are 3 cm from both AB and DC.

16.

A and B are two fixed points. Describe the locus of the points on this page that are equidistant from A and B.

17.

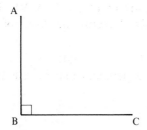

If $\widehat{ABC} = 90°$ describe the locus of points that are equidistant from AB and BC.

18. ABCD is a square of side 10 cm. Find the locus of points within the square equidistant from

a) AB and BC   b) AB and AD.

Is there any point that is equidistant from all three lines AB, BC and AD?
If so, where is it?

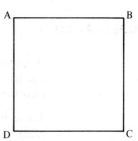

19.

A ─────────────────────────→──────────────── B

C ────────────────────→──────────────────── D

AB and CD are two parallel lines. Describe the locus of a point, between the two lines, that is always twice as far from AB as it is from CD.

20.

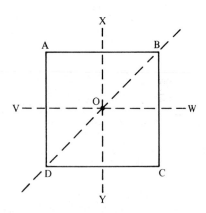

ABCD is a square, centre O, of side 4 cm. VW is parallel to AB and XY is parallel to AD. Describe the locus of A as the square is rotated about

a) XY   b) VW   c) DB

d) the axis through O perpendicular to ABCD.

SPECIAL LOCI

In questions 11 and 12 from exercise 24a the locus is a set of points traced out by a particular point, while in questions 15–19 the locus is a set of position points, all of which exist at the same time.

These loci have introduced us to the four most important loci in two dimensional work.

1. The locus of a point that moves in such a way that it is always at a fixed distance from a fixed point is called a circle. The fixed point is the centre of the circle, and the fixed distance is its radius.

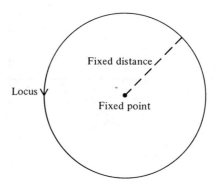

2. The locus of a point that moves in such a way that it is at a constant distance (d) from a line through two fixed points A and B, is the pair of straight lines drawn parallel to AB and distant d from it.

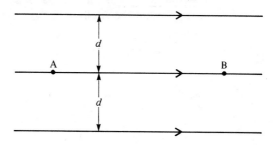

3. For points that are equidistant from two fixed points A and B, the locus is the perpendicular bisector of AB.

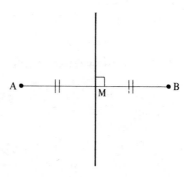

4. For points that are equidistant from two intersecting straight lines AXB and CXD, the locus is the pair of bisectors of the angles between the given lines. These bisectors are always at right angles to each other.

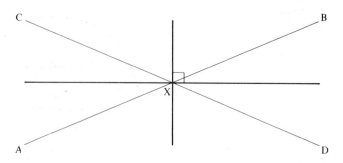

EXERCISE 24b **1.**

M is the midpoint of a chord AB of fixed length in a circle, centre O. Describe the locus of M as AB moves around the circle.

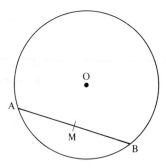

2.

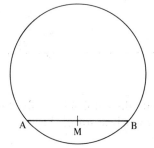

M is the midpoint of a chord AB. Describe the locus of the midpoints of the set of chords parallel to AB.

3.

A and B are two fixed points where AB = 6 cm. Describe the locus of a third point C if the area of triangle ABC is 12 cm².

4.

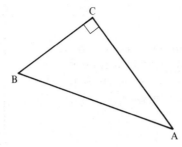

A and B are two fixed points.
Describe the locus of C if
$\widehat{ACB} = 90°$.

5. A is a fixed point. Describe the locus of the centres of circles which pass through A and have a radius of 5 cm.

6.

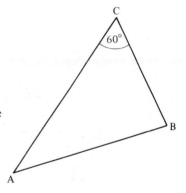

A and B are two fixed points.
Describe the locus of C if
$\widehat{ACB} = 60°$.

(Remember that angles in the same segment of a circle are equal.)

7.

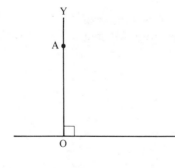

A is a point on OY such that
OA = 4 cm and B is a point on
OX. If $\widehat{XOY} = 90°$ describe the
locus of the midpoint of AB as B
moves along OX.

8.

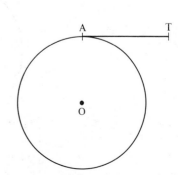

TA is of fixed length.
O is the centre of the circle and
$\widehat{OAT} = 90°$. Describe the locus
of T as A moves around the given
circle.

9.

Sketch the locus of   a) D   b) C
as the rectangle ABCD is rotated
through 90° clockwise about A in
the plane of the page.

10.

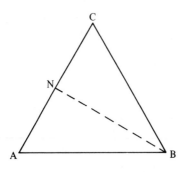

ABC is an equilateral triangle.
The triangle is rotated
clockwise about B until BC
becomes parallel to the lower
edge of the page. Sketch the
locus of

a) C
b) the foot, N, of the
   perpendicular from B to
   AC.

What angle has BA turned
through?

11.

Draw the locus of the centres of
circles that touch both AB and
BC.

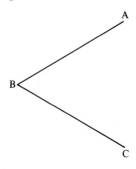

12.

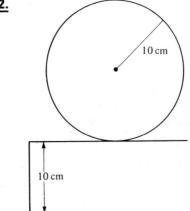

A wheel of radius 10 cm rolls
across a horizontal path, and
then down a step 10 cm deep,
before continuing to roll
horizontally. Sketch the locus
of the centre of the wheel.

13. Describe the locus of the centre of a coin of diameter 2 cm if it

    a) rolls around the inside of a circle with centre C and radius 5 cm
    b) rolls around the outside of a circle of radius 5 cm.

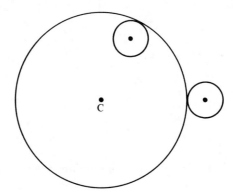

14. Describe the locus of the centre of a circle, of variable radius, which passes through two fixed points A and B.

15. Draw any triangle ABC such that $\widehat{ABC} = 90°$. Draw the locus of points equidistant from   a) A and B   b) B and C.

    What do you notice about the point of intersection of the two loci?

16.

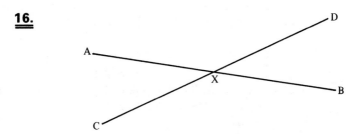

    A house P is to be built 100 m from a road AXB and 50 m from a road CXD. Show on a sketch the possible positions for P.

17.   X   Y

    X and Y represent two houses 100 m apart. Sketch the loci that will enable you to shade the area of land that is both nearer to X than to Y and is within 60 m of Y.

18.

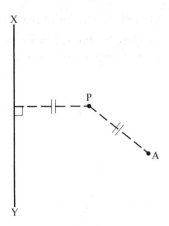

Plot the locus of a point P whose distance from a fixed point A is equal to its distance from a fixed line XY.

ACCURATE DRAWING

EXERCISE 24c Remember to make a rough sketch before you start each construction. Use suitable instruments including a *sharp* pencil.

1. Construct an angle $\widehat{ABC}$ of 60°. Construct the locus of points equidistant from AB and BC.

2. A and B are two points such that AB = 9 cm. Construct the locus of points equidistant from A and B.

3. A and B are two points such that AB = 8 cm.
   a) Construct the locus of points that are 3 cm from A.
   b) Construct the locus of points that are 2 cm from the line AB.
   c) In how many points do these loci intersect? How far is each point from B?

4. Two straight lines AXB and CXD intersect at X such that $\widehat{AXC} = 90°$.
   a) Construct the locus of points that are 4.5 cm from X.
   b) Construct the locus of points equidistant from AXB and CXD.
   c) In how many points do these loci intersect? How far is each point from X?

5. Draw a line AB that is 10 cm long. Construct the locus of a point P such that $\triangle ABP$ is isosceles with AP = BP.

6. Construct a triangle ABC in which AB = 10 cm, AC = 9 cm and BC = 8 cm.

   a) Construct the locus of points equidistant from A and B.

   b) Construct the locus of points equidistant from B and C.

   c) Describe the point of intersection, O, of the loci you have drawn in (a) and (b).

   d) Draw a circle, centre O, radius OA. (This is the *circumcircle* of ABC, O is the *circumcentre*.)

7. Draw a line AB that is 8 cm long. Construct the locus of a point C such that $A\widehat{C}B = 90°$ (C may lie on either side of AB).

8. a) Construct a rectangle ABCD such that AB = 12 cm and BC = 8 cm.

   b) Draw the locus of points equidistant from AB and BC.

   c) Draw the locus of points equidistant from A and B.

   d) Mark the point P that lies on the loci referred to in both (b) and (c). Measure PC.

9. a) Draw AB of length 12 cm. Construct the locus of a point P, above AB, such that $A\widehat{P}B = 90°$.

   b) Draw the locus of points that are 5 cm above AB.

   c) Mark the points P and Q that are 5 cm from AB, such that $A\widehat{P}B = A\widehat{Q}B = 90°$. Find the difference in length between AP and PB.

10. a) Construct a rectangle ABCD such that AB = 10 cm and BC = 8 cm.

   b) Draw the locus of points equidistant from AB and CD.

   c) Draw the locus of points, within the rectangle, that are 8 cm from C.

   d) Mark the point E, that is both equidistant from AB and CD, and 8 cm from C. Measure AE.

11. Draw a line AB that is 8 cm long. Construct the locus of a point P such that the area of $\triangle ABP$ is 24 cm^2.

12. a) Construct a triangle ABC in which AB = 12 cm, AC = 11 cm and BC = 8 cm.

   b) Draw the locus of points equidistant from AB and AC.

   c) Draw the locus of points equidistant from AC and CB.

   d) Mark, with I, the point of intersection of the two loci you have found in (b) and (c). What is special about the point I?

13. a) Construct a triangle ABC in which AB = 13 cm, $\widehat{ABC} = 45°$ and $\widehat{BAC} = 30°$.

b) Draw the locus of points that are 2.5 cm from BC.

c) Draw the locus of points that are 1.5 cm from AB.

d) Hence find the point, P, within the triangle, that is 2.5 cm from BC and 1.5 cm from AB. Measure AP.

Construct a triangle ABC in which AB = 9.5 cm, BC = 7 cm and $\widehat{ABC} = 60°$. Find the point D, within the triangle, that is 2 cm from AB and 5 cm from C. Measure BD.

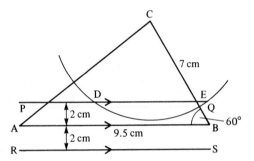

(Points that are 2 cm from AB lie on one or other of the two lines, PQ and RS, that are shown parallel to AB. Points that are 5 cm from C lie on the circle, centre C, radius 5 cm. This circle cuts PQ at D and E but cannot cut RS. Therefore RS need not be drawn in the accurate construction. From the sketch, D satisfies the given conditions but we cannot be certain whether E lies inside or outside the triangle until we do the construction.)

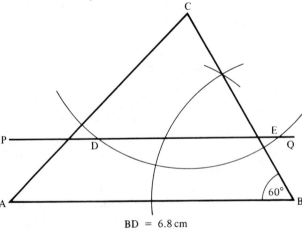

BD = 6.8 cm

14. ABCD is a rectangle with AB = 8 cm and BC = 5 cm. Construct this rectangle and find the point P which is 2 cm from AB and equidistant from AD and BC. Measure PB.

15. Construct a triangle ABC in which AB = 9.5 cm, BC = 8 cm and $\widehat{ABC} = 60°$. Find the point D, on the opposite side of AB from C, that is 7 cm from BC and 4.5 cm from AC. Measure CD.

16. Construct a rectangle ABCD with AB = 6.5 cm and AD = 8 cm. Find the point X which is 3 cm from AD and equidistant from AD and DC. Measure DX.

17. ABC is a triangle in which AB = 12 cm, BC = 9 cm and $\widehat{ABC} = 90°$. Construct this triangle and find a point D that is 4.5 cm from BC and equidistant from A and C. Measure AD.

LOCI INVOLVING REGIONS

A locus is the set of points that satisfies a given condition. If this condition involves an inequality, the set of points is a region rather than a line. For example, the locus of a point P that moves in such a way that it is always 10 cm from a fixed point A, is the *circumference* of a circle, centre A, radius 10 cm, whereas the locus of a point P that moves in such a way that it is always less than 10 cm from a fixed point A, is the region *within the circle* centre A, radius 10 cm.

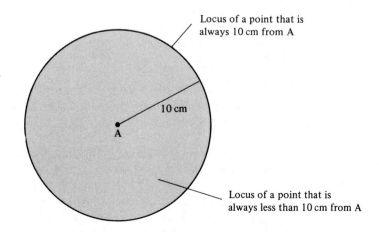

Locus of a point that is always 10 cm from A

10 cm

A

Locus of a point that is always less than 10 cm from A

EXERCISE 24d

A is a fixed point. Show, by shading in a suitable sketch, the locus of P such that AP < 5 cm.

(We begin by drawing the locus of P such that AP = 5 cm. This is a circle, centre A, radius 5 cm.)

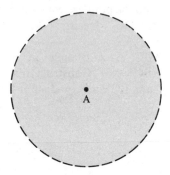

(If P is such that AP < 5 cm, P must be within the circle. P cannot lie on the circumference of the circle. We show this by using a dotted line. If the circle were to be included, it would be drawn as a solid line.)

1. A is a fixed point. Show, by shading in a suitable sketch, the locus of P such that AP ≤ 6 cm.

2. A is a fixed point. If 3 cm < AP < 6 cm, illustrate the locus of P.

3. ABC is a triangle. Illustrate the locus of P, within the triangle, such that BP < PC.

4. AB is a fixed line, 10 cm long. Illustrate the locus of P such that P is more than 3 cm from any point on AB but less than 6 cm from any point on AB.

5. AB is a fixed line, 10 cm long. Illustrate the locus of a point P such that $A\widehat{P}B < 90°$.

6. AB is a fixed line, 10 cm long. Illustrate the locus of a point P that is less than 7 cm from A and more than 8 cm from B.

7. Draw a line AB of length 6 cm. Draw accurately, on the same diagram, the locus of

a) the point P which moves so that AP = PB
b) the point Q such that $A\widehat{Q}B = 90°$
c) the point R which moves so that the area of triangle ARB is 6 cm².

Mark on your diagram a point X such that AX < XB, $A\widehat{X}B = 90°$, and the area of triangle AXB is 6 cm². Measure AX.

LOCI IN THREE DIMENSIONS

In three-dimensional work there are two important loci that concern us.

1. If a point moves so that it is always at a fixed distance from a fixed point its locus is a sphere.
   The fixed point is the centre of the sphere and the fixed distance is its radius.

2. The locus of points that are equidistant from two given points A and B is the plane bisecting AB at right angles.

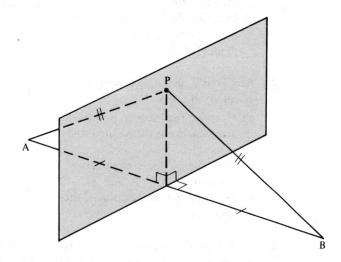

EXERCISE 24e   **1.** A is a fixed point and a point P moves in space so that AP = 6 cm. Describe the locus of P.

   2. A and B are two points 10 cm apart. P is a point in space such that PA = PB. Describe the locus of P.

3. a) A is a fixed point and a point X moves in space so that the length of AX is 5 cm. Describe the locus of X.

   b) B is a second fixed point so that AB = 8 cm. A point Y moves in space so that AY = YB. Describe the locus of Y.

   c) Describe the set of points of intersection of the two loci.

4. Describe the locus of a point in space that is always 10 cm from the surface of a sphere of radius 5 cm.

5. Describe the locus of a point in space that is always 5 cm from the surface of a sphere of radius 10 cm. What important difference is there between your answer to this question and your answer to question 4?

6. ABCD is a rectangle with AB = 12 cm and BC = 10 cm.

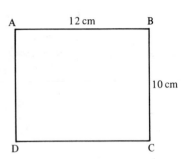

   a) Describe the locus of A as the rectangle is rotated about DC through 360°.

   b) Describe the locus of A as the rectangle is rotated about BC through 360°.

   c) Describe the locus of A as the rectangle is rotated about BD through 360°.

7. ABCD is a rectangle measuring 10 cm by 8 cm.

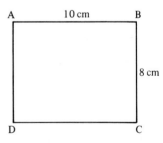

   a) Describe the locus of points that are 8 cm from the *plane* of the rectangle.

   b) Describe the locus of points equidistant from A and D.

   c) Describe the intersection of these two loci.

8. A, B and C are three points in space. Describe the locus of points equidistant from A and B and 10 cm from C.

9. Describe the locus of points equidistant from three given points A, B and C.

25 PLANS AND ELEVATIONS

PLANS AND ELEVATIONS

An architect presents drawings of a building partly in picture form and partly as diagrams.

The drawings need to show a *plan* of each floor, that is, a diagram of what is seen when looking down.

They also show *elevations*, that is, side and front views of the building and sometimes the back view as well.

This is a simple example.

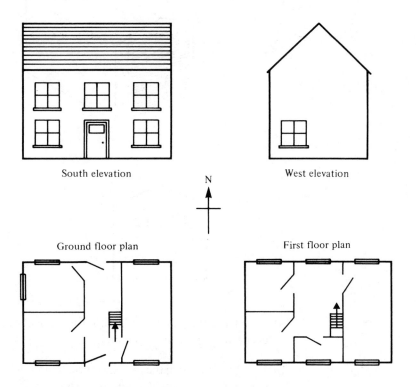

South elevation

N

West elevation

Ground floor plan

First floor plan

Notice that the heights of the two elevations agree and that the widths of the plans and elevations agree.

EXERCISE 25a **1.** a) How many windows are there on the east side of the house on the previous page?

b) Would the east elevation be the same width as the west elevation or the same width as the south elevation?

c) Draw the east elevation of the house. Make sure that it is the correct height and width.

2. Here is a drawing of a free-standing garage with one window. The rear elevation of the garage is also shown.

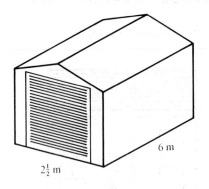

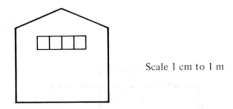

Scale 1 cm to 1 m

a) Sketch the plan and mark in the measurements, then draw the plan accurately on squared paper.

b) Use the drawing of the rear elevation to find the total height of the garage and the height of the side wall.

c) Sketch, and draw accurately, a side elevation of the garage.

3.

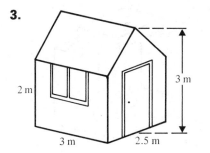
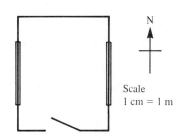

Scale
1 cm = 1 m

Above are drawings of a garden shed and the plan of the floor of the shed.

a) Draw the south elevation of the shed.

b) Draw the west elevation of the shed.

Notice that there is no information about the heights of windows or doors. Decide for yourself what they should be.

ARCHITECTS' DRAWINGS

Architects prepare drawings to show to their clients. For the builders, the drawings are on a large scale and show every detail and measurement.

(These working drawings are often called blueprints because they used to be copied by a method that gave white lines on a blue background.)

TECHNICAL DRAWING

Technical or engineering drawings are produced for the same reason as architectural ones, i.e. so that an object may be made exactly to the instructions.

Most drawings are now prepared on a computer using Computer Aided Design (CAD). They are very sophisticated and can show the view from any direction.

On the next page there is an example of a working drawing of a clamp. The drawing would probably be full-size but the measurements would be given as well.

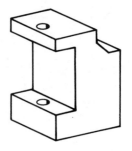

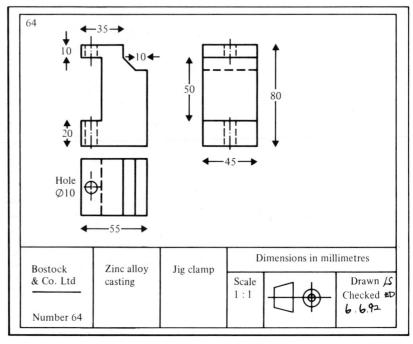

The person who prepares these drawings is called a draughtsman. There are rules for doing these drawings. In this chapter we will introduce a few of them.

If you look at a cube you will see something like this, which is a picture of the cube in perspective.

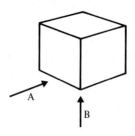

Technical drawings, however, are diagrams and not pictures.

For the elevation or view in the direction marked A we draw a square.

Elevation in direction A

For the plan (which is the view from above) we also draw a square.

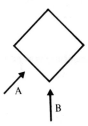

For the elevation in the direction marked B we see two sides but they do not look like squares. They are both foreshortened because we see them at an angle.

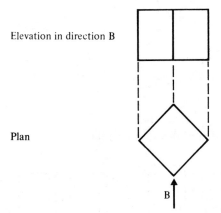

Elevation in direction B

Plan

Notice that the width of this elevation is the same as that of the plan and that its length is equal to that of a diagonal of the square.

EXERCISE 25b

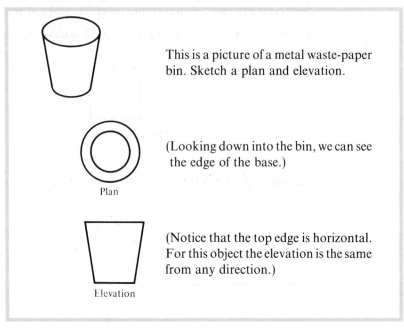

This is a picture of a metal waste-paper bin. Sketch a plan and elevation.

(Looking down into the bin, we can see the edge of the base.)

Plan

(Notice that the top edge is horizontal. For this object the elevation is the same from any direction.)

Elevation

1. This diagram shows part of a stone staircase.

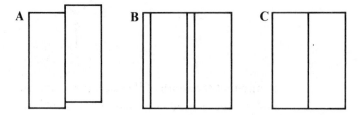

a) Which of the following diagrams is the plan?

A   B   C

b) Which of the following diagrams is the elevation from X?

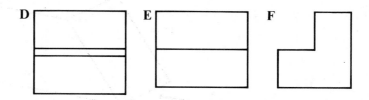

c) Which of the following drawings is the elevation from Y?

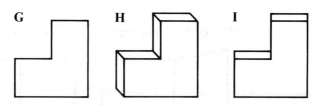

2.

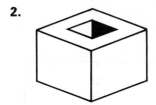

Sketch the plan of this object.

3.

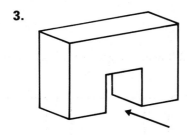

Sketch the elevation of this object in the direction of the arrow.

4.

a) Sketch a plan of this lampshade.

b) Sketch an elevation of the lampshade. Does it matter from which direction you choose to view it?

5.

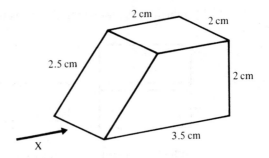

a) Which of the following diagrams is the plan of this object?

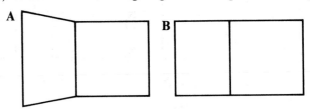

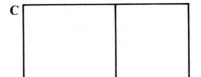

b) Which of the following diagrams is the elevation in the direction marked X?

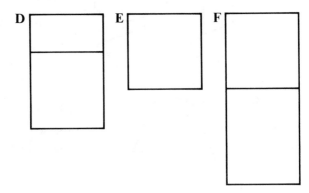

ACCURATE DRAWING

EXERCISE 25c

This is a square-based pyramid.

a) Sketch the plan of the pyramid and mark in the known measurements.

b) Which given measurement have you not used?

c) Draw the plan accurately.

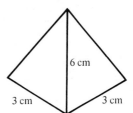

a) (Looking down from above, we see the edge of the square base with the top in the centre.)

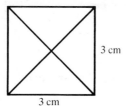

b) The length of the sloping edge, 6 cm, is not used.

c)

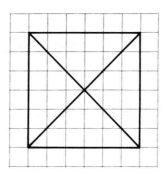

In each question, sketch the required diagrams and mark any known measurements.

Then draw each diagram accurately on squared paper.

1.

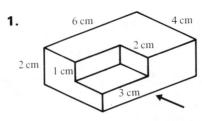

a) The plan.
b) The elevation in the direction of the arrow.

2.

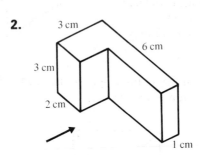

a) The plan.
b) The elevation in the direction of the arrow.

3. The height and depth of each step is 2 cm.

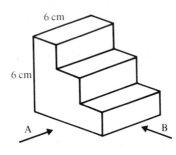

a) The plan.
b) The elevation in direction A.
c) The elevation in direction B.

4. The plan of a sphere of radius 3 cm.

5. This cylinder has a radius of 4 cm and a height of 6 cm.

a) The plan of the cylinder.
b) The elevation. (This is the same from any direction.)

HIDDEN LINES

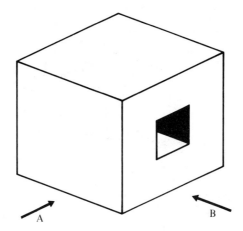

This cube has a hole through it. You might think that the plan of it is nothing more than a square, but to indicate the position of the hole we represent the hidden edges by broken lines.

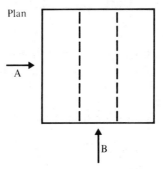

You can imagine that the object is made of clear plastic so that you see the hidden edges faintly.

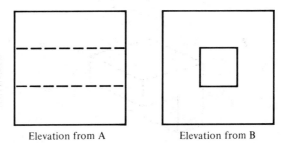

Elevation from A   Elevation from B

EXERCISE 25d In this exercise, use squared paper.

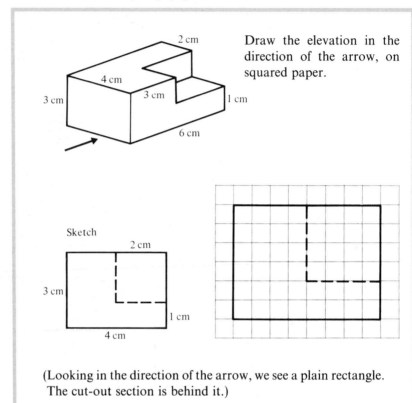

Draw the elevation in the direction of the arrow, on squared paper.

(Looking in the direction of the arrow, we see a plain rectangle. The cut-out section is behind it.)

In each question from 1 to 4, draw the plan.

1.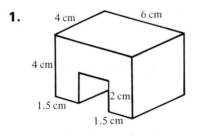

3.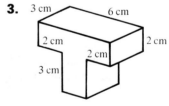

2.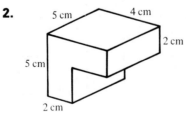

4.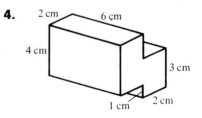

In each question from 5 to 8, draw the elevation in the direction of the arrow.

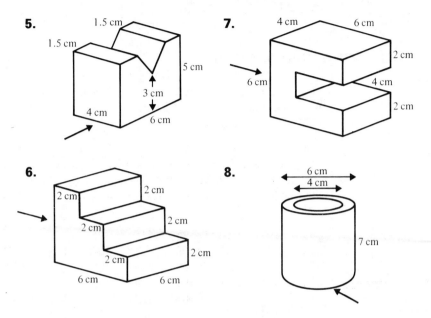

5.

1.5 cm

1.5 cm

5 cm

3 cm

4 cm

6 cm

7.

4 cm   6 cm

2 cm

6 cm   4 cm

2 cm

6.

2 cm   2 cm

2 cm   2 cm

2 cm   2 cm

6 cm   6 cm

8.

6 cm

4 cm

7 cm

In questions 9 and 10 draw the plan and the elevations in the directions of the arrows.

9.

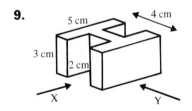

5 cm   4 cm

3 cm

2 cm

X   Y

All the pieces have a thickness of 1 cm.

10.

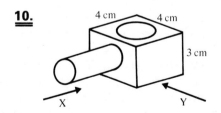

4 cm   4 cm

3 cm

X   Y

The hole, which goes right through this object, has a diameter of 3 cm.
The handle has a diameter of 2 cm. It is 4 cm long and is fixed to the centre of the face.

26 STATISTICS

COLLECTING INFORMATION

"Information Technology" is the name used for modern methods of dealing with information. Before information can be distributed, it has to be collected and organised.

The local health authority wanted some information about the heights of five-year-old children in its area. At their first school medical examination, the height of each child was recorded. The following figures are the recorded heights (in centimetres) of ninety five-year-olds from one infant school:

99	107	102	98	115	95	106	110	108	105
118	102	114	108	94	104	113	102	105	95
105	110	109	101	106	108	107	107	101	109
108	105	116	109	114	110	97	110	113	116
112	101	92	105	104	115	111	103	110	99
93	104	103	113	107	94	102	117	116	104
99	114	106	114	98	109	107	114	106	107
109	113	112	100	109	113	118	104	94	114
107	96	108	103	112	106	115	111	115	101

This set of figures was written down in the same order as the children came into the medical examination, so the heights are listed in a random order. Disorganised figures like these are called *raw data*. They need organising before we can make sense of them.

FREQUENCY TABLES

Because there are ninety heights recorded in the table above, it will be easier to work with them if we organise them in groups. Now if we use a number line to represent the heights of these children there is no point on the line which could not represent someone's height, i.e. information about height is continuous data. So the grouping we choose must not leave any "gaps" between the values included in consecutive groups.

Taking h cm to represent the height of any child, a suitable grouping is

$90 \leqslant h < 95 \quad 95 \leqslant h < 100 \quad 100 \leqslant h < 105 \quad 105 \leqslant h < 110 \quad 110 \leqslant h < 115 \quad 115 \leqslant h < 120$

Counting the number of heights in each group gives the following table which is called a frequency table:

Group	Tally	Frequency
$90 \leqslant h < 95$	卌	5
$95 \leqslant h < 100$	卌 ||||	9
$100 \leqslant h < 105$	卌 卌 卌 ||	17
$105 \leqslant h < 110$	卌 卌 卌 卌 卌 |||	28
$110 \leqslant h < 115$	卌 卌 卌 卌 |	21
$115 \leqslant h < 120$	卌 卌	10

Total: 90

When you make a frequency table from a set of raw data, work down the columns, making a tally mark for each value in the tally column next to the appropriate group. Do *not* go through the data looking for values that fit into the first group and the second group and so on.

BAR CHARTS

We saw in Book 1A that a frequency table can be illustrated by a bar chart.

The following bar chart illustrates the frequency table above:

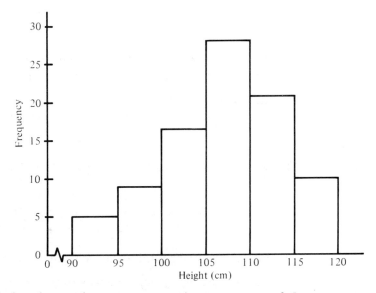

Notice that each group covers the same span of 5 cm so we make each bar the same width.

EXERCISE 26a **1.** The local fish and chip shop had 56 customers one Saturday evening. They spent the following amounts, in pence:

270	110	45	96	250	490	324	45
382	136	125	450	420	380	150	250
85	250	320	525	218	210	216	120
155	430	250	40	510	150	510	245
320	120	316	150	260	45	180	310
273	280	85	280	318	45	210	282
462	316	218	316	325	45	560	315

This data is *discrete*, i.e. each item is a whole number. This is because there is no number of pence between say, 99 p and 100 p. So we can use groups 0–99 p, 100–199 p, 200–299 p, 300–399 p, 400–499 p, 500–599 p, to make a frequency table and illustrate the data with a bar chart.

2. Fifty Junior School children joined the school's computer club. Their ages were recorded:

10	8	9	10	7	8	8	11	10	9
7	8	9	9	10	11	11	10	9	8
8	7	9	7	10	7	10	8	9	11
10	11	8	10	9	8	9	7	11	10
9	10	10	11	10	11	7	11	10	9

Make a frequency table showing the number of children of each age and illustrate this information with a bar chart.

Remember that this data is continuous, so we can use groups 7 years to less than 8 years, 8 years to less than 9 years and so on.

3. The weights of 30 bags of popcorn sold at a fête are given below:

69	83	75	65	68	68	73	70	80	79
70	76	63	86	69	65	66	74	66	68
70	60	67	74	65	65	67	88	81	63

First decide whether this data is discrete or continuous.

Choose your own groups and make a frequency table. Illustrate the data with a bar chart. Compare and discuss your bar chart with those of other members of your class.

HISTOGRAMS

The following frequency table was compiled from the weights of 100 people. This data is continuous because there is no number of kilograms which could not be someone's weight. In this example there are 15 people with weights from 50 kg up to, but not including, 60 kg. Anyone with a weight of 60 kg belongs in the second group and so on.

Weight (w kg)	$50 \leqslant w < 60$	$60 \leqslant w < 70$	$70 \leqslant w < 80$	$80 \leqslant w < 90$	$90 \leqslant w < 100$	$100 \leqslant w < 110$
Frequency	15	30	35	15	3	2

The bar chart below illustrates this frequency table.

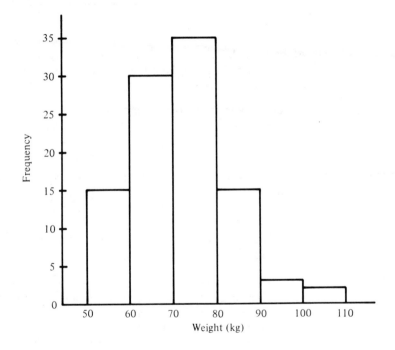

It is the *area* of a bar that gives the impression of the number of items in a group.

When a bar chart is constructed so that the *area* of each bar is proportional to the number of items in each group it is called a *histogram*.

The diagram above is a histogram: each group covers the same span so each bar has the same width. Hence the area of each bar is proportional to its height. In this case, therefore, the height of each bar is proportional to the number of items in the group.

Most bar charts showing frequencies are histograms but some are not. Consider the bar chart below. It shows the number of cans of fizzy drink sold by a manufacturer for each of the years 1982, 1983 and 1984:

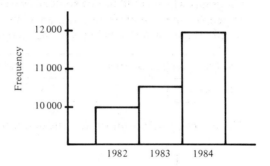

All bars are the same width but, because the frequency scale does not start at zero, the height and therefore the area, of the bar is not proportional to the frequency of the sales that year.

Hence this bar chart is *not* a histogram.

EXERCISE 26b   Draw a histogram to represent the frequency table in each question from 1 to 3.

1. This table shows the distribution of the ages of the people in the audience at a school concert.

Age (*n* years)	$0 \leqslant n < 20$	$20 \leqslant n < 40$	$40 \leqslant n < 60$	$60 \leqslant n < 80$	$80 \leqslant n < 100$
Frequency	73	34	27	10	6

2. This table gives the results of a survey into the weekly earnings of 100 sixteen-year-olds.

Weekly earnings (£ *p*)	$0 \leqslant p < 10$	$10 \leqslant p < 20$	$20 \leqslant p < 30$	$30 \leqslant p < 40$	$40 \leqslant p < 50$
Frequency	45	11	13	21	10

3. The table shows the distribution of the average marks of 40 children in the end-of-year examinations.

Average mark	1–20	21–40	41–60	61–80	81–100
Frequency	2	4	19	12	3

4. The histogram shows the distribution of the times taken by 50 children to get to school.

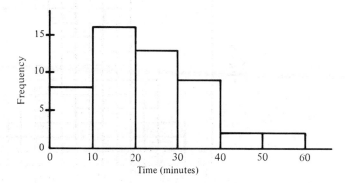

Construct a frequency table from this histogram.

5. The histogram is based on the number of hours that 30 children spent watching television on a particular Saturday.

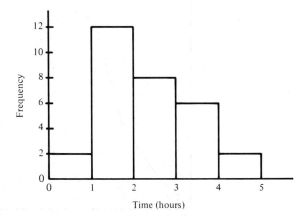

Construct a frequency table from this histogram.

FREQUENCY POLYGONS

This table gives the marks obtained by children in an examination:

Mark	1–10	11–20	21–30	31–40	41–50	51–60	61–70	71–80	81–90	91–100
Frequency	1	3	6	13	14	12	7	5	4	1

The histogram representing this data is given below.

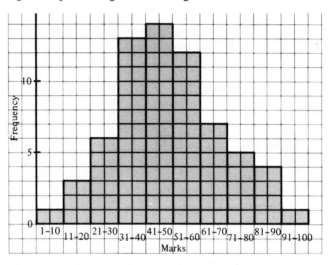

If, on a separate diagram, we plot the middle point at the top of each column of the histogram, and join these points in order, we have a *frequency polygon*. The frequency polygon for our histogram is given below.

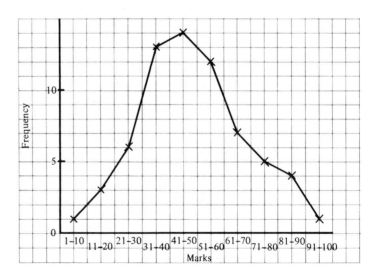

Sometimes the frequency polygon is superimposed on the histogram to give two different representations of the data in a single diagram.

The combined histogram and frequency polygon is shown below.

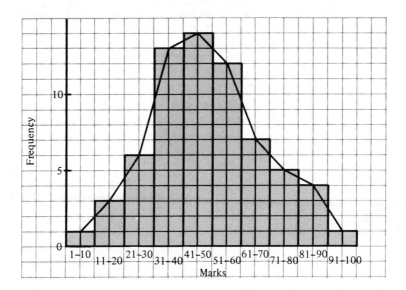

EXERCISE 26c In questions 1 and 2 represent the data given in the table by

a) a histogram   b) a frequency polygon

1. The table shows the results of a survey on the pocket money received by 100 ten-year-olds.

Weekly pocket money (p)	0–19	20–39	40–59	60–79	80–99	100–119
Frequency	5	15	27	33	16	4

2. The table shows the distribution of goals scored in the 46 football league matches played one Saturday.

Total number of goals scored	0	1	2	3	4	5	6	7
Frequency	7	10	9	3	9	4	1	3

3. Draw, on the same diagram, the histogram and frequency polygon for the data given in the following table which shows the distribution of goals scored by the twenty-six teams in a hockey league.

Number of goals scored	50–54	55–59	60–64	65–69	70–74	75–79	80–84
Frequency	3	9	4	6	0	2	2

4. Draw, on the same diagram, the histogram and frequency polygon for the data given in the following table, which shows the distribution of shoe sizes for the sixty teachers on a school staff.

Shoe size	3	4	5	6	7	8	9	10	11
Frequency	2	5	8	9	11	13	8	3	1

VARYING WIDTH BARS

Consider the following frequency table which shows the results of a survey on the weekly pay of 100 women:

Pay (£n)	$0 \leqslant n < 20$	$20 \leqslant n < 40$	$40 \leqslant n < 60$	$60 \leqslant n < 100$	$100 \leqslant n < 200$
Frequency	20	36	25	14	5

The spans of the groups are *not* equal.
the first three groups each cover a span of £20
the fourth group spans £40, twice the span of the first three
the fifth group spans £100, five times the span of the first three.

To illustrate this distribution on a histogram, we must make
the width of the first three bars equal
the width of the fourth bar twice that of the first three
the width of the fifth bar five times that of the first three.

Now consider the 14 incomes in the group £60 up to £100. If we suppose that these incomes are evenly spread throughout this group then there are
seven women with incomes in the group £60 to £80 and
seven women with incomes in the group £80 to £100.

The fourth bar is therefore seven units high for the whole group. Because the width of this bar is twice that of the others used, its *area* is proportional to the frequency of the group.

We can see from this example that if we have a bar twice as wide as the others used, the height of that bar is *half* the number of items in the group.

If we also assume that the incomes in the fifth group are evenly spread throughout the group then there is one woman with an income in each of the subgroups spanning £20. The fifth bar is therefore one unit high for the whole group. Again, because the width of this bar is five times that of the first three bars, its area is proportional to the frequency of the group.

Note that the wider bars are not sub-divided by vertical lines at the intermediate scale numbers, e.g. in the histogram below there are no vertical divisions at 80, 120, 140, 160 or 180.

Note also that we do not use "frequency" to label the vertical axis because it is the area of the bar which represents the frequency and not the height.

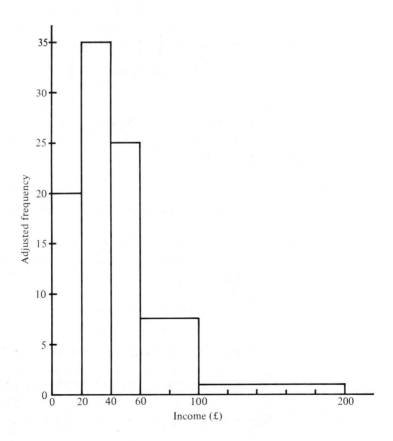

EXERCISE 26d

Draw a histogram to illustrate the following frequency table which shows the distribution of the ages of 300 people attending a school concert:

Age (n years)	$0 \leqslant n < 20$	$20 \leqslant n < 30$	$30 \leqslant n < 40$	$40 \leqslant n < 50$	$50 \leqslant n < 60$	$60 \leqslant n < 100$
Frequency	20	40	100	110	10	20

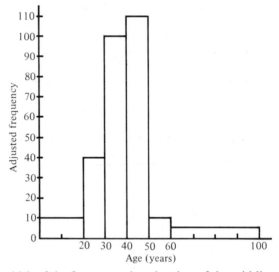

(The width of the first group is twice that of the middle groups, so its height is half of its frequency, i.e. $\frac{1}{2}$ of 20. The last group is four times the width of the first group, so the height of the bar is $\frac{1}{4}$ of 20 on the frequency scale.)

Draw a histogram to illustrate the frequency tables in questions 1 to 5.

1. The table gives the distribution of marks of 30 children in a test:

Mark	0–39	40–59	60–79	80–99
Frequency	8	8	10	4

2. The table gives the distribution of lengths of 50 blades of grass:

Length (h mm)	$40 \leqslant h < 50$	$50 \leqslant h < 60$	$60 \leqslant h < 70$	$70 \leqslant h < 80$	$80 \leqslant h < 110$
Frequency	6	8	12	12	12

3. The table shows the distribution of the times taken by 40 children to complete an obstacle race:

Time (t seconds)	$0 \leqslant t < 40$	$40 \leqslant t < 50$	$50 \leqslant t < 60$	$60 \leqslant t < 70$
Frequency	8	15	7	10

4. The table shows the distribution of weights of 30 bags of chips from a fish and chip shop:

Weight (w grams)	$0 \leqslant w < 50$	$50 \leqslant w < 60$	$60 \leqslant w < 70$	$70 \leqslant w < 80$
Frequency	4	8	14	4

5. The table gives the distribution of marks of 100 children in an end-of-term mathematics examination:

Mark	0–29	30–39	40–49	50–59	60–99
Frequency	10	15	25	34	16

Make a frequency table from the histogram below which shows the distribution of the ages of people boarding buses at the bus station between 0830 and 0900 one morning.

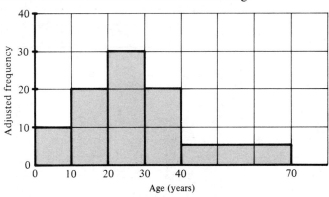

Age (n years)	$0 \leqslant n < 10$	$10 \leqslant n < 20$	$20 \leqslant n < 30$	$30 \leqslant n < 40$	$40 \leqslant n < 70$
Frequency	10	20	30	20	15

(The width of the last bar is three times that of the other bars, so the frequency in this group is three times the height of the bar.)

In questions 6 to 9, make a frequency table from the histogram:

6.

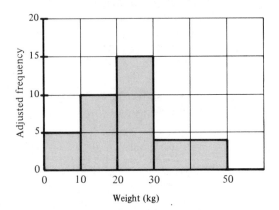

7.

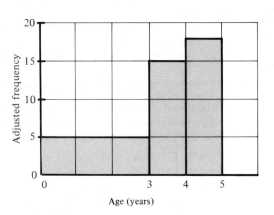

8.

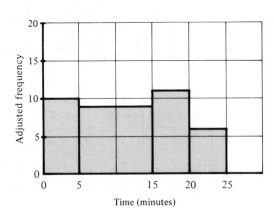

9.

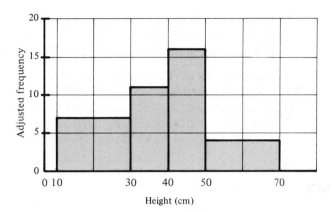

Height (cm)

MEAN, MODE AND MEDIAN

In Book 2A we saw that, when we have a set of numbers there are three different measures we can use to attempt to give a "typical member" that is representative of the set.

MEAN

The *mean* (arithmetic average) of a set of n numbers is the sum of the numbers divided by n.

The mean of the set 2, 6, 8, 8, 10, 10, 12, is

$$\frac{2 + 6 + 8 + 8 + 10 + 10 + 12}{7} = \frac{56}{7} = 8$$

The mean value of a set of numbers is the most frequently used form of average, so much so that the word "average" is often used for "mean value".

For example, if you were asked for your average mark in a set of examinations, you would total the marks and divide by the number of examinations, i.e. you would find the mean mark.

MODE

In a set of numbers, the mode is the number that occurs most often.
For example, for the set 2, 2, 4, 4, 4, 5, 6, 6, the mode is 4 as no other number occurs more than twice.

The mode is easier to find if the numbers are arranged in order of size.

If the numbers in a set are all different, there is no mode.
For example, the set 1, 2, 3, 5, 8, 10, has no mode.

If there are two (or more) numbers which equally occur most often, there are two (or more) modes.

For example, in the set 1, 2, 2, 3, 5, 5, 8, both 2 and 5 are modes.

The mode of a set of numbers is sometimes more useful than the mean and it is much easier to find. For example, if a shoe manufacturer records the size of each pair of shoes that is sold in different areas, it would be the mode of those sizes that he would be interested in (i.e. which size is the most common), because he would have to supply most of that size.

MEDIAN

If we arrange a set of numbers in order of size, the median is the number in the middle.

For example, for the seven numbers in the set 2, 4, 5, ⑦, 7, 8, 9, the median is 7.

When there is an even number of numbers in the set, the median is the mean of the two middle numbers.

For example, for the eight numbers in the set 2, 3, 4, 4, 5, 6, 7, 7, the median is the mean of 4 and 5, i.e. 4.5.

The median is sometimes a better representative number to use than the mean. Suppose, for example, that a pupil got the following marks for end of term examinations; 5, 52, 58, 59, 66.

The mean mark is 48 which is lower than all but one of these marks. This is because one very low mark has affected the mean and in this case the mean mark does not give a fair impression of performance. It would be better to use the median mark, which is 58, as representative of the set.

For a small set of numbers, say 15, it is easy to find the median and we can see that it is the 8th value, i.e. the $\left(\frac{15+1}{2}\right)$th value. From examples such as this we deduce that, for n numbers arranged in order of size, the median is the $\left(\frac{n+1}{2}\right)$ th number.

For example, for 59 numbers, the median is the $\left(\frac{59+1}{2}\right)$ th number, i.e. the 30th number.

For 60 numbers, the median is the $\left(\frac{60+1}{2}\right)$ th number, i.e. the $\left(30\frac{1}{2}\right)$th number. This means the average of the 30th and 31st numbers.

RANGE

The mean, mode or median of a set of data is a single number which can be taken to represent, each in a different way, the whole set. Another aspect of the data which we often need to assess is the spread of the separate items in the set. There are several ways of measuring spread. The simplest (though not particularly useful) way is to find the *range*.

The range of a set of data is the difference between the lowest and highest values.

EXERCISE 26e

A page from a novel by D. H. Lawrence was chosen at random and the number of letters in each of the first twenty words on that page was recorded:

3, 4, 5, 3, 7, 8, 3, 3, 6, 2, 4, 6, 4, 6, 3, 13, 4, 3, 3, 2

Find the mean, modal and median number of letters per word. Find also the range of these numbers.

Arranging the numbers in size order:

2, 2, 3, 3, 3, 3, 3, 3, 3, 4, 4, 4, 4, 5, 6, 6, 6, 7, 8, 13

10th 11th

The mean is $\frac{92}{20} = 4.6$

The mode is 3

The median is the value of the $\left(\frac{20 + 1}{2}\right)$ th number,

i.e. the $\left(10\frac{1}{2}\right)$ th number, which is the average of the 10th and 11th numbers

∴ the median is 4

The range is given by $(13 - 2)$ letters, i.e. 11 letters.

Find the mean, mode, median and range of the following sets of numbers. Remember to arrange them in order of size first. Give answers correct to three significant figures where necessary:

1. 3, 6, 2, 5, 9, 2, 4

2. 10, 8, 10, 16, 7, 9, 10, 8, 9

3. 13, 16, 12, 14, 19, 12, 14, 13

4. 1.6, 2.4, 3.9, 1.7, 1.6, 0.2, 1.3, 2.0

5. 4, 3, 4, 5, 2, 5, 4, 3

6. 5, 7, 9, 12, 10, 8, 7, 9, 7, 8, 12, 7

7. 0.8, 0.7, 0.6, 0.7, 0.8, 0.8, 0.9, 0.5

8. 1.3, 1.8, 1.7, 1.9, 1.4, 1.5, 1.3, 1.8, 1.2

9. Ten music students took a Grade 3 piano examination. They obtained the following marks: 106, 125, 132, 140, 108, 102, 75, 135, 146, 123. Find the mean and median marks. Which of these two representative measures would be most useful to the teacher who entered the students? (Give *brief* reasons; do not write an essay on the subject.)

10. A small firm employs ten people. The salaries of the employees are as follows: £30 000, £8000, £5000, £5000, £5000, £5000, £5000, £4000, £3000, £1500.

Find the mean, mode and median salary. Which of these three figures is a trade union official likely to be interested in, and why?

11. Thirty 15-year-olds were asked · how much pocket money they received each week and the following amounts (in pence) were recorded: 0, 0, 0, 50, 50, 50, 100, 100, 100, 100, 100, 100, 100, 150, 150, 200, 200, 200, 200, 200, 200, 200, 200, 200, 200, 250, 250, 250, 500, 1000.

Find the mean, mode and median amount.

If you were presenting your parents with an argument for an increase in pocket money, which of the three representative measures would you use and why?

12. The first eight customers at a supermarket one Saturday spent the following amounts: £25.10, £3.80, £20.50, £15.70, £38.40, £9.60, £46.20, £10.50. Find the mean and median amount spent.

13. The mean age of a family of five is 28 years 11 months. If the mean age of the parents is 46 years 2 months, find the mean age of the children. If the father's age is 48 years 7 months, how old is the mother?

14. In twelve completed innings, a batsman's mean score was 47.5. After a further innings his mean score fell to 44. How many runs did he score in his thirteenth innings?

FINDING THE MODE AND RANGE FROM A FREQUENCY TABLE

The frequency table shows the numbers of houses in a village that are occupied by different numbers of people:

Number of people living in one house	1	2	3	4	5	6
Frequency	10	8	15	25	12	4

It is clear that there are more houses with four people living in them than any other number, i.e. the mode is 4.

The smallest number of people living in any house is 1 and the largest number is 6 so the range of the number of occupants is $6 - 1$, i.e. 5.

EXERCISE 26f Write down the mode and the range for each of the following frequency tables or bar charts.

1. The number of newspapers bought in one week by the occupants of each house in a particular road was recorded. The results are shown in this frequency table:

No. of newspapers/week	2	3	4	5	6	7	8	9
Frequency	9	3	2	10	12	15	6	3

2.

Number of tickets bought per person for a football match	1	2	3	4	5	6	7
Frequency	250	200	100	50	10	3	1

3.

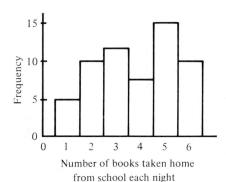

Number of books taken home
from school each night

FINDING THE MEAN FROM A FREQUENCY TABLE ━━━━━━━━━━

The pupils in class 9 were asked to state the number of children in their own family and the following frequency table was made:

Number of children per family	1	2	3	4	5
Frequency	7	15	5	2	1

Total number of families = 30

This information has not been grouped: all the numbers are here so we can total this set. We have seven families with one child giving seven children, fifteen with two children giving thirty children and so on, giving the total number of children as

$$(7 \times 1) + (15 \times 2) + (5 \times 3) + (2 \times 4) + (1 \times 5) = 65$$

There are 30 numbers in the set, so the mean is

$$\frac{65}{30} = 2.2 \text{ to } 1 \text{ d.p.}$$

i.e. there are, on average, 2.2 children per family.

To avoid unnecessary errors, this kind of calculation needs to be done systematically and it helps if the frequency table is written vertically. We can then add a column for the number of children in each group and sum the numbers in this column for the total number of children.

Number of children per family x	Frequency f	fx
1	7	7
2	15	30
3	5	15
4	2	8
5	1	5
	No. of families: 30	No. of children: 65

$\therefore$ the mean $= \dfrac{65}{30} = 2.2$ to 1 d.p.

EXERCISE 26g **1.** This table shows the results of counting the number of prickles per leaf on 50 holly leaves.

Number of prickles	1	2	3	4	5	6
Frequency	4	2	8	7	20	9

Find the mean number of prickles per leaf.

2. A six-sided dice was thrown 50 times. The table gives the number of times each score was obtained.

Score	1	2	3	4	5	6
Frequency	7	8	10	8	5	12

Find the mean score per throw.

3. Three coins were tossed together 30 times and the number of heads per throw was recorded.

Number of heads	0	1	2	3
Frequency	3	12	10	5

Find the mean number of heads per throw.

FINDING THE MEDIAN FROM A FREQUENCY TABLE

To find the median of a set of *items* we need to arrange them in order of size and pick out the one in the middle. This can easily be done if there are not many items and they can all be written down in order.

Now consider this frequency table which shows the distribution of the types of house currently on the list of a local estate agent.

Number of bedrooms	1	2	3	4	5	6
Frequency	6	7	16	12	3	1

The total number of houses is 45

If we want the median we must find the number of bedrooms in the 23rd house when they are arranged in order of size.

By beginning to add the frequencies we see that

There are 6 houses in the first column – not as many as 23
There are 6 + 7 = 13 houses in the first two columns
                                              – not as many as 23
There are 6 + 7 + 16 = 29 houses in the first three columns
                                              – more than 23

So the 23rd house is in the third column of the table,
i.e. the median number of bedrooms is 3.

Remember that if the total number of items is even, two of them are equally
"in the middle" and we take the mean of their values.

EXERCISE 26h **1.** Six sweet pea seeds were planted in each of twenty-five plant pots.
After germination the number of seedlings in each pot was counted
and the results are given in this table

Number of seedlings	1	2	3	4	5	6
Frequency	1	1	4	6	7	6

Find the median number of seedlings per pot.

2. Use Exercise 26g to find the median of the data given in
a) question 1   b) question 2   c) question 3.

FINDING THE MODE AND RANGE FROM A GROUPED FREQUENCY TABLE

This frequency table was made from information about the weights (in kg)
of 90 eleven-year-olds:

Weight (w kg)	$35 \leqslant w < 40$	$40 \leqslant w < 45$	$45 \leqslant w < 50$	$50 \leqslant w < 55$	$55 \leqslant w < 60$	$60 \leqslant w < 65$
Frequency	4	10	30	28	10	8

Because the weights have been placed in groups, we have lost some of the
detail; we do not know how many children have a weight of 46 kg or any
other particular weight. It is possible that more children have a weight of
52 kg than any other weight, but, as we just do not know, we cannot give
the mode as a single figure.

However, we *can* say that more children have a weight in the group
$45 \leqslant w < 50$ than in any other group. We call this the *modal group*.

To find the range we need the difference between the least weight and the greatest weight. We do not know exactly what the last weight is but it cannot be less than 35 kg and the greatest weight can be any value right up to 65 kg (although it cannot actually *be* 65 kg).

Therefore we *estimate* the range as

$$65\,kg - 35\,kg = 30\,kg$$

EXERCISE 26i Write down the modal group and estimate the range for each of the following frequency tables.

1.

Duration of a phone call (*t* minutes)	$0 \leqslant t < 5$	$5 \leqslant t < 10$	$10 \leqslant t < 15$	$15 \leqslant t < 20$
Frequency	20	30	10	5

2. This table shows the age groups of the employees at a small factory.

Age of employee (*n* years)	$18 \leqslant n < 28$	$28 \leqslant n < 38$	$38 \leqslant n < 48$	$48 \leqslant n < 58$	$58 \leqslant n < 68$
Frequency	46	59	41	37	17

3. This table shows the distribution of the number of morning customers at the Village Post Office in six months.

Number of customers	26–30	31–35	36–40	41–45	46–50
Frequency	20	31	54	35	16

FINDING THE MEAN FROM A GROUPED FREQUENCY TABLE

The pupils in one class were asked to count the number of items in their pockets and the following frequency table was drawn up:

Number of items	0–4	5–9	10–14	15–19	20–24
Frequency	6	11	6	4	3

We can see that eleven pupils had from 5 to 9 items in their pockets but we do not know the exact number that each individual pupil had. Therefore we cannot find exactly how many items the eleven pupils had altogether.

However, if we *assume* that the average number of items in that group is halfway between 5 and 9, i.e. $\frac{5+9}{2} = 7$, then we can estimate the total number of items in the group as $11 \times 7 = 77$.

Using the halfway value in the same way for the other groups we can find (approximately) the total number of items in the pockets of all 30 pupils.

(Note that a "halfway" value found in this way is not always a whole number.)

Number of items	Frequency f	Halfway values x	fx
0–4	6	2	12
5–9	11	7	77
10–14	6	12	72
15–19	4	17	68
20–24	3	22	66
	Total: 30		Total: 295

Therefore, the mean number of items is $\frac{295}{30} = 9.8$ to 1 d.p.

Remember that this calculation is based on the assumption that the average of each group is the "halfway" value in that group, so what we have found is an *estimate* of the mean value.

EXERCISE 26j

1. Fifty boxes of peaches were examined and the number of bad peaches in each box was recorded, with the following result:

No. of bad peaches per box	0–4	5–9	10–14	15–19
Frequency	34	11	4	1

Estimate the mean number of bad peaches per box.

2. Twenty tomato seeds were planted in a seed tray. Four weeks later the heights of the resulting plants were measured and the following frequency table was made:

Height (h cm)	$1 \leqslant h < 4$	$4 \leqslant h < 7$	$7 \leqslant h < 10$	$10 \leqslant h < 13$
Frequency	2	5	10	3

Estimate the mean height of the seedlings. (To estimate the average height of the seedlings in the first group we use $\frac{1}{2}(1+4)$, i.e. 2.5).

3. The table shows the result of a survey amongst 100 pupils on the amount of money each of them spent in the school tuck shop on one particular day:

Amount (pence)	0–24	25–49	50–74	75–99
Frequency	26	15	38	21

Find an estimate for the mean amount of money spent.

4. The histogram shows the result of an examination of 20 boxes of screws:

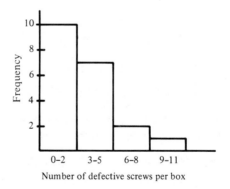

Number of defective screws per box

Make a frequency table and estimate the mean number of defective screws per box.

5. The table shows the distribution of heights of 50 adult females, measured to the nearest cm:

Height (h cm)	$145 \leqslant h < 150$	$150 \leqslant h < 155$	$155 \leqslant h < 160$	$160 \leqslant h < 165$	$165 \leqslant h < 170$	$170 \leqslant h < 175$
Frequency	1	3	21	18	5	2

Find the mean height.

FINDING THE MEDIAN FROM A GROUPED FREQUENCY TABLE

The frequency table below shows the distribution of the weights, in kilograms, of one hundred eleven-year-old children.

Weight (w)	$35 \leqslant w < 40$	$40 \leqslant w < 45$	$45 \leqslant w < 50$	$50 \leqslant w < 55$	$55 \leqslant w < 60$	$60 \leqslant w < 65$
Frequency	7	15	32	25	13	8

The total number of children is 100, so to find the median we must consider the weights of the 50th and 51st children when the weights are arranged in order. We add the frequencies in the columns until we find the column in which these two pupils are located.

$$7 + 15 = 22 \qquad \text{less than 50}$$
$$7 + 15 + 32 = 54 \quad \text{more than 51}$$

The median weight occurs in the third column of the table so it lies in the range $45 \leqslant w < 50$.

We could give an estimate of the median by taking the value at the middle of this range, i.e. $\frac{1}{2}(45 + 50)$, giving the median weight as $47\frac{1}{2}$ kg.

EXERCISE 26k Estimate the median value of the data given in questions 1 to 3 of Exercise 26j.

MIXED QUESTIONS

EXERCISE 26l Use the data given in each question from 1 to 4 to
(a) draw a histogram and superimpose a frequency polygon
(b) find the modal group
(c) state the maximum possible range
(d) find the group in which the median lies.

1. This table shows the distribution of the ages of the people attending a Pop Festival.

Age (n years)	$15 \leqslant n < 19$	$19 \leqslant n < 23$	$23 \leqslant n < 27$	$27 \leqslant n < 31$
Frequency	1900	2700	1600	3000

2. The distribution of the marks obtained by pupils sitting a test are given in the following table:

Mark	0–9	10–19	20–29	30–39	40–49	50–59
Frequency	9	13	27	43	28	20

3. A school organises a Grand Prize Draw to raise money to buy a mini-
    bus. Tickets are sold at 50 p per book and pupils are encouraged to
    sell as many as possible. The table below shows the distribution of the
    numbers of books sold by pupils in the school.

Number of books sold	0–5	6–10	11–15	16–20	21–25	26–30
Frequency	33	124	173	144	95	31

4. This table is based on a cricketer's scores in one season.

Score	0–19	20–39	40–59	60–79	80–99	100–119	120–139
Frequency	8	14	33	6	5	3	1

5. Ninety pupils in one year sat a mathematics examination and the
    following frequency table was drawn up from the results.

    Find the modal group and estimate the range and the median.

Mark	0–14	15–29	30–44	45–59	60–74	75–89	90–100
Frequency	7	5	6	34	22	12	4

6. This table shows the results of a survey among 70 pupils on the
    amount of money that each of them brought to school on a particular
    morning:

Amount (pence)	0–49	50–99	100–149	150–199
Frequency	11	35	15	9

    a) Estimate the mode and the median amount of money.
    b) Find the likely mean amount of money.

In questions 7 and 8 find a value for the mean of the frequency distribution.
(These tables were used earlier in Exercises 26b and 26d.)

7. This table shows the distribution of the ages of the people in the
    audience at a school concert.

Age (n years)	$0 \leqslant n < 20$	$20 \leqslant n < 40$	$40 \leqslant n < 60$	$60 \leqslant n < 80$	$80 \leqslant n < 100$
Frequency	73	34	27	10	6

8. The table gives the distribution of marks of 30 children in a test:

Mark	0–39	40–59	60–79	80–99
Frequency	8	8	10	4

9. Use the list of numbers on the first page of this chapter to find the mean height of the 90 five-year-olds used in that example. Then find the mean using the frequency table drawn up for those heights (page 437).

Compare your answers.

10. Use the information gathered from a project of your own to find the mean

a) from the raw data

b) from a grouped frequency distribution made from the raw data.

PROJECTS

EXERCISE 26h From a chosen project, collect the information (raw data), decide on the groups it can be divided into and make a frequency table. Illustrate your results with a histogram and find the mean value from the frequency table. Find also the mode or modal group and range of the data.

Suggestions for Class Projects

1. Heights of pupils in the class.

2. Weights of pupils in the class.

3. Costs of journeys to school this morning.

4. Times of journeys to school (in minutes).

5. Numbers of brothers and sisters per pupil.

6. Numbers of items in school bags.

7. Ages of pupils (in months).

8. Results of an experiment in science (where every pupil did the same experiment).

9. The prices of a 500 g packet of cornflakes (or any other item) from as many shops as possible.

Suggestions for Individual Projects

10. Choose a page of text from a book and record the number of letters per word (about 100 words is enough).

11. Use the same page of text and record the number of words per sentence.

12. Choose a completely different type of book from that used in question 10 and repeat questions 10 and 11.

13. Count the number of people per car passing a particular place in the evening rush hour.

14. Repeat question 13 at a different time of day.

15. From as many brochures as possible, find the cost of a two-week holiday in August to a particular place, say in Spain.

16. Use either a daily newspaper, or your school weather station if you have one, to record the daily rainfall in your area for a period of 30 days.

17. Use the class register to find the number of absences for each pupil in your class last term.